Reproduction, Larval Biology, and Recruitment of the Deep-Sea Benthos

Reproduction, Larval Biology, and Recruitment of the Deep-Sea Benthos

EDITED BY
Craig M. Young and Kevin J. Eckelbarger

COLUMBIA UNIVERSITY PRESS NEW YORK

Columbia University Press
New York Chichester, West Sussex

Library of Congress Cataloging-in-Publication Data
Reproduction, larval biology, and recruitment of the deep-sea benthos
/ edited by Craig M. Young and Kevin J. Eckelbarger.
p. cm.
Includes bibliographical references and index.
ISBN 0-231-08004-2 (alk. paper)
1. Deep-sea fauna—Reproduction. 2. Deep-sea fauna—Larvae.
3. Marine invertebrates—Reproduction. 4. Marine invertebrates—
Larvae. 5. Deep-sea ecology. 6. Benthos. I. Young, Craig M.
II. Eckelbarger, Kevin J.
QL125.5R47 1994
574.1′6′09162—dc20 93-38716
CIP

Casebound editions of Columbia University Press books are
printed on permanent and durable acid-free paper.

Printed in the United States of America
c 10 9 8 7 6 5 4 3 2 1

The cover photograph of a deep-sea protoconch is reprinted from an article by R. A. Lutz, P. Bouchet, D. Jablonski, R. D. Turner, and A. Warén with the permission of the *American Malacological Bulletin*.

CONTENTS

Preface ix

Contributors xiii

ONE A Tale of Two Dogmas: The Early History of Deep-Sea Reproductive Biology 1

Craig M. Young

TWO Cold-Water Echinoderms Break "Thorson's Rule" 26

John S. Pearse

THREE Adaptations for Reproduction Among Deep-Sea Benthic Molluscs: An Appraisal of the Existing Evidence 44

Rudolf S. Scheltema

FOUR Molluscan Life History Traits at Deep-Sea Hydrothermal Vents and Cold Methane/Sulfide Seeps 76

Richard G. Gustafson and Richard A. Lutz

FIVE Ontogenetic Migration and Dispersal of Deep-Sea Gastropod Larvae 98

Phillipe Bouchet and Anders Warén

SIX The Relationship between Development Pattern and Species Diversity in Deep-Sea Prosobranch Snails 119

Carol T. Stuart and Michael A. Rex

SEVEN Ultrastructural Features of Gonads and Gametes in Deep-Sea Invertebrates 137

Kevin J. Eckelbarger

EIGHT Environmental Control of Quasi-Continuous and Seasonal Reproduction in Deep-Sea Benthic Invertebrates 158

Paul A. Tyler, Lucia S. Campos-Creasy, and Lucy A. Giles

NINE The Biology of External Fertilization in Deep-Sea Echinoderms 179

Craig M. Young

TEN Implications of Mesoscale Flows for Dispersal of Deep-Sea Larvae 201

Lauren S. Mullineaux

ELEVEN Recruitment Ecology and Age Structure of Deep-Sea Invertebrate Populations 224

John D. Gage

TWELVE Life History Studies of Deep-Sea Benthic Infauna: Polychaeta, Aplacophora, and Cumacea from the Continental Slope Off Massachusetts 244

James A. Blake and Les Watling

THIRTEEN The Influence of Natural Organic Enhancement on Life Histories and Community Structure of Bathyal Polychaetes 261

Lisa A. Levin, Gail R. Plaia, and Cynthia L. Huggett

FOURTEEN The Advantages of Dispersal in a Patchy Environment: Effects of Disturbance in a Cellular Automaton Model 284

Ron J. Etter and Hal Caswell

FIFTEEN Ecological Patterns in the Deep-Sea Benthos: How Are They Related to Reproduction, Larval Biology, and Recruitment? 306

J. Frederick Grassle

Taxonomic Index 315

Author Index 322

Subject Index 329

PREFACE

Interest in the deep sea has exploded over the last fifteen years, primarily as a result of intense work at hydrothermal vent fields along the mid-oceanic ridges. Until recently, traditional sampling with trawls, dredges, and cores dominated the field. Technological advances, particularly manned submersibles, have now made it possible to not only sample the deep sea, but to conduct experiments there. Earlier oceanographers could only dream of what we can now accomplish. One such scientist was Walter Garstang, whose poem "An Oceanographer's Dream" was included in his posthumously published book of poems, *Larval Forms* (London: Blackwell, 1951). The poem begins

> I built myself a stately treasure house
> Beside the rolling ocean's swell,
> With cunning apparatus to disclose
> The gems below that dwell,—
>
> Those frail living gems "of purest ray serene"
> Oft frail, fantastic, strange and rare,
> Whose lives sheer mysteries have ever been,
> But need not be—with care.

Only forty years after Garstang's death, we have entered an era where, "with care" and by the use of "cunning apparatus," it is possible to study not only the identities and numbers of deep-sea creatures, but also how they live: their physiology, morphology, behavior, reproduction, and development. Descriptions of pattern are slowly giving way to experiments that test hypotheses about processes, and descriptions of gross morphology are being supplemented by studies at the cellular and subcellular level. It was in this exciting academic climate that the American Society of Zoologists convened a symposium having the same title as this book in Atlanta, Georgia, December 29–30, 1991. Fifteen scientists with a common interest in deep-sea reproduction presented papers outlining and circumscribing the present state of our knowledge on reproduction and larval biology of deep-sea organisms. The idea for the symposium developed through an unlikely yet ongoing research collaboration between the coeditors, an ecologist and a functional morphologist, when it became clear that only minimal effort has been made to understand the biological and ecological processes that control the reproduction of deep-sea organisms.

With the notable exception of the Scottish Marine Biological Association's deep-sea sampling program in the Rockall Trough and on the Hebridean Slope, few organized research efforts have been directed specifically at understanding the processes controlling reproduction and recruitment of deep-sea animals. Much of what we know has been obtained as by-products of sampling programs with other primary goals. This is surprising, as so much of our basic understanding of deep-sea ecology hinges on a better understanding of these processes. Many fundamental questions remain—mostly questions that have been addressed thoroughly for shallow-water species. How do animals in the deep sea control the timing of gametogenesis to ensure reproductive synchrony and successful fertilization? How are seeps, vents, seamounts, rock outcrops, and other isolated habitats colonized? What are the food sources of deep-sea planktotrophic larvae, and what role do larvae play in water column biological processes? Why are the gametes and gonads of some deep-sea animals modified from those of shallow-water species? How is fertilization accomplished in low density populations and without light as a spawning cue? How far do deep-sea larvae migrate vertically? How far do they disperse?

Like many symposium volumes, this one has the dual purpose of reviewing our current knowledge and pointing out the areas where new research is needed. Some of the chapters are extensive reviews, whereas others report the results of original recent research. Deep-sea studies of reproduction have long been shackled by a few unfounded generalizations and untested hypotheses. This volume breaks the chain of Thorson's Rule in the first two chapters, then goes on to emphasize the rich diversity of processes that characterize reproduction in the deep sea and the methods that can be used to study them. When one considers that the majority of the planet lies under more than 2 km of sea water, and yet nearly all our knowledge about the reproduction and larval biology of marine invertebrates is based on species that can be collected by researchers in wading boots, we have some exciting and rewarding work ahead. We predict and hope that the steadily rising interest in invertebrate reproduction and development, the use of manned submersibles, and the increased emphasis on in situ experimentation will spawn (excuse the pun) a new and multidisciplinary field of deep-sea reproductive biology. We hope that this volume will stimulate workers in many disciplines to direct their attention away from the intertidal zone and toward the abyss.

Many individuals contributed to the production of this volume. In particular, we thank our many colleagues who served as anonymous reviewers for the papers that appear here. The American Society of Zoologists provided travel funds for some overseas participants, and the National Science Foundation supported the travel of many others. The entire manuscript was proofread carefully and typeset for submission by Barbara Heglund. Dan McCarthy, Cate Hurlbut, Kumi Ekaratue, Mark Devin, and Patricia Richards attended to many details while the volume was being assembled.

CONTRIBUTORS

James A. Blake
Science Applications International Corporation
89 Water Street
Woods Hole, Massachusetts 02543

Phillipe Bouchet
Muséum national d'Histoire naturelle
Paris, France

Lucia S. Campos-Creasey
Department of Oceanography
The University
Southampton S09 5NH
United Kingdom

Hal Caswell
Woods Hole Oceanographic Institution
Woods Hole, Massachusetts 02543

Kevin J. Eckelbarger
Darling Marine Center
University of Maine
Walpole, Maine 04573

Ron J. Etter
Department of Biology
University of Massachusetts at Boston
Boston, Massachusetts 02125

John D. Gage
Scottish Marine Biological Association
Dunstaffnage Marine Laboratory
P.O. Box 3
Oban, Scotland

Lucy A. Giles
Department of Oceanography
The University
Southampton S09 5NH
United Kingdom

J. Frederick Grassle
Institute of Marine and Coastal Sciences
Rutgers University
New Brunswick, New Jersey 08903

Richard G. Gustafson
Institute of Marine and Coastal Sciences
Rutgers University
New Brunswick, New Jersey 08903

Cynthia L. Huggett
Department of Marine, Earth and Atmospheric Sciences
North Carolina State University
Raleigh, North Carolina 27695-8208

Lisa A. Levin
Marine Life Research Group
Scripps Institution of Oceanography
La Jolla, California 92093-0218

Richard A. Lutz
Institute of Marine and Coastal Sciences
Rutgers University
New Brunswick, New Jersey 08903

Lauren S. Mullineaux
Woods Hole Oceanographic Institution
Woods Hole, Massachusetts 02543

John S. Pearse
Institute of Marine Science and Department of Biology
University of California
Santa Cruz, California 95064

Gayle R. Plaia
Department of Marine, Earth and Atmospheric Sciences
North Carolina State University
Raleigh, North Carolina 27695-8208

Michael A. Rex
Department of Biology
University of Massachusetts at Boston
Boston, Massachusetts 02125

Rudolf S. Scheltema
Woods Hole Oceanographic Institution
Woods Hole, Massachusetts 02543

Carol T. Stuart
Department of Zoology
University of Massachusetts at Amherst
Amherst, Massachusetts 01003

Paul A. Tyler
Department of Oceanography
The University
Southampton S09 5NH
United Kingdom

Les Watling
Darling Marine Center
University of Maine
Walpole, Maine 04573

Anders Warèn
Naturhistoriska Riksmuseet
Stockholm, Sweden

Craig M. Young
Harbor Branch Oceanographic Institution
5600 U.S. 1 N
Ft. Pierce, Florida 34946

CHAPTER ONE

A Tale of Two Dogmas: The Early History of Deep-Sea Reproductive Biology

Craig M. Young

ABSTRACT

Writings about deep-sea reproduction have traditionally been placed in the context of two long-standing hypotheses, Orton's Rule and Thorson's Rule. The former, which predicts continuous reproduction in the isothermal regions of the deep sea, has now been disproven for many species in numerous animal phyla. The latter hypothesis, which predicts that deep-sea animals should brood, enjoyed widespread acceptance until very recently, when numerous exceptions have been documented. Thorson's Rule was originally articulated by several authors of the *Challenger Reports* about a half century before it was formalized by Thorson, but not all of the *Challenger* authors believed in the absence of deep-sea larvae. Indeed, a careful search of the literature reveals more than twenty pieces of evidence for deep-sea pelagic larvae published before Thorson's 1936 monograph. From the earliest days of deep-sea exploration, biologists could have concluded that reproduction is accomplished in the deep sea by a diversity of mechanisms.

"Since deep-sea data are few and difficult to obtain, each school of thought stands on a considerable body of opinion. What is only opinion and what is known for certain are not easy to sort out in the literature."

(Carney et al. 1983)

The study of deep-sea reproduction, embryology, and larval biology, like the overall field of deep-sea biology, is a relatively young science. Neither Aristotle nor the Renaissance masters expressed opinions about how animals reproduce at great depths. Indeed, thanks in part to the depauperate deep-water fauna of the Aegean Sea and despite fragments of evidence to the contrary (reviewed by Herdman 1923; Mills 1983), the British naturalist Edward Forbes was able to convince a generation of biologists as late as 1844 that conditions below 300 fathoms are not suitable for life (Forbes 1844). The roots of deep-sea reproductive biology may not be deep, but they are tangled. Some of the earliest observations and most influential ideas are found scattered in the expedition reports and taxonomic tomes published in the decades surrounding the turn of the present century. Frequently, however, generalizations were based on "naturalist's intuition" rather than hard facts. To this day, breakthroughs are often tied more closely to technological advances than to the development of ideas and theories.

Research in deep-sea biology accelerated with the expeditions of HMS *Lightning* in 1868, the four cruises of HMS *Porcupine* between 1869 and 1870 (Thomson 1874), and the circumnavigation by HMS *Challenger* between 1872 and 1876. Recent reviews discussing these and later oceanographic expeditions (Menzies et al. 1973; Mills 1983) make it clear that observational, explorational, and descriptive studies have dominated the field throughout its history. Even today, for example, there is an ongoing effort to explore and describe hydrothermal vents, back-arc basins, and mid-oceanic ridges by manned and unmanned submersibles. The testing of hypotheses in deep-sea research did not come into its own until about thirty years ago, and the application of true experimental approaches began even more recently. When the remarkable fauna of hydrothermal vents was discovered in 1977, technology was already available to begin process-oriented experimental work. Consequently, physiological, ecological, microbiological, and biochemical results flooded the literature very quickly. Such was not the case in the early days; some ideas and hypotheses about reproduction remained entrenched in the literature for more than a century before they were questioned because the technology required for testing them developed slowly. It was only natural that some of the untested hypotheses should become practically canonized by tradition.

Hoary Hypotheses

Ideas about deep-sea reproduction may be entrenched, but they are not diverse. In fact, a perusal of review articles that deal with the deep sea generally turns up only two major hypotheses, both of which have remained in the literature for a very long time, and both of which have now been tested repeatedly and found wanting. The first is that animals living in the isothermal and otherwise constant environment of the deep sea should reproduce continuously. This hypothesis follows logically from the idea first articulated by John Orton (1920) that seawater temperature controls breeding. Thorson (1946) referred to this idea as Orton's Rule, but a careful reading of Orton's paper makes it clear that the author regarded this as nothing more than a tentative and testable hypothesis as far as the deep sea was concerned. In true Popperian fashion (Popper 1959), Orton offered the means whereby the hypothesis could be (and later was) falsified:

> Systematic records of breeding of deep-sea animals may not be available for some time, but it would be well worth recording the sexual condition of all animals taken from these regions. . . . There can be no doubt that an oceanographical expedition to the Atlantic in winter would yield most valuable biological knowledge, providing work could be done. (Orton 1920:354)

Orton's Rule was tested nearly a half century later, when several teams of oceanographers began to evaluate deep-sea samples obtained in various seasons, thereby demonstrating noncontinuous breeding in a several invertebrate phyla (George and Menzies 1967, 1968; Schoener 1967; Rokop 1974, 1977; Tyler et al. 1982; Harrison 1988). It has now become clear that numerous seasonal processes exist in isothermal deep-sea environments (reviewed by Tyler 1988; Scheltema, this volume) and that factors other than temperature may be important in entraining reproductive seasons (reviewed by Giese and Kanatani 1988). Interestingly, the recent discovery that detrital material arrives in annual pulses on the deep-sea floor was first suggested by Henry Nottridge Moseley, one of the naturalists on the *Challenger* expedition in 1880:

> Life must be monotonous in the deep sea. There must be an entire absence of seasons, no day and night, no change of temperature. Possibly there is at some places a periodical variation in the supply of food falling from above which may give rise to a little annual excitement amongst the inhabitants.
> (Moseley 1880:592)

Because Orton's generalization was not debated much by the early deep-sea biologists and has been put to rest by modern ones, I will not consider its history further.

The most pervasive long-standing hypothesis about deep-sea reproduction is that deep-sea and polar animals should undergo direct development, bypassing the pelagic embryonic and larval stages so prevalent among shallow-water invertebrates. This idea has been named Thorson's Rule (Mileikovsky 1971) after the great Danish synthesizer Gunnar Thorson (figure 1.1A), whose dissertation research in East Greenland provided what appeared to be solid evidence in support of the idea (Thorson 1936; but see Pearse, this volume). Thorson articulated the rule plainly and forcefully in his 1936 and 1946 monographs and again in his oft-cited 1950 review. Where the deep sea was concerned, his hypothesis was based in part on fragmentary evidence and in part upon logic:

> It is quite natural that the deep-sea species must have a nonpelagic larval development. As the phytoplankton production, on account of the processes of assimilation, can only take place in the uppermost water layers, the larvae of the deep-sea species, if they are actually to pass through a pelagic larval development, must move through a water column often 2000–4000 m high or even more from the bottom to the food-producing surface layer, to cover the same distance after having completed their pelagic development, and sink to the bottom. It is obvious that under these circumstances the conditions for completing the pelagic larval development are extremely unfavourable; a great number of the larvae will be destroyed on the way . . . As already stated, it is therefore the opinion of biologists that the deep-sea animals complete their reproduction without any pelagic larval stage. (Thorson 1936:129–130)

Although Thorson's own data were all from the Arctic, not the deep sea, he nevertheless pressed this hypothesis on a scientific community willing to accept it virtually on faith. The idea actually antedated Thorson's formulation by at least a half century, even though considerable evidence for pelagic larval development was available from the earliest days. Thorson's Rule has been embraced time and time again (e.g., Madsen 1961; Scheltema 1972; Crisp 1986; Emlet et al. 1987); like the phoenix, it rises from ashes with virtually every text on marine biology or oceanography, though authors of recent books devoted to deep-sea biology have all been careful to point out the numerous exceptions to the rule (Menzies et al. 1973; Marshall 1979; Gage and Tyler 1991).

It is not my intention to drive yet another nail into the coffin of Thorson's Rule. Instead, I will consider the development of this idea in the eighty years or so before Thorson popularized it. This will be done by documenting some of the opinions of early biologists about the nature of deep-sea reproduction and by asking whether the data available to Thorson and other reviewers in the second half of the twentieth century supported the generalizations they made. In the process, I hope to remind biologists of some of the important contributions to our knowledge of deep-sea reproduction, larval biology, and embryology made during the early years of deep-sea exploration.

Thorson's Rule Before Thorson: The Early Evidence for Brooding

Sir John Murray (figure 1.1B), successor of Wyville Thomson (figure 1.1C), the organizer of the *Challenger* expedition, maintained a small office in Edinburgh (figure 1.1E). From there, he corresponded with the top biologists of the day, organizing their efforts with the *Challenger* collections and editing their contributions. As the major editor of the *Challenger Reports,* Murray influenced the course of science by selectively including certain interesting discoveries in those

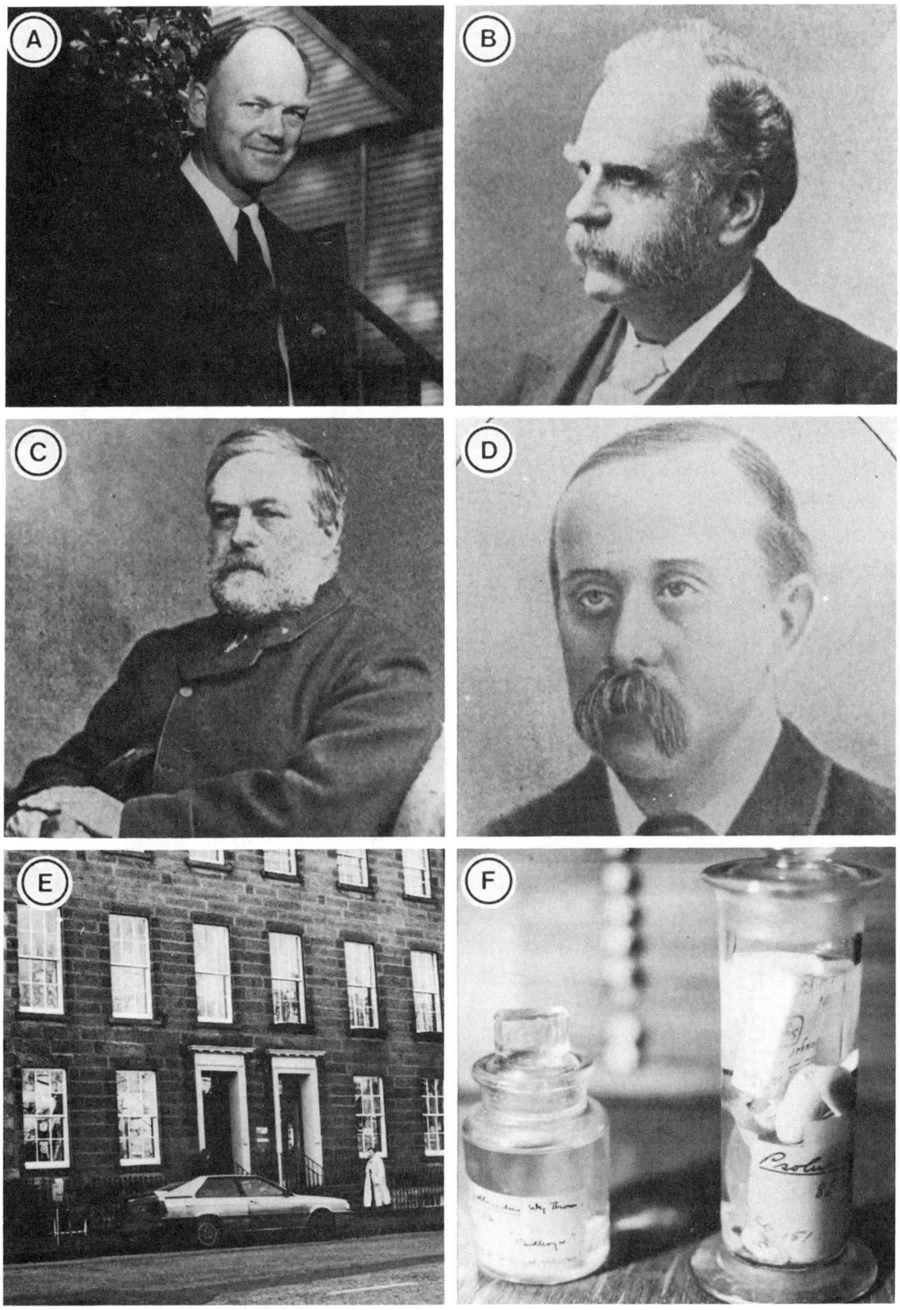

FIGURE 1.1. A: Gunnar Thorson in the 1950s (photo by Ruth Turner). B: Sir John Murray. C: Sir Charles Wyville Thomson. D: Henry Nottridge Moseley. E: The former *Challenger* office at 32 Queen Street, Edinburgh. It is now (1992) a real-estate office. F: Brooding sea cucumbers, *Psolus ephippifer,* collected by the scientists on HMS *Challenger*. These are now in the British Museum (Natural History).

volumes that would ultimately be the most read and quoted. Volume 1, *Narrative of the Cruise of HMS "Challenger"* (Tizard et al. 1885), was one such volume. It was not the first to appear (twelve volumes had earlier publication dates), but because this volume contains a lively travelogue of the voyage, and woodcuts that depict everything from the ship itself to a fanciful mermaid pulling a plankton net, *Narrative* is by far the most often cited in popular accounts, historical treatments, and even scientific papers. Most oceanographers know much more about the details of biology recounted in volume 1 than the details found in the forty-nine weightier taxonomic tomes.

Descriptions of brooding echinoderms figured prominently in volume 1, *Narrative,* and also in Wyville Thomson's introduction to the Zoology reports (Thomson 1880). Early on, Thomson planned to write about three of his own personal interests himself: the stalked crinoids, the hexactinellid sponges, and the brooding echinoderms (Herdman 1923). He managed to write up text only for the echinoderms before his death in 1882. The crinoids and hexactinellids were ultimately sent to other qualified systematists to work up. Thomson's account of the brooding echinoderms (Thomson 1878) was published by the Linnaean Society two years before the appearance of the first *Challenger Report* in 1880 and nearly a decade before Murray's volume 1 *Narrative* appeared. When editing volume 1 (Tizard et al. 1885), Murray quoted verbatim the major part of Thomson's (1878) beautifully illustrated paper on brooding echinoderms and the authors who wrote the systematic accounts of the various echinoderm classes added little to the original descriptions of brooding; instead, they quoted the works of Wyville Thomson again and again (Agassiz 1881; Lyman 1882; Theel 1882, 1886; Sladen 1889).

The emphasis on brooding echinoderms in the narrative report is not surprising; after finding that the deep-sea fauna was similar throughout the world oceans, Moseley (figure 1.1D) commented in 1880, "We got quite tired on the *Challenger* of dredging up the same monotonous animals wherever we went" (p. 547). It was undoubtedly exciting to obtain samples from which lifestyles and processes could be easily inferred.

The brooding Antarctic echinoderms made such an impression on some of the *Challenger* naturalists that they quickly formed the opinion that brooding dominates in the deep sea. Perhaps the most dogmatic early statement of Thorson's Rule comes from a speculative letter written by John Murray to the young William Herdman in 1894 and reproduced in Herdman's 1923 biography of his mentors: "These animals without pelagic free-swimming larvae also descended to the deep sea as the waters cooled. . . . Polar animals and deep-sea animals have all a direct development" (Herdman 1923:81).

When one delves beyond *Narrative,* it becomes apparent that Murray's generalization (Herdman 1923) was not justified by the data at hand, even for the Echinodermata (see Pearse, this volume). Because we tend to think of the *Challenger* voyage primarily as an effort to understand the deep sea, it is not commonly recognized that the brooders described by Thomson mostly occurred in shallow water or had eurybathic distributions. The relationship between egg size and developmental mode was incompletely understood in the days surrounding the turn of the century, and the interpretation of reproductive patterns was sometimes biased because only the adult stage was available for study. Whereas brooding is easily demonstrated by the presence of young, broadcast spawning could not at that time be inferred from morphology.

The *Challenger* holothurians, as described by Hjalmar Theel (figure 1.2A), included two brooders, *Cladodactyla crocea,* which lived in shallow land-locked fjords in Antarctica, and *Psolus ephippifer* (figure 1.1F), a eurybathic Antarctic species (Theel 1882, 1886). Neither of these species belongs to the Elasipoda, which is the dominant holothurian order in the deep sea; indeed, brooding among the Elasipods was first documented in a single species, *Oneirophanta mutabilis,* more than a half century later (Hansen 1968). Theel almost certainly would have found brooded embryos if they were present; his detailed descriptions of internal and external anatomy, including reproductive structures and gametes, are among the best from that era. However, absence of evidence did not prevent some early biologists from speculating about developmental patterns.

For example, even though neither Thomson (1878) nor Theel (1882) found evidence for brooding in elasipod holothurians, Thomson assumed that elasipods must brood nevertheless: "In some Echinoderms the embryo passes into the adult condition without any free larval stage, which seems to be the case with several shallow-water Holothurians, and doubtless even within the Elasipoda" (Tizard et al. 1885:281). He then went on to speculate about how lecithotrophic larvae would be distributed if they existed: "If these latter were subject to a more complicated metamorphosis, it is most probable that their larvae would not be able to live at the surface of the sea, but keep close to the floor of the ocean" (Tizard et al. 1885:281).

The American oceanographer and biologist Alexander Agassiz (figure 1.2B) wrote the systematic description of the *Challenger* Echinoidea. He described 62 species living deeper than 100 fathoms, only two of which were definite brooders, *Hemiaster* (now *Abatus*) *cavernosa* and *Goniocidaris canaliculata.* Both of these are Antarctic species that extend from very shallow water to bathyal depths. Since positive evidence of brooding was available for only 3 percent of the 62 deep-water species, it should have been impossible for Agassiz and his contemporaries to conclude that brooding *predominates* among deep-sea echinoids. However, Agassiz (1881) extended the available data with an interesting piece of circumstantial evidence. In examining the genital plates of echinoid tests, he discovered that some gonopores were larger than others and speculated that those with large gonopores must in fact release very large eggs. This conclusion is supported by a recent study by Emlet (1989). Agassiz reasoned that since all of the recently discovered brooders had large eggs, species with large eggs must all brood. In the language of the logician, this categorical syllogism suffers from the fallacy of "undistributed middles" (Hurley 1988). In form, the statement is equivalent to: "All brachiopods have shells and this animal has a shell, so this animal must be a brachiopod." The animal in question could be a brachiopod, of course, but it could also be a snail, a clam, an ostracod, or anything else with an exoskeleton. Likewise, animals with large eggs might be brooders, but they might also be pelagic lecithotrophs.

This fallacy of logic caused Agassiz and many subsequent workers to err repeatedly. For example, Agassiz stated of *Porocidaris elegans,* which belongs to a genus now suspected of pelagic lecithotrophy, "Large female genital openings probably indicate that *Porocidaris elegans,* like *Goniocidaris canaliculata,* is viviparous" (Agassiz 1881:42). He arrived at a similar conclusion with respect to the echinothuriids, which produce floating, pelagic lecithotrophic larvae: "Judging from the large size of the genital openings and the large size of the eggs in one of the species, this group of sea urchins is probably viviparous" (Agassiz 1881:73).

Agassiz was not the only author in history to mistakenly assume that large egg sizes are always indicative of brooding. At least in his early papers, Thorson apparently did not understand that large eggs could be produced by animals with pelagic development. A few excerpts from his 1936 paper will serve to illustrate this point:

> Earlier authors only discuss the fact that "eggs rich in yolk" and accordingly with direct development are more met with in the Arctic than in southern seas. (p. 77)

> The considerable egg size and the consequent nonpelagic development thus seems to be the genotypically determined mode of reproduction characteristic of these species. (p. 78)

> There is every probability that the big eggs and the subsequent non-pelagic development in the gastropods as well as the lamellibranchs and the shrimps is a method of reproduction characteristic of the species throughout their area of distribution. (p. 79)

Deep-sea biologists have generally followed the tradition of Agassiz and Thorson by improperly inferring brooding from large egg sizes. For example, F. Jensenius Madsen (1961), in his thoughtful monograph on the Porcellanasterid starfish, noted: "The ripe eggs measure 0.5 to 0.6 mm in diameter which confirms an expected nonpelagic development." As recently as 1984 Tyler and Gage, in an otherwise brilliant paper on the abyssal cidaroid and echinothuriid urchins of the Rockall Trough, made a similar statement: "The dominant reproductive strategy is production of a large egg and, hence, direct development with little dispersal."

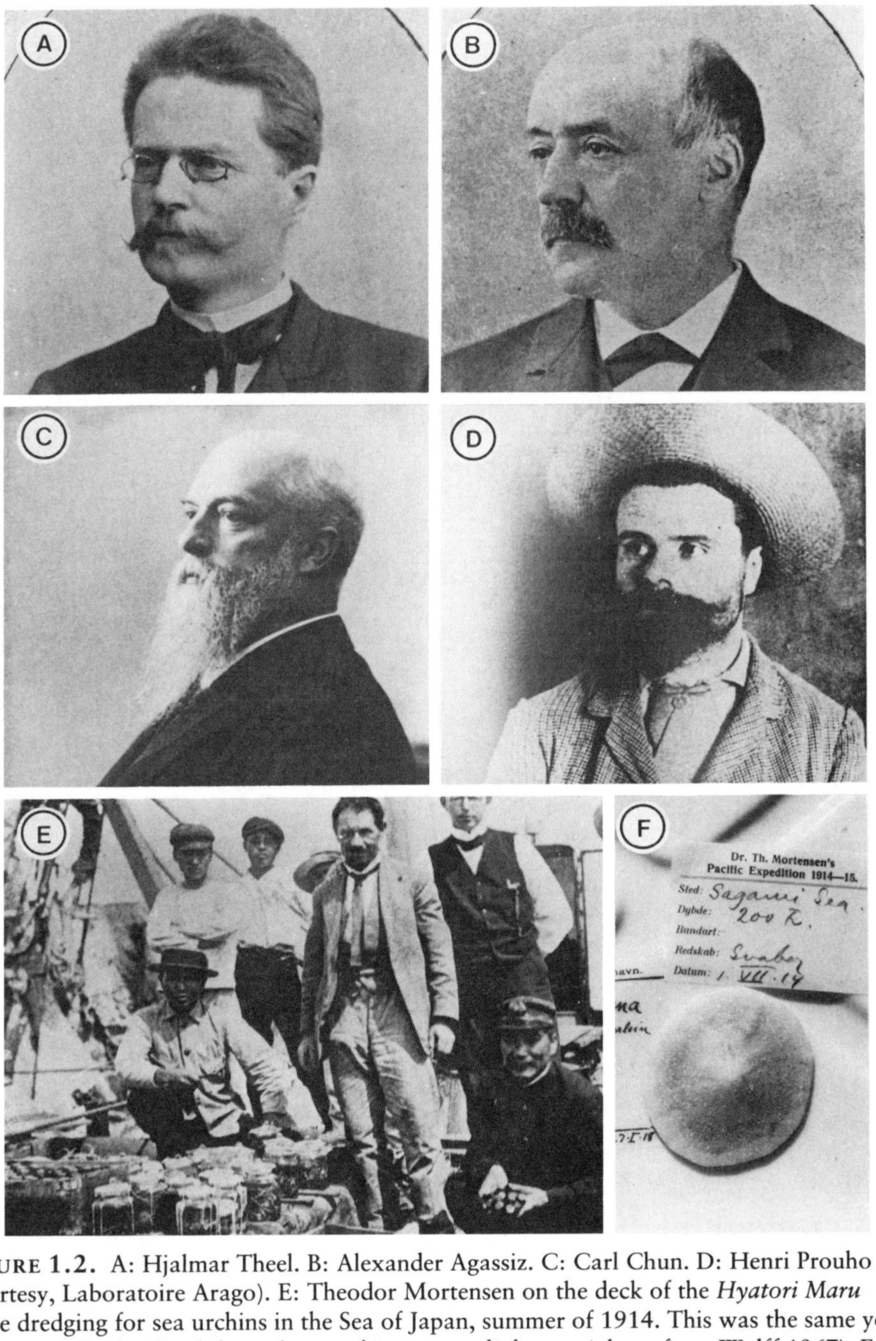

FIGURE 1.2. A: Hjalmar Theel. B: Alexander Agassiz. C: Carl Chun. D: Henri Prouho (courtesy, Laboratoire Arago). E: Theodor Mortensen on the deck of the *Hyatori Maru* while dredging for sea urchins in the Sea of Japan, summer of 1914. This was the same year he successfully fertilized the embryos of *Laganum diplopora* (photo from Wolff 1967). F: Mortensen's specimens of *Laganum diplopora* in the Zoologisk Museum, Copenhagen. These probably came from the same dredge haul as the individuals whose gonads were macerated for embryological studies.

As John Pearse (this volume) and Mileikovsky (1971) have shown convincingly, both polar and abyssal echinoderm faunas are dominated not by brooders but by pelagic lecithotrophs that produce large eggs. Moreover, large eggs do not necessarily imply low dispersal. Large yolky eggs of numerous species float at the surface, where they may disperse widely (e.g., the starfish *Mediaster aequalis* can delay metamorphosis for up to a year; Birkeland et al. 1971). Recent data indicate that the yolky eggs of most bathyal and abyssal echinoderms also float (Young and Cameron 1987; Cameron et al. 1988; personal observations).

The *Challenger* volume on asteroids (Sladen 1889) reports brooding in the same two species that Thomson (1878) reported: *Leptychaster kergulensis, Hymenaster pellucidus,* as well as a third species, *Hymenaster praecoquis*. These latter two species are members of the distinctive family Pterasteridae, a number of which retain broods under the supradorsal membrane.

Pelagic Larvae in the Early Deep-Sea Literature

As recently as 1971, Mileikovsky recognized only one definite example of pelagic larval development in an abyssal invertebrate. However, both circumstantial and direct evidence of deep-sea larvae was available at the time of the *Challenger* expedition and in the early decades of the twentieth century. Thorson (1936, 1946, 1950) was aware of some of this evidence, but was not convinced by it (Milekovsky 1971). In this section, I will review evidence for larval development in deep-sea benthic invertebrates that was available to Thorson in 1936.

Not all of the *Challenger* authors shared Murray's opinion (Herdman 1923) about the predominance of direct development. Two years after returning from the *Challenger* voyage, H. N. Moseley (figure 1.1D) wrote an eloquent three-part summary of the cruise findings for *Nature* (Moseley 1880) in which he alluded several times to pelagic development in the deep sea. For example, he noted the existence of eyed larvae in some deep-sea crustaceans which are blind as adults and interpreted this as evidence that larvae migrate to depths where vision might be possible. Of the deep-sea coral *Cryptohelia pudica,* which he described in greater detail the following year (Moseley 1881), he observed: "The fully developed larvae are worm-like, and when ready escape and swim away to found each a new colony. For corals, like us, travel in their youth and see the world, and become stationary like us only in later life" (Moseley 1880:546).

In his discussion of the plankton and nekton of the open sea, Moseley made yet another anthropomorphic allusion to deep-sea larvae: "It seems probable that some animals which live near the sea surface when young in all the pleasures of warmth and sunlight, sink when fully grown to lead a sluggish life on the cold and dismal bottom" (p. 571). He also cited work by Lütken, in which the young of deep-sea angler fishes were found in the guts of surface-living predators: "If this deep-sea fish really develops in the early stage at the surface, how do the eggs reach the top of the water? Possibly they rise slowly from the bottom. Perhaps some other deep-sea animals go through their early stages at the surface" (p. 571). Remarkably, these opinions have virtually never been cited in later literature, probably because they were overshadowed by Thomson's more memorable drawings of brooding echinoderms.

Echinodermata

In the late 1870s, Henri Prouho (figure 1.2C), who was later to gain a reputation as a great authority on bryozoans, was a graduate student at the newly built laboratory Arago at Banyuls sur Mer. For a dissertation project, Prouho chose to study the embryology of two species of sea urchins which could be collected by trawl from the Bay of Biscay. One of these, known at the time as *Dorocidaris papillata,* was in fact *Cidaris cidaris,* a eurybathic species that occurs not only in the Mediterranean, but also to a depth of 1800 m off the Atlantic continental slope from Norway to the Cape Verde Islands (Mortensen 1927). The echinopluteus larvae reared from gametes by Prouho looked similar to the larvae of other cidaroids described much later; the eggs were small and the larvae were transparent and planktotrophic (Prouho 1888).

A large and spectacular holothurian larva with a cosmopolitan distribution was described originally by the German oceanographer Carl Chun (figure 1.2D) and given the name *Auricularia nudibranchiata* (Chun 1887). Both Chun and Mortensen (1921) provided some morphological evidence that this was the larva of a deep-sea Elasipod, though MacBride (1920) took exception and attributed it to a shallow-water synaptid because of the presence of wheel ossicles. Later authors have sided with MacBride (reviewed by Smiley et al. 1991). The question of identity remains unresolved, but Smiley et al. (1991) have proposed *A. nudibranchiata* may be the larva of an abyssal synaptid. Using closing nets off Naples in 1886, Chun (1887) also noted the presence of many other kinds of invertebrate larvae to depths of at least 1400 m (Mills 1983).

More than three decades after Prouho reared the first bathyal echinoid larvae, embryos, and larvae of two other bathyal species were also cultured. In 1912, Cresswell Shearer and his colleagues obtained ripe specimens of *Echinus acutus* from commercial trawlers in Plymouth. This species, which lives on the continental slope at depths as great as 1280 meters (Mortensen 1927), yielded small eggs that developed into planktotrophic larvae that differed in only minor respects from the congeneric shallow-water species *Echinus esculentus* (Shearer et al. 1914). Then in 1914, the Danish biologist Theodor Mortensen (figure 1.2E) dredged the clypeasteroid *Laganum diplopora* (figure 1.2F) from 800 m in the Sagami sea, macerated gonads, and reared the resulting embryos into perfect 2-armed planktotrophic plutei (Mortensen 1921). Although he was able to keep them alive for only a few days before departing Japan and abandoning the culture, Mortensen did note that these deep-sea larvae resembled those of shallow-water relatives and also that they tended to rest on the bottom of the culture vessel. In his discussion of deep-sea reproduction, Mortensen (1921) called attention to the holothurian larvae described by Chun and to other larvae collected during the plankton expedition of 1889 (Mortensen 1898), all of which he considered as evidence that pelagic larval development occurs in the deep sea. Mortensen (1921) speculated on the origin of these larvae:

> Were they carried out from the coastal waters by currents, or did they rise from the bottom? So long as we do not know the parental origin of the larvae found in the open sea the question cannot be answered definitely. Most probably they come from both sources. The question of whether any deep-sea Echinoderms have truly pelagic larvae must, after what we now know, be answered in the affirmative. *Laganum diplopora* has been directly proved to have pelagic larvae, and it is almost certain that also *Pedicellaster 6-radiatus* has pelagic larvae (*Stellophaera mirabilis*); judging from the character of the eggs many other deep-sea forms must have pelagic larvae (though a great percentage of them have large eggs and therefore certainly must have an abbreviated development.
> (Mortensen 1921:248–249)

Note that Mortensen equated large eggs with "abbreviated development," not with brooding. Despite Mortensen's published opinion that "many deep-sea forms must have pelagic larvae" Thorson (who was a graduate student in the same department where Mortensen was a senior faculty member) attributed the opposite point of view to Mortensen just fifteen years later: "Dr. Mortensen, who has thoroughly studied these phenomena as regards the echinoderms, has informed me that with a single exception, the deep-sea echinoderms must, according to the experience gained so far, be assumed to have a nonpelagic larval development, and the same seems to be the case with the other invertebrate groups" (Thorson 1936:129).

Whatever his real opinion about deep-sea larvae, Mortensen was troubled by the seemingly impossible vertical migration that would be required of a planktotrophic larva originating in the deep sea:

> The occurrence of larvae of deep-sea forms at the surface of the ocean is another problem not very easily explained (due allowance being made for the fact that we have as yet not definitely ascertained a single case of a larva of a deep-sea form found at the surface; we do not know exactly the depth at which *Stellophaera mirabilis* was taken, and we have not yet definitely proved *Auricularia nudibranchiata* to belong to a deep-sea Holothurian). It seems impossible that the larvae could rise all the way by means of their ciliary movements. One cannot help suggesting that vertical currents may have something to do with the transport of the larvae from the deeper layers towards the surface and the

> inverse. . . . Upon the whole, we are hardly in possession of facts enough even for a definite formulation of the problems connected with the deep-sea larvae. We can only see this much that continued researches will be sure to bring important results. (Mortensen 1921:249–250)

Although he did not realize it at the time, Mortensen (1898) also described the larva of a deep-sea ophiuroid. This species, which had been collected in large numbers during Victor Hensen's plankton expedition, could not be assigned to a known adult at the time, so Mortensen (1898) called the larva *Ophiopluteus ramosus*. Semenova et al. (1964) later suggested that *O. ramosus* was the larva of a bathyal ophiuroid, and Tyler and Gage (1982) finally showed it to be *Ophiocten gracilis,* a bathyal species that occurs abundantly throughout the Northeast Atlantic.

Gonads and gametes of many echinoderms were illustrated and described in the *Challenger Reports*. For example: Herbert Carpenter (figure 1.3A) detailed the structure of genital pinnules of deep-sea and shallow-water crinoids (Carpenter 1884) and Theel (1882, 1886) illustrated the positions of the gonads and gonopores and the structure of the male and female genital tubules in elasipod and other deep-sea holothurians. In some holothurians, Theel found genital papillae which he called gonadal processes. In others, he illustrated multiple gonopores, and even drew the eggs in scaled drawings that allow us to infer their probable lecithotrophic development (Theel 1886). The anatomy illustrated in most of these species suggests free spawning rather than brooding.

Brachiopoda and bryozoa

The years between 1910 and 1915 were important ones for deep-sea larval biology. While Mortensen and Shearer were rearing bathyal echinoid larvae in the Pacific and Atlantic Oceans respectively, deep-sea brachiopod larvae were being discovered in the Indian Ocean. The Scottish naturalist Henry Ashworth (figure 1.3B), who worked in the same department at Edinburgh as Forbes, Thomson, and Murray, collected these larvae from the plankton while cruising not on a research ship, but on a commercial passenger vessel (Orient Lines) to Australia. He secured permission from the master of the vessel to take plankton samples from the seawater pump, which drew water from 18 feet below the surface and dumped it into large tanks on deck. From these tanks, he obtained larvae of several species of inarticulate brachiopods in both the Red Sea and the Indian Ocean (Ashworth 1915). One of these, which he described in detail, was the larva of *Pelagodiscus atlanticus,* which was already known to occur at depths ranging from 200 fathoms to 2425 fathoms. Although these larvae came from the surface plankton, other inarticulate brachiopod larvae of uncertain affinity had already been collected years earlier at 1000–1200 meters during the plankton expedition (Simroth 1897) and at 3000 meters depth by the German South Polar expedition (Eichler 1911). The larvae of inarticulate brachipods are planktotrophic and spend up to several weeks in the plankton (Long and Stricker 1991).

The *Challenger Report* dealing with brachiopods (Davidson 1880) says nothing about reproductive processes, and only the shells are illustrated. The same applies to bryozoans, except that many of the illustrations show ovicells similar to those in which the embryos are brooded prior to release of lecithotrophic coronate larvae (Busk 1884, 1886).

Mollusca

Beginning in about 1888, Prince Albert I of Monaco used a diversity of approaches on his oceanographic ships *l'Hirondelle, Princess Alice I,* and *Princess Alice II* to collect deep-water plankton and benthos in the North and equatorial Atlantic and in the Mediterranean Sea (Mills 1983). Many of the gastropods collected by dredging in deep water had intact protoconchs, which were illustrated by Dautzenberg and Fisher (1896, 1897) and others. Examination of the protoconch size and morphology of these animals suggests that many have planktonic development. One of the *Challenger Reports* dealing with gastropods (Watson 1886) also illustrated protoconchs of some deep-sea gastropods. Surprisingly, Pelseneer's monograph on the anatomy of deep-sea gastropods (1888) contained very little information about reproduction.

Although the relationship between developmental mode and protoconch size was not understood at the time, Thorson was a major proponent of the "shell apex theory" which relied on protoconch form to infer developmental mode. He certainly could have considered these published illustrations of protoconchs when speculating about reproduction in the deep sea. The deep-sea malacology literature from around the turn of the century was large (Clarke 1962) and it seems likely that developmental modes could have been inferred from protoconchs and prodissoconchs for many of the bathyal species illustrated. However, as Bouchet (this volume) has noted, the larval shells of many abyssal species are generally eroded away.

Arthropoda

Both direct and circumstantial evidence for larval development in deep-sea crustaceans was presented in the *Challenger Reports* and in other papers written by the *Challenger* naturalists. Rudolf von Willemöes-Suhm (figure 1.3C) died during the expedition, but not before sending home a description of a blind megalopa larva collected from a depth of 1675 m (Willemöes-Suhm 1876). Other crustacean larvae were also obtained from deep water around the turn of the century. Selbie (1914) described an unusual, large larva which was later shown to be the young form of the deep-sea eryonid *Polycheles* sp. (Gurney 1939). These animals swim in the water column at great depths, but resemble the adults so closely that Gurney (1939), without having access to earlier stages, speculated that development might be direct. As early as 1897, Milne-Edwards

FIGURE 1.3. A: P. H. Carpenter. B: Henry Ashworth. C: Rudolf von Willemöes-Suhm. D: P. P. C. Hoek. E: Richard Hertwig. F: William Herdman.

and Bouvier demonstrated that galatheid-like shrimp in the family Uroptychidae undergo direct development from small eggs. In his treatment of the deep-sea Anomura, Henderson (1888) made the following observation about the reproduction of galatheid crabs:

> In the majority of the deep-water galatheids—with the exception of those belonging to the genus *Munida*—the eggs carried by the female are few in number and of remarkably large size. It may be inferred from this that the deep sea enemies are fewer, and the chances of each individual egg undergoing its full development therefore relatively greater, the result being diminished production on the part of the parents; while the large size of the ova perhaps indicates a protracted embryonic existence.

The presumed relationships among larval mortality, fecundity and egg size articulated by Henderson are reminiscent of Thorson's (1950) discussion of larval wastage, and also of one premise of Vance's (1973) well-known model.

Dolfein (1904) examined patterns of egg size among deep-sea crustaceans. He noted that adult crabs with reduced eyes often produce large eggs and bypass the larval stage, whereas species with normal eyes generally have small eggs and normal larval development. Some species with small eggs (and presumably planktonic larval development) were found in very deep water.

Cirripedes in the *Challenger* collection were worked up by P. P. C. Hoek (1883, 1884), who had interests in anatomy and ecology as well as systematics. Within the ubiquitous deep-sea genus *Scalpellum,* Hoek (figure 1.3D) discovered dioecious and androdioecious species (Charnov 1987) in which complemental males colonized the region near the oviducal opening of much larger females or hermaphrodites. In some cases, Hoek found and illustrated the cyprid larvae which metamorphosed into these complemental males. Although he was not able to determine if cyprids actually disperse in the water column, the larvae he pictured look similar in most respects to the cyprids of other barnacles which most certainly do, including scalpellids from shallower fjordic waters (Svane 1986). Hoek observed both nauplius and cyprid stages of a specimen of *Scalpellum stroemii* dredged from about 1000 m depth during the 1882 cruise of HMS *Triton.* He noted:

> On opening it, I found it contained eggs. They were large and not extremely numerous. On studying them with the microscope I found they had passed already the Nauplius-stage, and had arrived at the Cypris-stage. The exuviae of the Nauplius-larva still adhered to the covering of the Cypris, though it was not easy to make out which parts had developed from the Nauplius-appendages.
>
> (Hoek 1884:75)

Because the nauplius exuvium was still attached to the cypris in the mantle cavity of the female, Hoek assumed that the naupliar stage does not disperse in the plankton. It is also possible, however, that the cyprid stage is very short lived.

Hoek also described the *Challenger* pycnogonids and provided the first thorough description of reproduction in this group. His systematic treatment includes observations on numerous deep-sea species, including the giant abyssal pycnogonids in the genus *Collosendis,* but he unfortunately chose a shallow-water species of *Nymphon* for his exhaustive description of embryology (Hoek 1881), assuming that "in general the deep-sea animals will exhibit the same anatomical structure, and pass through the same development as the littoral or shallow-water forms" (p. 100).

Cnidaria

H. N. Moseley's descriptions of the morphology of deep-sea hydrocorals include numerous illustrations of the planula larvae prior to release (Moseley 1881). For example, he illustrated in detail various stages of embryogenesis and planula development in the gonophores of *Errina labiata* and *Cryptohelia pudica,* as well as spermatogenesis in *Astylus subviridis* and *Sporadophora dichotoma.* His illustrations make it clear that many of the deep-sea hydrocorals reproduce in virtually the same fashion as shallow-water relatives. Although the text (Moseley 1881) does not indicate whether planulae crawl away or swim, Moseley (1880) expressed the opinion elsewhere that the larvae were pelagic. Little information is provided in the *Challenger Reports* on the reproduction of deep-water scleractinians or octocorals (Moseley 1881), but monographs on hydrozoans (Allmann 1883) and pennatulids (Kolliker 1880) contain illustrations of gametes.

An important contribution to our understanding of anthozoan reproduction came from the first of two monographs on Actiniarians by Richard Hertwig (1882). Using well-preserved specimens of *Corallimorphus rigidus* collected from 1375–1950 fathoms, Hertwig (figure 1.3E) observed and illustrated large yolky eggs residing in the entoderm and migrating into the mesoglea. This migration of the ova was also observed in the symbiotic anemone *Calliactis parasitica* from relatively shallow water. It is now known to be a general phenomenon in the cnidaria, though its discovery is often attributed to later German scientists (Campbell 1974).

Ascidiacea

Deep-sea ascidians from the *Challenger* expedition were monographed by William Herdman (figure 1.3F) in 1882 and 1886, though one of the most interesting species, *Hypobythius calycodes* was originally described in a paper by Moseley (1876). Remarkably, Moseley's careful observations of gonads and gametes in this species included a drawing of sperm with eccentric mitochondria. Although the significance of this was not known at the time (Moseley described it as "a small transparent vesicle attached to one side of the head"), the sperm of virtually all ascidians are now known to have this feature. *Hypobythius caly-*

codes has remarkably large eggs (1.5 mm) for a solitary ascidian. It is not known whether this species reproduces by pelagic lecithotrophy or by brooding.

Herdman's (1882) monograph on the simple ascidians included a careful drawing of the oviduct of a ripe individual of *Bathyoncus mirabilis*. The scale of the drawing suggests that this species produces eggs similar in size to those of shallow-water solitary ascidians and releases eggs and sperm near the excurrent siphon, where they are probably spawned to undergo a pelagic lecithotrophic larval development not unlike that of most other solitary ascidians. Similar anatomical evidence for free spawning was given for several other species as well.

Among the compound ascidians, Herdman (1886) found larvae in *Cystodytes draschii* dredged from 400 fathoms, *Polyclinum molle* (600 fathoms), *Amaroucium recumbeus* (245 fathoms), *Leptoclinum tenue* (to 600 fathoms) and *Coelocormus huxleyi* (600 fathoms). In the latter species, tadpoles were embedded in a section of the test containing no adult zooids. Ascidian tadpoles can possess either two pigmented sensory structures (ocellus, statocyst), a single statocyst, or a compound sensory organ known as a statolith (reviewed by Svane and Young 1989). Of the colonial ascidian larvae observed by Herdman, two had both an ocellus and a statocyst, and two had only a single sensory organ, probably a statocyst.

OTHER PHYLA

The *Challenger Reports* contain isolated observations about reproduction in several smaller phyla, but they also fail to give us much information about some large taxa, including polychaetes, bivalves, and brachiopods.

Hubrecht's (1887) treatment of nemerteans provides much detail about gonad structure, as virtually all of the worms in the monograph were described from paraffin sections. However, because he did not give collection depths for the specimens described, most of the information is not easily placed in a deep-sea context.

A female specimen of the echiurid, *Bonellia sumhi,* dredged from a depth of 1340 fathoms off Nova Scotia, had numerous dwarf males attached and contained eggs in the "uterus" (Selenka 1885). The extreme sexual dimorphism in this genus had been described earlier for a shallow-water species (reviewed by Stephen and Edmonds 1972) by Kowalevsky and others. The discovery of a deep-sea species with dwarf males showed that reproduction was similar in this genus regardless of depth, and suggests the occurrence of larvae.

The calcareous sponges described from the *Challenger* collection by Poléjaeff (1883) included five species that contained amphiblastula larvae. However, the deepest of these, *Sycon arcticum,* came from just 100 fathoms; reproduction was not described for any abyssal species. The monograph on Hexactinellida (Schulze 1887) emphasized spicules and contained no useful information on the reproductive process.

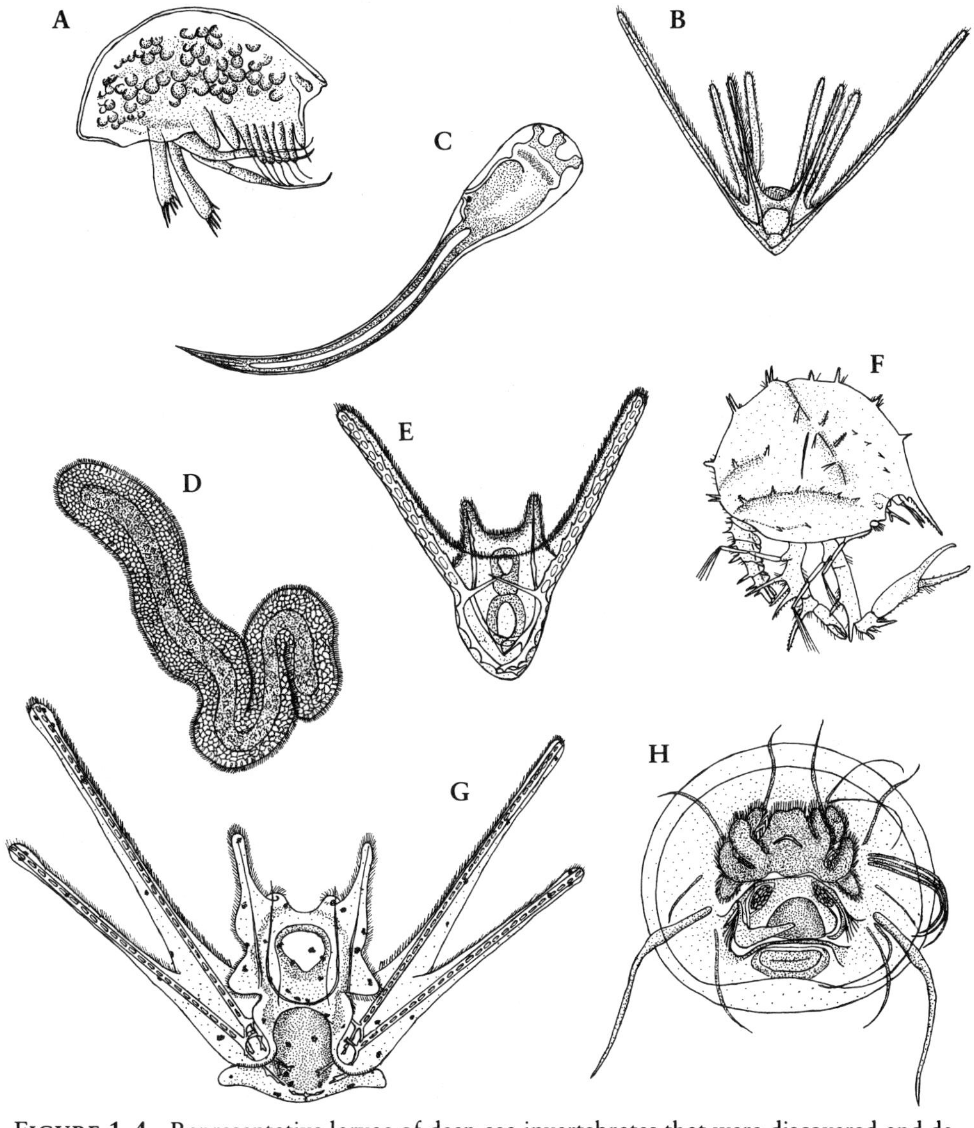

FIGURE 1.4. Representative larvae of deep-sea invertebrates that were discovered and described before 1936. A: Cypris larva of the barnacle *Scalpellum stroemi* (after Hoek 1883). B: "*Ophiopluteus ramosus,*" the larva of *Ophiocten gracilis* (after Mortensen 1889). C: Tadpole larva of the ascidian *Cystodytes draschii* (after Herdman 1886). D: Planula larva of *Cryptohelia affinis* (after Moseley 1880). E: Echinopluteus larva of the echinoid *Lagauum diplopora* (after Mortensen 1921). F: Larva of *Polycheles sp.* (after Seibie 1914). G: Echinopluteus larva of the echinoid *Cidaris cidaris* (after Prouho 1988). H: Larva of the inarticulate brachiopod *Pelagodiscus atlanticus* (after Ashworth 1915).

Summary

It is clear that much evidence for pelagic larvae was available to Thorson when he formulated "Thorson's Rule" in 1936. Admittedly, much of it is circumstantial, some is based on opinion, and some is found in obscure places. Nevertheless, even a perusal of the *Challenger Reports* would have placed reasonable doubts on the strong generalization that Thorson (1950) made. Figure 1.4 illustrates some of the larvae of deep-sea animals that were observed and drawn prior to 1936. To summarize, pre-Thorson evidence for pelagic larval development in the deep sea includes:

1. Planktotrophic larvae of three bathyal echinoids, *Cidaris cidaris, Laganum diplopora,* and *Echinus acutus* reared in the laboratory (Prouho 1888; Shearer et al. 1914; Mortensen 1921).
2. Planktonic echinoderm larvae collected in the plankton, *Auricularia nudibranchiata, Stellophaera mirabilis,* and *Ophiopluteus ramosus* suspected of originating in the deep sea (Chun 1887; Mortensen 1898, 1921).
3. A number of deep-sea echinoderms with small eggs, suggesting planktotrophic development (Mortensen 1921).
4. Anatomical evidence for free spawning in elasipod holothurians (Theel 1886).
5. Larvae of the deep-sea inarticulate brachiopod *Pelagodiscus atlanticus* captured in shallow water (Ashworth 1915).
6. Larvae of unidentified articulate brachiopods captured at bathyal and abyssal depths (Simroth 1897; Eichler 1911).
7. Deep-sea Cheilostome bryozoans with ovicells, suggesting reproduction by coronate larvae (Busk 1884).
8. Protoconchs of deep-sea gastropods illustrated in taxonomic works (Watson 1886; Dautzenberg and Fisher 1896) suggest planktonic development.
9. A blind megalopa larva collected from a depth of 1675 m (Willemöes-Suhm 1876). Eyed crustacean larvae produced by blind adults, suggesting vertical migration (Moseley 1880).
10. Post-larva of eryonid crab *Polycheles* is described (Selbie 1914).
11. Male cyprids and nauplii of deep-sea barnacles in the genus *Scalpellum* (Hoek 1883).
12. Planula larvae of various hydrocorals (Moseley 1881) and hydrozoans (Allman 1883) illustrated.
13. Anatomical evidence and small eggs indicate free spawning and external development in solitary ascidians (Herdman 1882).
14. Larvae are observed in at least 5 species of abyssal compound ascidians (Herdman 1886). Larvae of two species are probably blind.
15. Dwarf males in abyssal echiurid suggest larval development (Selenka 1885).
16. Amphiblastulae found in calcareous sponges from bathyal depths (Poléjaeff 1883).

A major problem of interpretation that led to the common acceptance of Thorson's Rule was the assumption that large eggs are always indicative of brooding. In actual fact, brooders generally (though not always) do have large eggs, but many animals with large eggs produce pelagic lecithotrophic larvae (Pearse, this volume). It seems unlikely that several generations of scientists would have made this error independently. The staying power of Thorson's Rule in the community of deep-sea biologists can instead be attributed to tradition. We tend not to question the opinions of prominent figures, like Thorson, who cast long shadows in a particular field. Since Thorson essentially "wrote the book" (Thorson 1946) on larval ecology, many biologists have probably assumed that his facts were correct and that his literature searches were thorough.

In this review I have emphasized the contributions of the *Challenger Reports* because of their general accessibility and their historical significance. I have also devoted a disproportionate amount of space to the Echinodermata because they were the organisms that provided the original impetus for the great dogma that deep-sea animals brood. Many additional observations on reproduction in the deep sea are scattered through the English, Swedish, Dutch, German, French and Danish expedition reports, but space does not permit their inclusion in this brief review. If one were to assemble *all* of the existing information about deep-sea reproductive patterns in one place, it would certainly paint a patchwork mosaic of reproductive diversity rather than a monotonous, general pattern to which all species conform.

Acknowledgments

This work would not have been possible without access to excellent libraries staffed with helpful librarians. In particular, I thank the staffs of the M.B.L. library in Woods Hole, the library of the Natural History Museum in London, and the Harbor Branch Oceanographic library in Florida. Kristen Metzger and Marilyn Morris of H.B.O.I. always responded cheerfully when I requested an interlibrary loan of yet another obscure paper. In London, Brian Gardner and Roland Emson helped me find portraits and obituaries, Andrew Smith provided a space in his office where I could copy photos, Nigel Merritt pointed me in the direction of the *Challenger* archives, and Gordon Patterson allowed me to see, handle, and photograph some of the brooding echinoderms collected during the *Challenger* expedition. Access to Mortensen's specimens was provided by Margit Jensen at the Zoologisk Museum in Copenhagen. Torben Wolff, Tony Rice, Bob Hessler, Paul Tyler, and John Pearse provided historical insights and stimulating conversation about the development of ideas. John Gage directed me to important landmarks in Edinburgh and provided access to the library at Dunstaffnage Marine Laboratory in Oban. Professor Jarl-Ove Stromberg provided access to the library and pleasant working conditions at the Kristineberg Marine Biological Station in Sweden. I thank Cate Hurlbut for commenting on a draft manuscript, Barbara Heglund and Dan McCarthy for editorial assistance, and Patricia Richards for help with library research. Travel for this research was supported by NATO collaborative research grant CRG-900628 and by NSF grants OCE-8916264 and OCE-9116560. This is Harbor Branch Contribution Number 1009.

REFERENCES

Agassiz, A. 1881. Echinoidea. *Rep. Sci. Res. Voy. H.M.S. Challenger* Zoology 3(9):1–321, 65 pls.

Allmann, G. J. 1883. Hydroida, Part 1. *Rep. Sci. Res. Voy. H.M.S. Challenger* Zoology 7(20):1–54, 20 pls.

Ashworth, J. H. 1915. On larvae of *Lingula* and *Pelagodiscus*. *Trans. Roy. Soc. Edinburgh* 51:45–70.

Birkeland, C. E., F. S. Chia, and R. R. Strathmann. 1971. Development, substratum selection, delay of metamorphosis and growth in the seastar *Mediaster aequalis* Stimpson. *Biol. Bull.* 141:99–108.

Busk, G. 1884. Polyzoa, Part 1. The Cheilostomata. *Rep. Sci. Res. Voy. H.M.S. Challenger* Zoology 10(30):1–47, 6 pls.

Busk, G. 1886. Polyzoa, Part 2. The Cyclostomata, Ctenostomata, and Pedicellinea. *Rep. Sci. Res. Voy. H.M.S. Challenger* Zoology 17(l):1–47, 10 pls.

Cameron, J. L., F. S. McEuen, and C. M. Young. 1988. Floating lecithotrophic eggs from the bathyal echinothuriid sea urchin *Araeosoma fenestratum*. In R. D. Burke, P. V. Mladenov, P. Lambert, and R. L. Parsley, eds., *Echinoderm Biology*, pp. 177–80. Rotterdam: Balkema.

Campbell, R. D. 1974. Cnidaria. In A. C. Giese and J. S. Pearse, eds., *Reproduction of Marine Invertebrates*, vol. 1: *Acoelomate and Pseudocoelomate Metazoans*, pp. 133–199. New York: Academic Press.

Carney, R. S., R. L. Haedrich, and G. T. Rowe. 1983. Zonation of fauna in the deep sea. In G. T. Rowe, ed., *The Sea*, 8:371–398. New York: Wiley-Interscience.

Carpenter, P. H. 1884. Crinoidea, Part 1. *Rep. Sci. Res. Voy. H.M.S. Challenger* Zoology 11(31):1–442, 69 pls.

Charnov, E. L. 1987. Sexuality and hermaphroditism in barnacles: a natural selection approach. In A. J. Southward, ed., *Barnacle Biology*, pp. 89–103. Rotterdam: Balkema.

Chun, C. 1887. *Die pelagische Thierwelt in grosseren Meerestiefen und ihre Beziehungen zu der Oberflachenfauna*. Bibliotheca Zoologica, No. 1. Cassel: Theodor Fischer, 72 pp.

Clarke, A. H., Jr. 1962. Annotated list and bibliography of the abyssal marine molluscs of the world. *Nat. Mus. Can. Bull.* 181:1–114.

Crisp, D. J. 1986. A comparison between the reproduction of high and low latitude barnacles, including *Balanus balanoides* and *Tetraclita pacifica*. In M. J. Thompson, R. Sarojini, and R. Nagabhushanam, eds., *Biology of Benthic Marine Invertebrates*, pp. 69–84. New Delhi: Oxford and IBH Publishing.

Dautzenberg, P. and H. Fisher. 1896. Dragages effectés par "L'Hirondelle" et par la "Princesse-Alice," 1888–1895. Mollusques gastéropodes. *Mémoires de la Société Zool. de France* 9:395–498.

Dautzenberg, P. and H. Fisher. 1897. Dragages effectés par "L'Hirondelle" et par la "Princesse-Alice," 1888–1896. *Mémoires de la Société Zool. de France* 10:139–234.

Davidson, T. 1880. Brachiopoda. *Rep. Sci. Res. Voy. H.M.S. Challenger* Zoology 1(1):1–67, 4 pls.

Doflein, F. 1904. Brachyura. Wiss. Ergebn. "Valdivia." 6:1–314.

Eichler, P. 1911. Die Brachiopoden der deutschen Südpolar-Exped. *D. Südp.-Exped.* Bd. xii. Zool. iv. Bd., pp. 384–385.

Emlet, R. B. 1989. Apical skeletons of sea urchins (Echinodermata: Echinoidea): Two methods for inferring mode of larval development. *Paleobiology* 15:223–254.

Emlet, R. B., L. R. McEdward, and R. R. Strathmann. 1987. Echinoderm larval ecology viewed from the egg. In M. Jangoux and J. M. Lawrence, eds., *Echinoderm Studies 2*, pp. 55–136. Rotterdam: Balkema.

Forbes, E. 1844. Report on the Mollusca and Radiata of the Aegean Sea, and their distribution, considered as bearing on geology. *Report (1843) to the 13th meeting of the British Association for the Advancement of Science,* pp. 30–193.

Gage, J. D. and P. A. Tyler. 1991. *Deep-Sea Biology: A Natural History of Organisms at the Deep-Sea Floor.* Cambridge: Cambridge University Press, 504 pp.

George, R. Y. and R. J. Menzies. 1967. Indication of cyclic reproductive activity in abyssal organisms. *Nature* 215:878.

George, R. Y. and R. J. Menzies. 1968. Further evidence for seasonal breeding cycles in deep sea. *Nature* 220:80–81.

Giese, A. C. and H. Kanatani. 1988. Maturation and spawning. In A. C. Giese, J. S. Pearse, and V. B. Pearse, eds., *Reproduction of Marine Invertebrates,* vol. 8: *General Aspects: Seeking Unity from Diversity,* pp. 251–329. Palo Alto and Pacific Grove, Calif.: Blackwell/Boxwood.

Gurney, R. 1939. *Bibliography of the Larvae of Decapod Crustacea.* London: Ray Society, 306 pp.

Hansen, B. 1968. Brood protection in the deep-sea holothurian *Oneirophanta mutabilis* Theel. *Nature* 217:1062–1063.

Harrison, K. 1988. Seasonal reproduction in deep-sea Crustacea (Isopoda: Asellota). *J. Nat. Hist.* 22:175–197.

Henderson, J. R. 1888. Anomura. *Rep. Sci. Res. Voy. H.M.S. Challenger* Zoology 27(59):1–221, 21 pls.

Herdman, W. A. 1882. Tunicata, Part 1. *Rep. Sci. Res. Voy. H.M.S. Challenger* Zoology 6(17):1–296, 37 pls.

Herdman, W. A. 1886. Tunicata, Part 2. *Rep. Sci. Res. Voy. H.M.S. Challenger* Zoology 14(38):1–432, 49 pls.

Herdman, W. A. 1923. *Founders of Oceanography and Their Work.* London: Edward Arnold, 340 pp.

Hertwig, R. 1882. Actiniaria. *Rep. Sci. Res. Voy. H.M.S. Challenger* Zoology 6(15):1–136, 14 pls.

Hoek, P. P. C. 1881. Pycnogonida. *Rep. Sci. Res. Voy. H.M.S. Challenger* Zoology 3(10):1–167, 21 pls.

Hoek, P. P. C. 1883. Cirripedia (Systematic Part). *Rep. Sci. Res. Voy. H.M.S. Challenger* Zoology 8(25):1–169, 13 pls.

Hoek, P. P. C. 1884. Cirripedia (Anatomic Part). *Rep. Sci. Res. Voy. H.M.S. Challenger* Zoology 10(27):1–47, 6 pls.

Hubrecht, A. A. W. 1887. Nemertea. *Rep. Sci. Res. Voy. H.M.S. Challenger* Zoology 19(54):1–150, 16 pls.

Hurley, P. J. 1988. *A Concise Introduction to Logic.* Belmont, Calif.: Wadsworth.

Kölliker, A. von. 1880. Pennatulida. *Rep. Sci. Res. Voy. H.M.S. Challenger* Zoology 1(2):1–41, 11 pls.

Long, J. A. and S. A. Stricker. 1991. Brachiopods. In A. C. Giese, J. S. Pearse, and V. B. Pearse, eds., *Reproduction of Marine Invertebrates,* vol. 6: *Echinoderms and Lophophorates,* pp. 47–84. Pacific Grove, Calif.: Boxwood Press.

Lyman, T. 1882. Ophiuroidea. *Rep. Sci. Res. Voy. H.M.S. Challenger* Zoology 5(14):1–386, 48 pls.

MacBride, E. W. 1920. Echinodermata (Part II) and Enteropneusta. Larvae of Echinodermata and Enteropneusta. *British Antarctic ("Terra Nova") Expedition, 1910. Natural History Report, Zoology* 4:83–94. British Museum (Natural History), London.

Madsen, F. J. 1961. On the zoogeography and origin of the abyssal fauna. *Galathea Report* 4:177–218.

Marshall, N. B. 1979. *Developments in Deep-Sea Biology.* Poole, Dorsett: Blandford Press.

Menzies, R. J., R. Y. George, and G. T. Rowe. 1973. *Abyssal Environment and Ecology of the World Oceans*. New York: Wiley-Interscience, 488 pp.

Mileikovsky, S. A. 1971. Types of larval development in marine bottom invertebrates, their distribution and ecological significance. *Mar. Biol.* 10:193–213.

Mills, E. L. 1983. Problems of deep-sea biology: an historical perspective. In G. Rowe, ed., *The Sea,* 8:1–77. New York: Wiley-Interscience.

Milne-Edwards, H. and E. L. Bouvier. 1897. Descriptions des Crustacés de la Famille des Galathéidé recueillis pendant l'Expédition. *Mem. Mus. Comp. Zool. Harvard* 19:1–141.

Mortensen, T. 1898. Die Echinodermenlarven der Plankton-Expedition bebst einer systematischen Revision der bisher bekannten Echinodermenlarven. *Ergebn. Plankton-Exped.* 2j:1–120.

Mortensen, T. 1921. *Studies of the Development and Larval Forms of Echinoderms*. Copenhagen: G. E. C. Gad, 266 pp.

Mortensen, T. 1927. *Handbook of the Echinoderms of the British Isles*. Oxford: Oxford University Press, 471 pp.

Moseley, H. N. 1876. On deep-sea ascidians &c. *Linn. Soc. Trans.* ser. ii. 1:287.

Moseley, H. N. 1880. Deep-sea dredgings and life in the deep sea. *Nature* 21:543–547, 569–572, 591–593.

Moseley, H. N. 1881. Hydroid, Alcyonarian, and Madreporarian Corals. *Rep. Sci. Res. Voy. H.M.S. Challenger* Zoology 2(7):1–248, 32 pls.

Orton, J. H. 1920. Sea temperature, breeding, and distribution in marine animals. *J. Mar. Biol. Assoc. U. K.* 12:339–366.

Pelseneer, P. 1888. Anatomy of the deep-sea Mollusca. *Rep. Sci. Res. Voy. H.M.S. Challenger*. Zoology 27(74):1–42, 4 pls.

Poléjaeff, N. 1883. Calcarea. *Rep. Sci. Res. Voy. H.M.S. Challenger* Zoology 7(24):1–76, 9 pls.

Popper, K. R. 1959. *The Logic of Scientific Discovery*. New York: Basic Books, 479 pp.

Prouho, H. 1888. Recherches sur le *Dorocidaris papillata* et quelques autres échinides de la Méditerranée. *Arch. Zool. expér. et génér.* (2) v. 1888:213–380.

Rokop, F. J. 1974. Reproductive patterns in deep-sea benthos. *Science* 186:743–745.

Rokop, F. J. 1977. Seasonal reproduction of the brachiopod *Frieleia halli* and the scaphopod *Cadulus californicus* at bathyal depths in the deep sea. *Marine Biology* 43:237–246.

Scheltema, R. S. 1972. Reproduction and dispersal of bottom dwelling deep-sea invertebrates: A speculative summary. In R. W. Bauer, ed., *Barobiology and Experimental Biology of the Deep Sea,* pp. 58–68. Chapel Hill: University of North Carolina Press.

Schoener, A. 1967. Evidence for reproductive periodicity in the deep-sea. *Ecology* 49:81–87.

Schulze, F. E. 1887. Hexactinellida. *Rep. Sci. Res. Voy. H.M.S. Challenger* Zoology 21(53):1–513, 104 pls.

Selbie, C. M. 1914. The Decapoda Reptantia of the coasts of Ireland. Part I: Palinura, Astacura, and Amomura (except Paguridea). *Sci. Invest. Fish. Br. Ireland.* 1914 i, 116 pp.

Selenka, E. 1885. Gephyrea. *Rep. Sci. Res. Voy. H.M.S. Challenger* Zoology 13(36):1–24, 4 pls.

Semenova, T. N., S. A. Mileikovsky, and K. N. Nesis. 1964. The morphology, distribution, and seasonal incidence of the ophiuroid larva of *Ophiocten sericium* (Forbes) in the northwest Atlantic, Norwegian Sea, and Barents Sea. *Okeanologiya* 4:669–683 (in Russian).

Shearer, C., W. DeMorgan, and H. M. Fuchs. 1914. On the experimental hybridization of echinoids. *Phil. Trans. Roy. Soc. B.* 204:255–362, 7 pls.

Simroth, H. 1897. Die Brachiopoden der Plankton-Exped. *Ergebn. Plankton-Exped.* Bd. ii, pp. 1–19.

Sladen, W. P. 1889. Asteroidea. *Rep. Sci. Res. Voy. H.M.S. Challenger* Zoology 30(51):1–893, 118 pls.

Smiley, S., F. S. McEuen, C. Chaffee, and S. Krishnan. 1991. Echinodermata: Holothuroidea. In A. C. Giese, J. S. Pearse, and V. B. Pearse, eds., *Reproduction of Marine Invertebrates,* vol. 6: *Echinoderms and Lophophorates,* pp. 663–750. Pacific Grove, Calif.: Boxwood Press.

Stephen, A. C. and S. J. Edmonds. 1972. *The phyla sipuncula and echiura.* Trustees of the British Museum (Natural History), London, 528 pp.

Svane, I. 1986. Sex determination in *Scalpellum scalpellum* (Cirripedia: Thoracida: Lepadomorpha), a hermaphroditic goose barnacle with dwarf males. *Mar. Biol.* 90:249–253.

Svane, I. and C. M. Young. 1989. The ecology and behaviour of ascidian larvae. *Oceanogr. Mar. Biol. Ann. Rev.* 27:45–90.

Theel, H. 1882. Holothuroidea, Part 1. *Rep. Sci. Res. Voy. H.M.S. Challenger* Zoology 4(13):1–176, 46 pls.

Theel, H. 1886. Holothuroidea, Part 2. *Rep. Sci. Res. Voy. H.M.S. Challenger* Zoology 14(39):1–290, 16 pls.

Thomson, C. W. 1874. *The Depths of the Sea.* London: Macmillan, 527 pp.

Thomson, C. W. 1878. Notice of some peculiarities in the mode of propagation of certain echinoderms of the Southern Sea. *J. Linn. Soc. Zool.* 13:55–79.

Thomson, C. W. 1880. General introduction to the zoological series of reports. *Rep. Sci. Res. Voy. H.M.S. Challenger* Zoology 1:1–62.

Thorson, G. 1936. The larval development, growth, and metabolism of Arctic marine bottom invertebrates. *Medd. Grnland* 100:1–55.

Thorson, G. 1946. Reproduction and larval development of Danish marine bottom invertebrates. *Medd. Komn. Danm. Fisk. of Havunders., Ser. Plankton* 4:1–523.

Thorson, G. 1950. Reproductive and larval ecology of marine bottom invertebrates. *Biol. Rev.* 25:1–45.

Tizard, T. H., H. N. Moseley, J. Y. Buchanan, and J. Murray. 1885. Narrative of the cruise of H.M.S. *Challenger,* with a general account of the scientific results of the expedition. *Rep. Sci. Res. Voy. H.M.S. Challenger* Narrative I:1–1110, 75 pls.

Tyler, P. A. 1988. Seasonality in the deep sea. *Oceanogr. Mar. Biol. Ann. Rev.* 26:227–258.

Tyler, P. A. and J. D. Gage. 1982. *Ophiopluteus ramosus,* the larval form of *Ophiocten gracilis* (Echinodermata: Ophiuroidea) from the Rockall Trough. *J. Mar. Biol. Assoc. U. K.* 62:485–486.

Tyler, P. A. and J. D. Gage. 1984. The reproductive biology of echinothurid and cidarid sea urchins from the deep sea (Rockall Trough, northeast Atlantic Ocean). *Mar. Biol.* 80:63–74.

Tyler, P. A., A. Grant, S. L. Pain, and J. D. Gage. 1982. Is annual reproduction in deep-sea echinoderms a response to variability in their environment? *Nature* 300:747–749.

Vance, R. R. 1973. On reproductive strategies in marine benthic invertebrates. *Am. Nat.* 107:339–352.

Watson, R. B. 1886. Scaphopoda and Gasteropoda. *Rep. Sci. Res. Voy. H.M.S. Challenger* Zoology XV(xlii):1–756, 53 pls.

Willemöes-Suhm, R. von. 1876. Preliminary report to Prof. Wyville Thomson, F.R.S., director of the civilian scientific staff, on observations made during the earlier part of the voyage of H.M.S. *Challenger. Proc. Roy. Soc. Lond.* 24:569–585.

Wolff, T. 1967. *Danish Expeditions on the Seven Seas.* Copenhagen: Rhodos.

Young, C. M. and J. L. Cameron. 1987. Laboratory and *in situ* flotation rates of lecithotrophic eggs from the bathyal echinoid *Phormosoma placenta. Deep-Sea Res.* 34:1629–1639.

CHAPTER TWO

Cold-Water Echinoderms Break "Thorson's Rule"

John S. Pearse

ABSTRACT

A major paradigm in marine biology links polar and deep-sea habitats with nonpelagic development by benthic marine invertebrates. The idea originated in the last century and was developed by Gunnar Thorson in the first half of this century to become known as "Thorson's Rule." This hypothesis never had strong general supporting evidence. It arose from misinterpretations of available observations and persisted because it could be so readily "explained" in terms of the supposed hazards to pelagic larval life in cold-water habitats. It is false for many groups, including echinoderms; over 70 percent of the echinoderms of northeast Greenland, the deep-sea of the northeast Atlantic, and the high Antarctic produce pelagic larvae. Although most of these larvae are nonfeeding lecithotrophs, some of the most abundant species in polar and probably deep seas produce planktotrophic larvae similar to those in temperate and tropical regions. These conclusions highlight a growing recognition that simple generalizations about complex ecological phenomena are likely to be both inadequate and misleading. Thorson's Rule should be laid to rest, and Thorson should be remembered for his stimulating hypotheses that generated so many contributions in marine biology.

> *"Natural history tells us unequivocally that we are foolish to look for general answers to specific questions about how organisms perform. . . . At the organismic level the questions are apt to be general and the answers specific."* (Bartholomew 1986:328)

The idea that faunas of the deep and polar seas are similar arose nearly 150 years ago when Sir James Clark Ross noted that animals from the Antarctic coast resembled those he had collected during earlier Arctic expeditions. Ross (1847) suggested that the polar faunas might be continuous with those across the floor of the deep sea, provided sea temperatures remained low—a condition later confirmed by the great oceanographic expeditions of the second half of the nineteenth century (Buchan 1895). This idea, at least in general terms, gained wide credence by the beginning of the twentieth century and remains to the present; the morphological similarity among arctic, deep-sea, and antarctic species is recognized as the result of similar adaptations and perhaps common ancestry (Murray 1895; Lipps and Hickman 1982; Gage and Tyler 1991). These similarities mirror major parallels in environmental conditions. Both deep and polar seas generally remain within a few degrees of freezing, are devoid of most sunlight much or all of the time, and depend mainly on primary production elsewhere (surface layers for the deep-sea, offshore, and subpolar regions for polar shores).

Consequently, it is not unreasonable to suppose that biological processes also might be similar for deep-sea and polar species. One such process is reproduction. Thomson (1878, 1885) reported finding several echinoderms in subantarctic waters that were "peculiar" because they brooded their young and had no larval stage. These observations, and the apparent lack of larvae in polar plankton, led Murray (1894:81) to write "Polar animals and deep-sea animals have all a direct development (so also fresh-water animals)." This notion was picked up by later workers (e.g., Giard 1905; Mortensen 1913, 1921), and Thorson (1936), reinforced by his own studies in east Greenland, similarly concluded that (p. 128) "in some areas outside the Arctic, too, a nonpelagic larval development is the rule. . . . [these] are 1) the deep oceans, 2) the antarctic seas, and 3) areas with brackish water (and fresh water)." He further proposed that nonpelagic development among Arctic, Antarctic, and deep-sea faunas was a response to the same set of environmental conditions: low temperature and poor food conditions for pelagic larvae. Thorson restated these generalizations repeatedly in his 1946 and 1950 reviews with little additional information; in 1961 he wrote (p. 456) "As is well known, nearly all invertebrate species living in Arctic and Antarctic coastal zones as well as all inhabitants of the deep sea seem to have a nonpelagic larval development." Mileikovsky (1971) coined this generalization "Thorson's Rule," and as recently as 1986 Crisp considered it "the one valid generalization" concerning patterns of marine invertebrate life histories. Vance (1973) was the first to attempt to explain this "known geographic trend in developmental patterns" by mathematical models, largely supporting Thorson's intuition, and others refined and extended his efforts (e.g., Christiansen and Fenchel 1979; Grant 1983; Roughgarden 1989).

Thorson (1936) recognized four types of sexual reproduction by marine invertebrates: pelagic planktotrophy, pelagic lecithotrophy, nonpelagic lecithotrophy, and viviparity which he considered did "not differ essentially with" nonpelagic lecithotrophy. These categories have been modified by several subsequent workers (summarized by Turner et al. 1986). Shuto (1974) and Jablonski and Lutz (1983), for example, distinguished only planktotrophy and lecithotrophy (= nonplanktotrophy for Jablonski and Lutz 1983). In both cases, both pelagic and nonpelagic lecithotrophy were combined, in part, because nonfeeding development in gastropods is indicated by large eggs with stored materials, and egg size (or resulting gastropod larval shell size) can be easily determined from preserved specimens. Turner et al. (1986) also divided developmental mode first into planktotrophy and lecithotrophy (= nonplanktotrophy), and then only partly distinguished pelagic development in several subcategories of lecithotrophy.

Thorson (1936) himself may have contributed to the ambiguous mixing of pelagic and nonpelagic lecithotrophic modes, when he mistakenly followed Mortensen (1910, 1927a) in using egg size to determine whether development was pelagic or not. Nevertheless, Thorson (1936, 1946, 1950, 1961) explicitly and consistently distinguished the two nonequivalent dichotomies in modes of development: pelagic versus nonpelagic and feeding (planktotrophic) versus nonfeeding (lecithotrophic or nonplanktotrophic). Furthermore, although he considered nonpelagic development to be always nonfeeding, he recognized both pelagic planktotrophic and pelagic lecithotrophic development. He believed that pelagic lecithotrophic larvae were generally of minor importance worldwide and "absent from the arctic and antarctic coastal waters and probably also from the deep sea" (Thorson 1946:437).

Roughgarden (1989) explained Thorson's implied gradient in pelagic lecithotrophic development, increasing from the poles to the equator, as the result of either more intense benthic predation in the tropics, selecting against benthic development, or increased pressure for dispersal in the tropics. It is now apparent that no explanation is needed; Thorson was wrong both in general and in particular regarding polar and deep-sea echinoderms, in which pelagic lecithotrophic development is the major mode of development.

In this paper I re-examine Thorson's data on echinoderms and then review studies done over the past several decades. Thorson's generalization regarding mode of development is shown to be invalid for polar and deep-sea echinoderms, as is probably also the case for other groups in these habitats (e.g., deep-sea bivalves and gastropods, Lightfoot et al. 1979; Rex et al. 1979; Rex and Warén 1982; Jablonski and Lutz 1983). Its demise may be disappointing to those of us who have used it as a convenient backdrop for both empirical and theoretical studies. However, recognizing its failure may permit us to deal more rigorously with complex biological patterns that reflect an accumulation of species-specific natural histories.

Echinoderms are especially suitable for such an evaluation because they are widespread, diverse, abundant, extensively studied, and still considered good examples of Thorson's Rule (e.g., Lawrence 1987). Much of the work on them has been reviewed recently (deep sea: Tyler 1986, 1988; Gage and Tyler 1991; Antarctic: Bosch and Pearse 1990; Pearse et al. 1991a). Indeed, some of the earliest examples in support of Thorson's generalization were echinoderms—and some of the first cracks in it, in both polar and deep-sea habitats, came from studies on echinoderms.

Thorson's Data Re-Examined

Thorson (1936) reported the mode of reproduction for 24 of the approximately 40 species of echinoderms known from the fjords of northeast Greenland. One, the asteroid *Stephanasterias albula,* reproduces solely by asexual fission (see Mladenov et al. 1986) while the others are sexual. Of these 23 species, Thorson recognized 5 that produced pelagic larvae, 4 of which are also planktotrophic. Three of these species, the ophiuroids *Ophiocten sericeum* and *Ophiura robusta,* and the echinoid *Strongylocentrotus droebachiensis* were "extremely common as bottom animals in the fjords" (p. 20), and their pluteus larvae were found in the plankton during the summer. Thus some of the commonest echinoderms in the northeast Greenland fjords were recognized by Thorson as exceptions to what later became Thorson's Rule. Thorson (1936) discounted the relevance of these species because they all have "an arctic-boreal or still wider distribution" (p. 24) while he believed that most of the remaining species had "a 'purely' arctic distribution," thereby somehow being more representative of the Arctic.

Thorson (1936) recognized only one species in northeast Greenland with pelagic lecithotrophic development, the holothuroid *Cucumaria frondosa*; he downplayed its relevance not only because it has a widespread arctic-boreal distribution but also apparently because in northeast Greenland it was known only for the outer coast. Thorson listed 12 other species that produced large eggs (up to 800 μm diameter—his published measurements above 1 mm were obviously off by a factor of 10). These he considered to have nonpelagic, direct development.

It now seems more likely that all 12 northeast Greenland species that Thorson (1936) said had nonpelagic, direct development have long-lived, pelagic lecithotrophic larvae. None of the species is known to brood embryos, and benthic lecithotrophic development is rare in echinoderms. Most species of echinoderms with eggs ranging between about 300 and 1000 μm in diameter produce either brooded embryos or pelagic lecithotrophic larvae that remain in the plankton for weeks to months (Emlet et al. 1987; Pearse and Cameron 1991). One of Thorson's 12 species, the asteroid *Crossaster papposus* with egg diameters of 750–800 μm, already was known to have pelagic lecithotrophic larvae; Gemmill (1920) found that when reared at 10°C, they swam near the surface of culture dishes for 19 days before settling—they probably take much longer to develop at polar temperatures. Two other species, the asteroids *Ctenodiscus crispatus* and *Asterias lincki,* also are now known to produce pelagic lecithotrophic larvae (Falk-Petersen 1982), as most likely does another species, the ophiuroid *Ophiacantha bidentata* (Tyler and Gage 1982). A fifth species, the echinoid *Pourtalesia jeffreysi,* with an egg size of only about 225 μm (Harvey and Gage 1984) to 320–380 μm (Thorson 1936), may even produce a planktotrophic larva (Emlet et al. 1987; Pearse and Cameron 1991).

Of the 6 east Greenland species Thorson (1936) listed as brooders, two are comatulid crinoids that were assumed to brood externally (Mortensen 1910). This is almost certainly incorrect; these two species, like most other comatulids, probably produce pelagic lecithotrophic larvae that sometimes settle on or near adults (Holland 1991). Another of Thorson's brooding species, the asteroid *Pteraster militaris,* is now known to produce pelagic lecithotrophic larvae as well as to brood embryos (McClary and Mladenov 1988), while the other pterasterid on Thorson's list, *Hymenaster pellucidus,* is now known to brood (Clark and Downey 1992), unlike its deep-sea congener *H. membranaceus* (Gage and Tyler 1991). Moreover, Thorson's third asteroid, *Henricia sanguinolenta,* sometimes releases eggs that develop pelagically (Mortensen 1927a); it also may represent a species complex with brooding and non-brooding forms, as has been found for the *Henricia* complex of the northeast Pacific (Strathmann 1987). Regarding distribution, 3 of the 4 confirmed brooding asteroids Thorson listed are widespread arctic-boreal species, and the fourth, *Leptasterias groenlandica,* belongs to a diverse arctic-boreal genus in which all the species brood. There is no reason to believe that brooding or nonpelagic development is at all characteristic of "purely" arctic species of echinoderms as claimed by Thorson.

Mode of Development in the Antarctic

Like the east Greenland echinoderms studied by Thorson (1936), some of the most abundant and conspicuous echinoderms in McMurdo Sound, Antarctica, also produce pelagic planktotrophic larvae: the asteroids *Odontaster validus, O. meridionalis,* and *Porania antarctica* (Pearse and Bosch 1986; Bosch 1988, 1989), the echinoid *Sterechinus neumayeri* (Bosch et al. 1987), and the ophiuroid *Ophionotus victoriae* (I. Bosch and R. Britton, unpublished observations of egg size, about 180 μm diameter). These species are abundant in shallow water around the Antarctic continent but are barely represented in subantarctic waters; they are typical antarctic species.

The mode of development has been determined for 14 of the 16 shallow-water asteroids known in McMurdo Sound (Bosch and Pearse 1990): 3 have pelagic planktotrophic larvae; 8 have pelagic lecithotrophic larvae; and only 3 have nonpelagic development (2 brooders and 1 with an unusual demersal, lecithotrophic larva [Bosch 1989]). Another, *Cuenotaster involutus,* with buoyant eggs about 1 mm in diameter, certainly produces pelagic lecithotrophic larvae (Pearse, unpublished). The pelagic lecithotrophic larvae of one species, *Acodontaster hodgsoni,* took 3.5 months to settle and metamorphose, which is probably typical for species with lecithotrophic larvae (Bosch and Pearse 1990).

Echinoids are much less diverse than the asteroids in McMurdo Sound. In addition to *Sterechinus neumayeri,* with pelagic planktotrophic larvae, there are 2 species of brooding spatangoids (Pearse and McClintock 1990) and a brooding cidaroid (*Ctenocidaris perrieri,* I. Bosch and R. Britton, unpublished). All antarctic species of these latter 2 echinoid orders are known to brood (Emlet et al. 1987), and as Emlet (1990) argues, this may reflect phylogenetic and historic constraints; there are few brooding echinoids in either the arctic or deep-sea. Although a number of deep-water, antarctic crinoids are notable because they brood their embryos (Holland 1991), the one shallow-water crinoid found in McMurdo Sound produces lecithotrophic larvae, as do most other crinoids (McClintock and Pearse 1987). Little is known about reproduction of the ophiuroids or holothuroids in McMurdo Sound. However, on the basis of egg size (about 350 μm diameter), the large ophiuroid *Ophiosparta gigas* probably produces pelagic lecithotrophic larvae (I. Bosch and R. Britton, unpublished observations).

Mode of Development in the Deep Sea

The extraordinarily comprehensive work by Paul Tyler, John Gage, and their colleagues on reproduction of deep-sea echinoderms of the northeast Atlantic, particularly in the Rockall Trough, parallels our work in McMurdo Sound. Two asteroids (*Dytaster grandis* (= *D. insignis)* and *Plutonaster bifrons*), 2 ophiuroids (*Ophiura ljungmani* and *Ophiocten gracilis*), and 4 echinoids (*Echinus affinis, E. alexandri, E. acutus,* and *Cidaris cidaris*) produce eggs about 110–120 μm in diameter that most likely develop into pelagic planktotrophic larvae (Gage and Tyler 1981; Tyler et al. 1982; Tyler and Gage 1984a, b). Two more echinoids, holasteroids with eggs about 225–345 μm in diameter, also probably produce pelagic larvae, perhaps planktotrophs (Harvey and Gage 1984). Another 9 species of asteroids and 6 species of echinoids produce eggs 850–1500 μm in diameter that probably develop into pelagic lecithotrophic larvae (Tyler et al. 1982, Tyler and Gage 1984b). None of these species has been found to brood embryos (Gage and Tyler 1991), including *Hymenaster membranaceus,* which Clark and Downey (1992) consider to be synonymous with *H. pellucidus,* a known brooder.. The holothuroid *Oneirophanta mutabilis* was reported to be an intra-ovarian brooder off Central America (Hansen 1968), but not in the northeast Atlantic (Tyler and Billett 1988). Although brooding has not been found in echinoderms of the Rockall Trough area, some deep-sea species are known to brood, including the echinoids *Hypsiechinus coronatus* of the far-north Atlantic (Mortensen 1903) and *Aporocidaris milleri* of the Pacific (Mortensen 1927b).

Tyler et al. (1982) interpreted egg sizes as did Mortensen (1927a) and Thorson (1936), and suggested that species with eggs about 1 mm in diameter would have "direct development omitting a larval stage." This interpretation was maintained by Tyler and Gage (1984b) for additional species of echinoids, by Tyler and Billett (1988) for several species of holothuroids, and by Tyler (1986, 1988) in his reviews of reproduction of deep-sea echinoderms. However, Gage and Tyler (1991) did recognize that large eggs might produce pelagic larvae when they noted the in situ demonstration of floating lecithotrophic eggs of several echinoids at bathyal depths by Young and Cameron (1987) and Cameron et al. (1988).

It now seems likely that most deep-sea echinoderms with large yolky eggs produce pelagic lecithotrophic larvae that live for months in the plankton, and nonpelagic development, including brooding, is as unusual in the deep-sea as it is elsewhere. That is the case for nearly all antarctic asteroids that similarly produce large eggs (Bosch and Pearse 1990), and as argued above, for most of Thorson's (1936) arctic species. Only a few species of echinoderms are known to produce demersal lecithotrophic larvae: the asteroids *Asterina gibbosa* (Emson and Crump 1979), *A. minor* (Komatsu et al. 1979), and the small form of *Porania antarctica* (Bosch 1989); these are very small asteroids with low fecundities. Other species with nonbuoyant eggs brood their embryos (Emlet et al. 1987, Pearse and Cameron 1991), and brooding should be seen in collected specimens, especially from cold water where brooding can take over six months (Bosch and Pearse 1990). Of course, until live eggs are examined, it cannot be certain that they are buoyant; however, because demersal, nonbrooded eggs seem so exceptional, it is probably best to assume that large lecithotrophic eggs are pelagic until shown otherwise.

It might be argued that the lecithotrophic larvae of deep-sea and polar species should be found in the plankton if they are pelagic. Murray (1895) believed polar species did not have pelagic larvae at all because larvae had not been found in polar plankton collections, yet Thorson (1936) found plenty of pluteus larvae in the plankton of northeast Greenland. Similarly, I used the absence of larvae of the antarctic asteroid *Odontaster validus* from a year-long plankton survey as evidence that the larvae were demersal and even proposed that deep-sea species would produce similar demersal larvae (Pearse 1969), only to find that these larvae are in the plankton, often in considerable abundance (Pearse and Bosch 1986, Bosch et al. 1990). In fact, factors regulating the patterns of distribution and abundance of larvae are poorly known, and larvae can remain within water masses in which they were spawned to drift near the bottom or be widely dispersed (Young and Chia 1987). Deep-sea plankton hauls have failed to collect larvae in areas of the northeast Atlantic with known populations of deep-sea echinoids and asteroids that produce planktotrophic larvae seasonally (Tyler 1986). The chances of collecting lecithotrophic larvae, which are produced in low numbers year around, seems even more remote.

Pelagic Development Prevails in Polar and Deep-Sea Echinoderms

Figure 2.1 compares egg sizes and probable developmental modes known for echinoderms from shallow water of northeast Greenland fjords, Rockall Trough of the deep northeast Atlantic, and shallow water in McMurdo Sound of the high Antarctic, with those in shallow water of Monterey Bay, central California. Nonpelagic, nonfeeding development, largely by brooding and viviparous forms, is found in about 15 to 30 percent of the species at all three shallow-water localities, but is absent from the deep-sea locality. On the other hand, pelagic lecithotrophic development predominates for all three cold-water localities, while pelagic planktotrophy is found in half of the species in the temperate locality. This latter pattern is what theoretical models predict: lecithotrophic

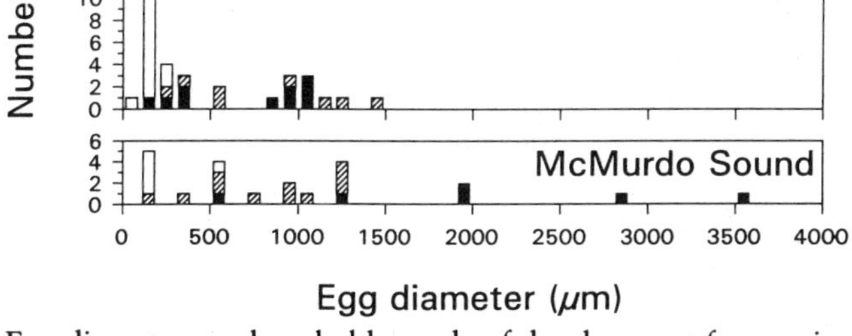

FIGURE 2.1. Egg diameters and probable mode of development for species of echinoderms from four localities. Open: pelagic planktotrophs; hatcheed: pelagic lecithotrophs; solid: benthic larvae of embryos (1 free-living demersal, 550 μm, McMurdo Sound; 4 viviparous, ≤ 400 μm, Monterey Bay; all others: brooders. See appendix for data.

development should be favored when larval food conditions are poor and development is slow, as is the case in cold-water environments (Vance 1973; Christiansen and Fenchel 1979; Roughgarden 1989); Thorson (1950) used the same conditions for his intuitive explanation of the supposed lack of pelagic larvae in polar and deep-sea habitats. Indeed, Thorson's explanation was for the wrong pattern: the contrast is not pelagic versus nonpelagic as he proposed, but feeding versus nonfeeding as recognized by theoreticians.

The question remains why so many cold-water species produce large, yolk-rich eggs that develop into long-lived pelagic lecithotrophic larvae. Lecithotrophic larvae of antarctic asteroids, at least, are very buoyant and must rise to the surface, which may be freezing or melting depending on wind and air temperature. When near the sea surface, the larvae could experience high levels of UV radiation. Deep-sea lecithotrophic larvae that are buoyant may rise thousands of meters, perhaps even crossing abrupt thermoclines. These hazards have been evoked to explain why cold-water species should have nonpelagic larvae. Grant (1983) recognized this problem with long-lived lecithotrophic larvae in general in his parting question (p. 554) "How do species with long-lived lecithotrophic larvae overcome the combined effect of lowered fecundity and high mortality during the larval period?"

The only apparent advantage of having pelagic lecithotrophic larvae seems to be dispersal, which increases outbreeding and chances of colonizing new habitats. Are these processes that important? Christiansen and Fenchel (1979) believed so, stating (p. 279) that the existence of lecithotrophic larvae "can only be explained as an adaptation to become more effective colonizers." Palmer and Strathmann (1981) and Strathmann (1985) argued persuasively against such importance for planktotrophic larvae, which Ghiselin (1987) and Roughgarden (1989), recognizing the problem with pelagic lecithotrophic larvae, only partially countered.

Regardless of the predominance of species with pelagic lecithotrophy in cold-water environments, about 20 to 25 percent of the echinoderm species—among the most abundant in these habitats—produce large numbers of small eggs that develop into pelagic planktotrophic larvae (about the same proportion as those that brood in polar habitats). Moreover, despite the oligotrophic nature of cold-water environments, there is no evidence that larvae in the Antarctic, at least, are food limited (Olson et al. 1987; Bosch et al. 1990; Pearse et al. 1991b). Certainly the fact that lecithotrophic development can occur in eggs only 200 μm in diameter suggests that the material in much larger eggs is not for larval nutrition. Consequently the intuitive and theoretical explanation for the predominance of lecithotrophic development in cold-water habitats—food limitation—also is probably wrong. Other factors, such as rate of development at low temperatures (Emlet et al. 1987) or juvenile size at metamorphosis and habitat complexity, including potential larval predators (Ghiselin 1987; Pearse et al. 1991a), may be more important for favoring pelagic lecithotrophy in cold-water environments. If we follow Bartholomew's (1986) wisdom, all these explanations, plus others, may apply, depending on the history and ecology of the particular species involved.

Acknowledgments

I appreciate the thoughtful comments, criticism, and encouragement from L. V. Basch, I. Bosch, A. Clarke, V. B. Pearse, P. A. Tyler, C. M. Young, and two anonymous reviewers. Both reviewers objected to my title, suggesting that I was setting up a straw man. I suggest they re-read Thorson's papers. My aim is to call sharp attention to a persistent false paradigm in the hope that it might finally be put aside. I also wish to pay homage to Gunnar Thorson for intuitively developing so well major global views in marine biology that shaped my own research efforts, and allowed me to examine them by working in spectacular settings all over the world. Research done in the Antarctic and in California supported by grants from the Division of Polar Programs, National Science Foundation (DPP 83–17082 and DPP 88–18354).

References

Bartholomew, G. A. 1986. The role of natural history in contemporary biology. *BioScience* 36:324–329.

Bosch, I. M. 1988. Reproduction and development of asteroids and an echinoid in shallow waters of McMurdo Sound, Antarctica. Ph.D. dissertation. University of California, Santa Cruz.

Bosch, I. M. 1989. Contrasting modes of reproduction in two antarctic asteroids in the genus *Porania,* with a description of unusual feeding and nonfeeding larval types. *Biol. Bull.* 177:77–82.

Bosch, I. M. and J. S. Pearse. 1990. Developmental types of shallow-water asteroids of McMurdo Sóund, Antarctica. *Mar. Biol.* 104:41–46.

Bosch, I. M., K. A. Beauchamp, M. E. Steele, and J. S. Pearse. 1987. Development, metamorphosis, and seasonal abundance of embryos and larvae of the antarctic sea urchin *Sterechinus neumayeri. Biol. Bull.* 173:126–135.

Bosch, I. M., J. S. Pearse, and L. V. Basch. 1990. Particulate food and growth of planktotrophic sea star larvae in McMurdo Sound, Antarctica. *Antarctic J. U.S.* 25:210–212.

Buchan, A. 1895. Report on oceanic circulation. *Rpt. Sci. Res., Voyage of H.M.S. Challenger, A summary of the scientific results, Appendix (Physics and Chemistry,* Part VIII), pp. 1–38.

Cameron, J. L., F. S. McEuen, and C. M. Young. 1988. Floating lecithotrophic eggs from the bathyal echinothuriid sea urchin *Araeosoma fenestratum.* In R. D. Burke, P. V. Mladenov, P. Lambert, and R. L. Parsley, eds., *Echinoderm Biology,* pp. 177–180. Rotterdam: A. A. Balkema.

Christiansen, F. B. and T. M. Fenchel. 1979. Evolution of marine invertebrate reproductive patterns. *Theor. Pop. Biol.* 16:267–282.

Clark, A. M. and M. E. Downey. 1992. *Starfishes of the Atlantic.* London: Chapman and Hall.

Crisp, D. J. 1986. A comparison between the reproduction of high and low latitude barnacles, including *Balanus balanoides* and *Tetraclita pacifica.* In M. J. Thompson, R. Sarojini, and R. Nagabhushanam, eds., *Biology of Benthic Marine Invertebrates,* pp. 69–84. New Delhi: Oxford and IBH Publishing.

Emlet, R. B. 1990. World patterns of developmental mode in echinoid echinoderms. In M. Hoshi and O. Yamashita, eds., *Advances in Invertebrate Reproduction 5,* pp. 329–335. Amsterdam: Elsevier Science.

Emlet, R. B., L. R. McEdward, and R. R. Strathmann. 1987. Echinoderm larval ecology viewed from the egg. In M. Jangoux and J. M. Lawrence, eds., *Echinoderm Studies,* vol. 2, pp. 55–136. Rotterdam: A. A. Balkema.

Emson, R. H. and R. G. Crump. 1979. Description of a new species of *Asterina* (Asteroidea), with an account of its ecology. *J. Mar. Biol. Ass. U. K.* 59:77–94.

Falk-Petersen, I. B. 1982. Breeding season and egg morphology of echinoderms in Balsfjoren, northern Norway. *Sarsia* 67:215–221.

Gage, J. D. and P. A. Tyler. 1981. Nonviable seasonal settlement of larvae of the upper bathyal brittlestar *Ophiocten gracilis* in the Rockall Trough Abyssal. *Mar. Biol.* 64:153–161.

Gage, J. D. and P. A. Tyler. 1991. *Deep-Sea Biology: A Natural History of Organisms at the Deep-Sea Floor.* Cambridge: Cambridge University Press.

Gemmill, J. F. 1920. The development of the starfish *Crossaster papposus,* Muller and Troschel. *Quart. J. Micro. Sci.* 64:155–190.

Ghiselin, M. T. 1987. Evolutionary aspects of marine invertebrate reproduction. In A. C. Giese, J. S. Pearse, and V. B. Pearse, eds., *Reproduction of Marine Invertebrates,* vol. 9: *General Aspects: Seeking Unity in Diversity,* pp. 609–665. Pacific Grove, Calif.: Boxwood Press.

Giard, A. 1905. La Poecilogonie. *6th Internt. Congr. Zool. Berne,* pp. 617–646.

Grant, A. 1983. On the evolution of brood protection in marine benthic invertebrates. *Amer. Nat.* 122:549–555.

Hansen, B. 1968. Brood protection in the deep-sea holothurian *Oneirophanta mutabilis* Theel. *Nature* 217:1062–1063.

Harvey, R. and J. D. Gage. 1984. Observations on the reproduction and postlarval morphology of pourtalesiid sea urchins in the Rockall Trough area (N.E. Atlantic). *Mar. Biol.* 82:181–190.

Holland, N. D. 1991. Echinodermata: Crinoidea. In A. C. Giese, J. S. Pearse, and V. B. Pearse, eds., *Reproduction of Marine Invertebrates,* vol. 4: *Echinoderms and Lophophorates,* pp. 247–299. Pacific Grove, Calif.: Boxwood Press.

Jablonski, D. and R. A. Lutz. 1983. Larval ecology of marine benthic invertebrates: Paleobiological implications. *Biol. Rev.* 58:21–89.

Komatsu, M., Y. T. Kano, H. Yoshizawa, S. Akabane, and C. Oguro. 1979. Reproduction and development of the hermaphroditic sea-star, *Asterina minor* Hayashi. *Biol. Bull.* 157:258–274.

Lawrence, J. 1987. *A Functional Biology of Echinoderms.* Baltimore: Johns Hopkins University Press.

Lightfoot, R., P. A. Tyler, and J. D. Gage. 1979. Seasonal reproduction of deep-sea bivalves and brittlestars. *Deep-Sea Res.* 26:967–973.

Lipps, J. H. and C. S. Hickman. 1982. Origin, age, and evolution of antarctic and deep-sea faunas. In W. G. Ernst and J. G. Morin eds., *The Environment of the Deep Sea,* pp. 324–356. Englewood Cliffs: Prentice-Hall.

McClary, D. J. and P. V. Mladenov. 1988. Brood and broadcast: A novel mode of reproduction in the sea star *Pteraster militaris.* In R. D. Burke, P. V. Mladenov, P. Lambert, and R. L. Parsley, eds., *Echinoderm Biology,* pp. 163–168. Rotterdam: A. A. Balkema.

McClintock, J. B. and J. S. Pearse. 1987. Reproductive biology of the common antarctic crinoid *Promachocrinus kerguelensis* (Echinodermata: Crinoidea). *Mar. Biol.* 96:151–155.

Mileikovsky, S. A. 1971. Types of larval development in marine bottom invertebrates, their distribution and ecological significance: A re-evaluation. *Mar. Biol.* 10:193–213.

Mladenov, P. V., S. F. Carson, and C. W. Walker. 1986. Reproductive ecology of an obligately fissiparous population of the sea star *Stephanasterias albula* (Stimpson). *J. Exp. Mar. Biol. Ecol.* 96:155–175.

Morris, R. H., D. P. Abbott, and E. C. Haderlie, eds. 1980. *Intertidal Invertebrates of California.* Stanford, Calif.: Stanford University Press.

Mortensen, T. 1903. Echinoidea. 1. *Danish Ingolf-Expedition* 4(1):1–193

Mortensen, T. 1910. Report on the echinoderms collected by the Danmark-Expedition at North East Greenland. *Medd. Grnland* 45:239–302.

Mortensen, T. 1913. Die Echinodermenlarven der Deutschen Sud-Polar Expedition 1901–1903. Deutsch. *Sudpol. Exped.* 14 (Zool. 6):69–111.

Mortensen, T. 1921. *Studies of the Development and Larval Forms of Echinoderms.* Copenhagen: G. E. C. Gad.

Mortensen, T. 1927a. *Handbook of the Echinoderms of the British Isles.* London: Oxford University Press.

Mortensen, T. 1927b. Report on the Echinoidea collected by the Albatross, Philippine Archipelago and adjacent regions, part 1. *Bull. U.S. Nat. Mus.* 100, 6 (4).

Murray, J. 1894. Letter partially quoted in W. A. Herdman 1923. *Founders of Oceanography and Their Work,* p. 81. London: Edward Arnold.

Murray, J. 1895. General observations on the distribution of marine organisms. *Rpt. Sci. Res., Voyage of the H.M.S. Challenger, A summary of the scientific results,* pp. 1431–1462.

Olson, R. R., I. Bosch, and J. S. Pearse. 1987. The antarctic larval food limitation hypothesis examined for the asteroid *Odontaster validus. Limnol. Oceanogr.* 32:686–690.

Palmer, A. R. and R. R. Strathmann. 1981. Scale of dispersal in varying environments and its implications for life histories of marine invertebrates. *Oecologia* 48:308–318.

Pearse, J. S. 1969. Slow developing demersal embryos and larvae of the antarctic sea star *Odontaster validus. Mar. Biol.* 3:110–116.

Pearse, J. S. and I. Bosch. 1986. Are the feeding larvae of the commonest antarctic asteroid really demersal? *Bull. Mar. Sci.* 39:477–484.

Pearse, J. S. and R. A. Cameron. 1991. Echinodermata: Echinoidea. In A. C. Giese, J. S. Pearse, and V. B. Pearse, eds., *Reproduction of Marine Invertebrates,* vol. 4: *Echinoderms and Lophophorates,* pp. 513–662. Pacific Grove, Calif.: Boxwood Press.

Pearse, J. S. and J. B. McClintock. 1990. A comparison of reproduction by the brooding spatangoid echinoids *Abatus shackletoni* and *A. nimrodi* in McMurdo Sound, Antarctica. *Invert. Repro. Develop.* 17:181–191.

Pearse, J. S., J. B. McClintock, and I. Bosch. 1991a. Reproduction of antarctic benthic invertebrates: Tempos, modes, and timing. *Amer. Zool.* 31:65–80.

Pearse, J. S., I. Bosch, V. B. Pearse, and L. V. Basch. 1991b. Differences in feeding on algae and bacteria by temperate and antarctic sea star larvae. *Antarctic J. U.S.* In press.

Rex, M. A. and A. Warén. 1982. Planktotrophic development in deep-sea prosobranch snails from the western North Atlantic. *Deep-Sea Res.* 29A:171–184.

Rex, M. A., C. A. Van Ummersen, and R. D. Turner. 1979. Reproductive pattern in the abyssal snail *Benthonella tenella* (Jeffreys). In S. E. Stancyk, ed., *Reproductive Ecology of Marine Invertebrates,* pp. 173–188. Columbia: University of South Carolina Press.

Ross, J. C. 1847. *A Voyage of Discovery and Research in the Southern and Antarctic*

Regions during the years 1839–43, Volume I. London: J. Murray (reprinted 1969, New York: Augustus M. Kelley).

Roughgarden, J. 1989. The evolution of marine life cycles. In M. W. Feldman, ed., *Mathematical Evolutionary Theory,* pp. 270–300. Princeton, N.J.: Princeton University Press.

Shuto, T. 1974. Larval ecology of prosobranch gastropods and its bearing on biogeography and paleontology. *Lethaia* 7:239–256.

Smiley S., F. S. McEuen, C. Chaffee, and S. Krishnan. 1991. Echinodermata: Holothuroidea. In A. C. Giese, J. S. Pearse, and V. B. Pearse, eds., *Reproduction of Marine Invertebrates,* vol. 4: *Echinoderms and Lophophorates,* pp. 663–750. Pacific Grove, Calif.: Boxwood Press.

Strathmann, M. F. 1987. Phylum Echinodermata: Class Asteroidea. In M. F. Strathmann, ed., *Reproduction and Development of Marine Invertebrates of the Northern Pacific Coast,* pp. 535–555. Seattle: University of Washington Press.

Strathmann, R. R. 1985. Feeding and nonfeeding larval development and life history evolution in marine invertebrates. *Ann. Rev. Ecol. Syst.* 16:339–362.

Thomson, C. W. 1878. Notice of some peculiarities on the mode of propagation of certain echinoderms of the southern seas. *J. Linn. Soc. Zool.* 13:55–79.

Thomson, C. W. 1885. Notes on the reproduction of certain echinoderms from the southern ocean. *Rpt. Sci. Res., Voyage of the H.M.S. Challenger,* vol. 1, *Narrative,* pp. 379–396.

Thorson, G. 1936. The larval development, growth, and metabolism of arctic marine bottom invertebrates compared with those of other seas. *Medd. Grnland* 100(6):1–155.

Thorson, G. 1946. Reproduction and larval development of Danish marine invertebrates, with special reference to the planktonic larvae in the sound (Øresund). *Medd. Komm. Dan. Fisk. Havunder. Ser. Plankton* 4(1):1–523.

Thorson, G. 1950. Reproductive and larval ecology of marine bottom invertebrates. *Biol. Rev.* 25:1–45.

Thorson, G. 1961. Length of pelagic larval life in marine bottom invertebrates as related to larval transport by ocean currents. In M. Sears, ed., *Oceanography,* pp. 455–474. Washington, D.C.: Amer. Assoc. Advanc. Science (publ. No. 67).

Turner, R. D., J. A. Pechenik, and C. B. Calloway. 1986. The language of benthic marine invertebrate development patterns: problems and needs. In M.-F. Thompson, R. Sarojini, and R. Nagabhushanam, eds., *Biology of Benthic Marine Organisms,* pp. 227–235. New Delhi: Oxford and IBH Publishing.

Tyler, P. A. 1986. Studies of a benthic time-series: Reproductive biology of benthic invertebrates in the Rockall Trough. *Proc. Roy. Soc. Edinburgh* 88B:175–190.

Tyler, P. A. 1988. Seasonality in the deep-sea. *Oceanogr. Mar. Biol. Ann. Rev.* 18:125–153.

Tyler, P. A. and D. S. M. Billett. 1988. The reproductive ecology of elasipodid holothurians from the NE Atlantic. *Biol. Oceanogr.* 5:273–296.

Tyler, P. A. and J. D. Gage. 1982. The reproductive biology of *Ophiacantha bidentata* (Echinodermata: Ophiuroidea) from the Rockall Trough. *J. Mar. Biol. Ass. U. K.* 62:45–55.

Tyler, P. A. and J. D. Gage. 1984a. Seasonal reproduction of *Echinus affinis* (Echinodermata: Echinoidea) in the Rockall Trough, northeast Atlantic Ocean. *Deep-Sea Res.* 31:387–402.

Tyler, P. A. and J. D. Gage. 1984b. The reproductive biology of echinothuriid and cidarid sea urchins from the deep-sea (Rockall Trough, northeast Atlantic Ocean). *Mar. Biol.* 80:63–74.

Tyler, P. A., A. Grant, S. L. Pain, and J. D. Gage. 1982. Is annual reproduction in deep-sea echinoderms a response to variability in their environment? *Nature (London)* 300:747–749.

Vance, R. R. 1973. On reproductive strategies in marine benthic invertebrates. *Am. Nat.* 107:339–352.

Young, C. M. and J. L. Cameron. 1987. Laboratory and *in situ* flotation rates of lecithotrophic eggs from the bathyal echinoid *Phormosoma placenta. Deep-Sea Res.* 34:1629–1639.

Young, C. M. and F.-S. Chia. 1987. Abundance and distribution of pelagic larvae as influenced by predation, behavior, and hydrographic factors. In A. C. Giese, J. S. Pearse, and V. B. Pearse, eds., *Reproduction of Marine Invertebrates,* vol. 9: *General Aspects: Seeking Unity in Diversity,* pp. 385–463. Pacific Grove, Calif.: Boxwood Press.

Appendix

Reproductive mode and egg diameters of species used in figure 2.1.

Northeast Greenland fjords, shallow water. N = 23; 10 asteroids, 7 ophiuroids, 2 echinoids, 2 holothuroids, 2 crinoids. Total species of echinoiderms in area, about 40. From Thorson (1936) unless indicated otherwise.

Planktotrophs (4; 17%)	Egg diameter (μm)
Ophiocten sericeum	170
Ophiura robusta	170
Ophiura sarsi	120[a]
Strongylocentrotus droebachiensis	160[b]

Pelagic lecithotrophs (15; 65%)	
Pontaster tenuispinus	600–650
Bathybiaster vexillifer	900[c]
Ctenodiscus crispatus	600–700
Poraniomorpha tumida	500–520
Asterias lincki	680
Crossaster papposus	800[b]
Ophiopleura borealis	550–580
Ophiopus arcticus	400
Ophiacantha bidentata	800
Ophioscolex glacialis	460–510
Elpidia glacialis	250–280 (immature?)
Cucumaria frondosa	650[a]
Pourtalesia jeffreysi	360–380
Heliometra glacialis	200[d]
Hathrometra prolixa	200[d]

Nonpelagic brooders (4; 17%)	
Pteraster militaris	1400[a]
Henricia sanguinolenta	1000[b]
Hymenaster pellucidus	500–6000 (immature?)
Leptasterias groenlandica	900[b]

[a]Falk-Petersen 1992. [b]Emlet et al. 1987. [c]Tyler 1986. [d]Estimated from Holland 1991.

Rockall Trough, Northeast Atlantic, deep sea. N = 37; 13 asteroids, 4 ophiuroids, 13 echinoids, 7 holothuroids. Probably most of the species in Rockall Trough. From Tyler (1986) unless indicated otherwise.

Planktotrophs (8; 22%)	Egg diameter (μm)
Plutonaster bifrons	110
Dytaster grandis (= *D. insignis*)	120
Ophiura ljungmani	110

Ophiocten gracilis	110
Cidaris cidaris	110
Echinus affinis	110
Echinus alexandri	110
Echinus acutus	110

Pelagic lecithotrophs (29; 78%)	
Bathybiaster vexillifer	900
Psilaster andromeda	900
Benthopecten simplex	900
Pontaster tenuispinus	850
Pectinaster filholi	850
Pseudarchaster parelii	900
Paragonaster subtilis	900
Hymenaster membranaceus	1100
Zoroaster fulgens	900
Brisinga endecacnemos	1250
Brisingella coronata	1250
Ophiacantha bidentata	600
Ophiomusium lymani	460
Poriocidaris purpurata	1500
Phormosoma placenta	1100
Hygrosoma petersi	1150
Sperosoma grimaldi	1100
Araeosoma fenestratum	1250
Calveriosoma histrix	1250
Echinosigra phiale	250–335[a]
Pourtalesia jeffreysi	225[a]
Pourtalesia miranda	250–345[a]
Ypsilothuria talismani	350[a]
Oneirophanta mutabilis	950[b]
Deima validum	700[b]
Benthodytes sordida	1200[b]
Psychropotes longicauda	>3000[b]
Psychropotes depressa	1800[b]
Psychropotes semperiana	3000[b]

[a]Harvey and Gage 1984. [b]Tyler and Billett 1988.

Monterey Bay, Central California, shallow water. N = 36; 16 asteroids, 7 ophiuroids, 3 echinoids, 10 holothuroids. Probably most of the species in shallow Monterey Bay. From Morris et al. (1980), Emlet et al. (1987), and Strathmann (1987), unless indicated otherwise.

Planktotrophs (18; 50%)	Egg diameter (μm)
Luidia foliolate	170
Astropecten californicus	170[a]
Dermasterias imbricata	200

Asterina miniata	170
Orthasterias koehleri	150
Pisaster brevispinus	160–170
Pisaster giganteus	165[a]
Pisaster ochraceus	150–160
Pycnopodia helianthoides	120
Ophiopholis aculeata	105
Ophiothrix spiculata	110
Ophiopteris papillosa	100
Amphiodia occidentalis	100
Strongylocentrotus purpuratus	80
Strongylocentrotus franciscanus	130–140
Dendraster excentricus	110–125
Stichopus californicus	185–210
Stichopus parvimensis	200[a]

Pelagic lecithotrophs (8; 22%)	
Mediaster aequalis	1100
Pteraster tesselatus	1460
Solaster dawsoni	950–1000
Henricia leviuscula	900–1500
Gorgonocephalus eucnemis	220
Cucumaria miniata	490–575
Cucumaria piperata	480–600
Eupentacta quinquesemita	370–415

Nonpelagic brooders/viviparous(v) (10; 28%)	
Leptasterias hexactis	900
Leptasterias pusilla	900[a]
Henricia species	700–1000
Amphipholis squamata (v)	105[b]
Ophioplocus esmarki (v)	330[b]
Cucumaria curata	1000
Cucumaria pseudocurata	1000
Pachythyone rubra (v)	250[c]
Lissothuria nutriens	1000
Leptosynapta albicans (v)	240–400[a]

[a]Estimated. [b]Rumrill and Pearse 1985. [c]Smiley et al. 1991.

McMurdo Sound, Antarctica, shallow water. N = 22; 15 asteroids, 2 ophiuroids, 4 echinoids, 1 crinoid. Most species of asteroids, echinoids, and crinoids recorded for shallow McMurdo Sound. From Bosch and Pearse (1990), Pearse et al. (1991a), and unpublished (see text).

Planktotrophs (5; 23%)	Egg diameter (μm)
Odontaster validus	170
Odontaster meridionalis	190

Porania antarctica (large form)	550
Sterechinus neumayeri	125
Ophionotus victoriae	185

Pelagic lecithotrophs (11; 50%)	
Bathybiaster loripes	930
Psilaster charcoti	950
Acodontaster conspicuus	700
Acodontaster elongatus	540
Acodontaster hodgsoni	550
Lophaster gaini	1280
Cuenotaster involutus	1000
Macroptychaster accrescens	1280
Perknaster fuscus	1200
Ophiosparta gigas	350
Promachocrinus kerguelensis	180

Nonpelagic lecithotrophs—demersal(d)/brooders (6; 27%)	
Porania antarctica (small form) (d)	550
Diplasterias brucei	2800
Notasterias armata	3500
Abatus shackletoni	1280
Abatus nimrodi	1970
Ctenocidaris perrieri	1950

CHAPTER THREE

Adaptations for Reproduction Among Deep-Sea Benthic Molluscs: An Appraisal of the Existing Evidence

Rudolf S. Scheltema

Abstract

Adaptations for reproduction among deep-sea Mollusca are fully as varied as those found among their shoal-water counterparts. Such diversity in adaptations for reproduction can be related to a corresponding wide diversity in the deep-sea benthic environment, until recently largely unexpected and mostly unknown. Among most deep-sea molluscs of those that have been studied there seems to be no obvious evidence for a coupling between periodicity of reproduction and primary productivity at the sea-surface, nor any evidence of a seasonal periodicity in reproduction. Fertility among deep-sea bivalve molluscs varies widely (2–30,000 eggs at once), but lifetime fecundity cannot be precisely determined since neither the rate of gametogenesis nor the age or length of life can be readily ascertained for most iteroparous species. The mode of development among deep-sea molluscs may be direct, without a planktonic stage, or indirect, either with a short nonfeeding (lecithotrophic) larva or alternatively a planktotrophic veliger stage feeding as it is passively dispersed by ocean currents. "Ontogenetic vertical migration" has been proposed for planktotrophic molluscan larvae. Although some indirect evidence has been advanced, up to now the veligers of only two species of eurybathal deep-sea gastropods are known to actually occur in

This paper is respectfully dedicated to my friend and colleague, Howard L. Sanders, on his seventy-first birthday, in recognition of his important contributions to the study of the deep-sea benthos.

the near-surface waters. The mode of development will have important consequences for (1) colonization or recruitment, (2) genetic exchange between populations, and (3) the biogeography of deep-sea molluscs, but studies on such aspects of molluscan life-history have only just begun.

"Have you walked through the depths of the ocean or dived to the floor of the sea?" Job 38:16

Less than two decades ago, most marine biologists characterized the deep sea as a physically monotonous environment with a nearly constant temperature between 1° and 3°C, varying only a fraction of a degree at any one location and with a salinity fluctuating at most only a few tenths of a part per thousand. Such physical changes are so small as to be considered of no particular physiological or ecological consequence. Because of such a constancy as well as the continuous and total darkness, the abyssal depths were regarded as an unchanging and timeless environment. Almost nothing was known about the periodicity or rate of particulate flux from the surface or about the effect of currents at the bottom boundary layer. The deep sea was regarded as an exceedingly stable and predictable environment and physical and biological processes were believed to proceed at a very slow rate. Now, twenty years later, new discoveries and observations require a reconsideration and revision of these earlier held beliefs. The purpose here is to appraise recent advances in the study of the deep-sea environment and to relate this new knowledge to what is known about adaptations for reproduction among deep-sea benthic molluscs.

Three kinds of adaptation for reproduction are considered. These are (1) temporal patterns of reproduction, i.e., the timing of gametogenesis and spawning; (2) fecundity or fertility, i.e., the number of gametes produced by an individual over its lifetime; and (3) mode of development, viz. (a) direct, without a vagile stage, development occurring either within the pallium or gill lamellae of the adult or otherwise in an encapsulating structure or gelatinous egg mass, or (b) indirect, including a planktonic larval stage. "Mixed" development with early stages protected within a brood cavity or egg capsule followed by a feeding planktonic larval stage is also encountered.

Temporal Patterns of Reproduction Among Deep-Sea Benthic Molluscs

Orton (1920) believed that "sea temperature must be the influence of paramount importance controlling [the periodicity of] breeding in marine animals" and concluded therefore, that "breeding should be continuous in homothermal regions." Accordingly, it was speculated that reproduction in the deep sea would be continuous rather than periodic. Indeed, if a regular seasonal periodicity of reproduction is to occur in the deep sea, it necessarily must be determined by a second-order phenomenon, such as the seasonal change in flux of particulate organic matter onto the bottom, which is a consequence of the annual cycle of primary productivity at the sea surface.

PERIODIC EVENTS IN THE DEEP-SEA BENTHIC ENVIRONMENT

The physical monotony and timelessness in the deep-sea envisioned only twenty years ago was related to the belief that events at the sea surface would not be perceived by organisms on the ocean floor (vide Menzies 1965). It was thought that there occurred no cyclic events on the bottom to which periodic biological activity could be cued. Not only was there no way to set a clock, but there was no obvious adaptive value for a bottom organism to "measure" time in a constant environment.

Among the first direct evidence to show a relationship between surface primary productivity and seasonal variation of particulate flux to great ocean depths was that of Deuser and Ross (1980). Using traps at 3200 m depth anchored 1000 m off the bottom, they demonstrated that seasonal changes in the accumulation of particulate matter was correlated with primary productivity of surface waters. Sinking rate of particulate matter has been found to range from less than 1 to more than 1000 m day^{-1}, but 100 m day^{-1} is not uncommon (McCave 1975; Hinga et al. 1979; Lampitt 1985). Wiebe et al. (1979) observed that diel vertical movement of organisms such as salps also contribute importantly to large-scale downward transport of organic matter. However, pellets are consumed or degraded in the water column (Lampitt et al. 1990), and it is estimated that up to 90 percent of the material that actually leaves the surface (i.e., about 10 percent of surface productivity) is actually consumed before reaching the bottom of the ocean. Thus, only about ca. 1 percent of the surface primary productivity arrives at the bottom (Gage and Tyler 1991:273).

Most measurements of particulate flux have been made with traps well off the bottom so as not to confound vertical flux from above with that of horizontal transport of fine suspended particulate matter within the nephlos layer along the bottom. Much of the existing data cannot reveal either the amount or periodicity of particulate matter that actually reaches the bottom boundary layer (i.e., to within a meter or less above the ocean floor), where benthic organisms actually live. Thus, existing measurements made with sediment traps could be largely irrelevant to the understanding of what happens on or directly above the bottom. Gage (1991) notes that measurements of particulate flux to the bottom-boundary-layer make certain tacit assumptions, namely that "the collectors provide a good estimate of the particles entering from above, and that the measurements are comparable despite time of year" and finally that "small particulate organic matter entering the BBL [bottom-boundary-layer] is the primary source of food for the organisms there." Gage concludes that differences in collector efficiency indicated by flume studies show that various designs of sediment traps are subject to significant bias and depend upon the flow regime (Butman et al. 1986).

A number of ambient conditions at the bottom-boundary-layer may confound or effectively mask any seasonal "signal" of particulate flux to bottom organisms. Intermittent diurnal tidally determined or seasonally varying horizontal advection along the bottom may effectively blur patterns of vertical flux from above. Vangriesheim and Khripounoff (1990), using a nephelometer, found semi-diurnal tidal oscillation at 0.5 and 120 m off the bottom at Moriadezek Terrace (Bay of Biscay) and demonstrated the need for observations at various time scales. Blaume et al. (1991) concluded from observations on the continental rise at 2500 m depth in the Barent Sea that biological and physical resuspension contribute importantly to increase in particulate and dissolved organic matter in the nephloid layer and that "lateral particulate flux far exceeds the vertical flux even at low velocities." The foregoing boundary layer effects are most important when their spatial and temporal scales are large relative to the vertical particle pulse of particulate matter.

In oligotrophic seas remote from continents and also in tropical regions, the standing crop of phytoplankton at the surface may be very small indeed, up to ten thousandfold less than that of coastal surface waters, and any seasonal differences in particulate flux to the bottom boundary layer may be so small as to be almost imperceptible. In contrast, where great ocean depths occur in proximity to continents (e.g., deep trenches and canyons) or over the continental rise, seasonal flux of particulate matter to the bottom can be expected to be very much greater owing to much higher primary productivity at the sea surface. Under such conditions the likelihood for a seasonal "signal" to the bottom-boundary-layer may be very much enhanced. Billet et al. (1983) used photographs of the ocean floor to observe a seasonal pulse of "phytodetrital" material at bathyal and abyssal depths (1370–3000 m) in Porcupine Sea Bight in the Northeast Atlantic (50°N, 13°W). Between mid-June and mid-July "dramatic changes [occurred] in the appearance of the sea bed. This 'detrital carpet' became resuspended whenever currents 1 m above the bottom exceeded 7 cm sec^{-1}" (Lampitt 1985).

Activity of some epi- and infaunal deposit-feeding benthic species may contribute to the bioturbation of the bottom sediment, and to the resuspension of particulate material into the nephlos layer, thereby confounding the seasonal "signal" of organic flux at or near the bottom. However, other infaunal species contribute to the stability of bottom sediments (Meadows and Tait 1985). Any effective seasonal cue for reproduction for filter feeding species (e.g., Pectinidae) necessarily must be perceived from just above the bottom, perhaps within the first half meter of the water column.

To summarize, wherever the primary productivity at the surface is high and markedly periodic, as occurs near continental land masses, there is a possibility that some deep sea benthic species will perceive seasonal changes in the amount of particulate organic material to reach the bottom. Under regions where surface productivity is very low, it is less likely that seasonal differences in particulate flux will be discerned. Resuspension and horizontal advection in some regions may confound any "signal" from the ocean surface.

Evidence for Periodicity of Reproduction

Three temporal patterns of reproduction can be distinguished among benthic molluscs: (1) continuous, nonperiodic reproduction where gametes are formed continuously throughout the reproductive maturity of adult individuals; (2) intermittently reoccurring reproduction where reproduction is noncontinuous but no demonstrable regularity in the periodicity of reproduction is evident; and (3) regularly reoccurring or seasonal reproduction where reproduction occurs at regular intervals and is seemingly keyed to seasonal environmental changes (e.g., temperature, Thorson 1946:424; Barnes and Barnes 1958). The questions that now arise are: (1) do deep-sea benthic molluscs show any seasonal periodicity in reproduction? and (2) can periodic changes in flux of particulate matter be correlated with such periodicity where it occurs?

The most convincing data for reproductive periodicity in deep-sea molluscs comes from direct observations of oogenesis using histological preparations. The occurrence of oocytes within the ovary can be established from periodic, sequential deep-sea samples taken from the same location over the course of a year, or better, several years. Such evidence will establish: (1) whether gametogenesis is periodic or continuous and, if periodic, what time of year oogenesis and spermatogenesis occur; and (2) whether gametogenesis is synchronous among all individuals within a population and therefore whether it is under environmental or endogenous control.

An intensive sampling regime is not always possible owing to fiscal and logistic constraints. Less definitive evidence however, sometimes can be obtained from a single deep-sea sample. If oogenesis is asynchronous, with all different stages of gametogenesis appearing simultaneously within the same ovary, then it may be inferred that reproduction is continuous and that oocytes are constantly formed and released in relatively small numbers. On the other hand, synchronous oogenesis suggests that spawning occurs periodically, but it reveals nothing about either the frequency at which spawning occurs or the interval between successive times of spawning. However, asynchronous gametogenesis, as determined from a single sample, cannot be definite proof that oogenesis is continuous. A closer examination plotting the size-frequencies of eggs may reveal a bimodal distribution suggesting periodic oogenesis (vide Grant and Tyler 1983).

Among some sublittoral species it has been shown that completion of gametogenesis and spawning may be widely separated in time. For example, in the intertidal gastropod *Nassarius obsoletus* the completion of gametogenesis precedes spawning by as much as five months (Scheltema 1967). At present there is no direct evidence for periodicity of spawning among deep-sea molluscs.

Data from the literature on temporal patterns of reproduction among benthic deep sea molluscan species are summarized in table 3.1. Among the protobranch bivalves, temporal patterns of reproduction are known for eleven species. In all instances the evidence comes from samples collected at various seasons of the year but not always from the same populations. Nine of the species reproduce continuously. Only *Ledella pustulosa* and *Yoldiella jeffreysi* show any evidence of periodicity (Gage et al. 1986). Reproduction in *Ledella pustulosa* apparently occurs "in January" (Gage et al. 1986) or "in the spring"

TABLE 3.1 *Temporal Patterns of Reproduction in Benthic Deep-Sea Mollusca*

Species	Pattern	Authority
BIVALVIA		
Protobranchia		
Nuculoidea bushae	continuous †	Scheltema unpubl.
Nuculoma granulosa	continuous †	Scheltema unpubl.
Deminucula atacellana[a]	continuous †	Scheltema 1972, Davis 1985, Gage et al. 1986
Brevinucula verrilli	continuous †	Scheltema unpubl.
Nucula darella	continuous †	Rokop 1974
Nucula pontonia	continuous †	Rokop 1974
Tindara cervola	continuous †	Rokop 1974
Malletia cuneata	continuous †	Rokop 1974, Giles et al. 1991
Ledella pustulosa	seasonal periodicity[b] †	Lightfoot et al. 1979, Gage et al. 1986, Giles et al. 1991
Yoldiella jeffreysi	seasonal periodicity[c] †	Lightfoot et al. 1979, Gage et al. 1986
Bathyarca sp.	continuous †	Rokop 1974
Eulamellibranchia		
Calyptogena magnifica	continuous ‡	Berg 1985
Bathymodiolus thermophilus	continuous ‡	Berg 1985
Gastropoda		
Colus jeffreyianus	continuous †	Colman et al. 1986
Calliotropis ottoi	continuous ‡	Colman & Tyler 1988
Benthonella tenulla	continuous ‡	Rex et al. 1979
Scaphopoda		
Cadulus californicus	seasonal periodicity †	Rokop 1974, 1977

[a] [= *Nucula cancellata,* see Schenck 1939]
[b] Reproduces in January
[c] Reproduces in October–November
† Determined by sequential sampling
‡ Inferred from asynchrony of oogenesis

(Giles et al. 1991) whereas *Yoldiella jeffreysi* reproduces "in October-November" (Gage et al. 1986). In neither species is there an obvious close correlation between the time of reproduction and the period of maximum flux of particulate matter onto the bottom, which in the northeast Atlantic, where the observations on *Ledella pustulosa* and *Yoldiella jeffreysi* were made, happens according to the observations of Lampitt (1985) between the end of May and mid-July (table 3.1) or, according to Billett et al. (1983), from the end of April to July.

To clarify this apparent paradox, Giles et al. (1991) examined seasonal differences in the diet of *Ledella pustulosa* and that of *Malletia cuneata*. The latter is a species in which gametogenesis is continuous throughout the year. Apparently *Ledella pustulosa* shows a seasonal increase in number of unicellular forms ingested (mostly coccolithophores, diatoms, and spores) which, according to the authors, is "coincident with the phytodetrital flux" but there is "no obvious seasonal variation [in ingestion] in *M. cuneata*." The exact significance of these results is elusive. Perhaps certain combinations of food are more nutritious than others but the precise causal relationship between reproductive periodicity and seasonal changes in availability of various particular kinds of food organisms will be difficult to establish for each individual bivalve species.

A second kind of evidence that may be used to infer periodic reproduction is the analysis of size-frequency distribution. Reoccurring distinctly separated peaks in size class result from intensive periods of reproduction, but few published examples of length-frequency analysis exist for deep-sea molluscs. Generally deep-sea bivalves exhibit a broad unimodel distribution with sexually mature individuals numerically predominating (Hutchings and Haedrich 1984, author's unpublished data). Such a size distribution is attributed to "rapid growth rates in early life, heavy predation on young individuals, and size-related mechanisms reducing niche overlap" (Hutchings and Haedrich 1984), but it could also be accounted for by a low and continuous recruitment. The method of size-frequency distribution has been successfully used with some non-molluscan taxa, but has not been useful in understanding the periodicity of reproduction among bivalve molluscs.

Tyler (1988) asks why species respond differently to periodic changes in the bottom environment. One answer may be that seasonal differences in the flux of particulate matter may be largely irrelevant to many deep-sea molluscan species. Among mollusc species only secondarily affected are (1) species such as *Calyptogena* sp. whose nutrition is based upon chemosynthetic primary production of symbiotic bacteria associated with deep-sea hydrothermal vents (Cavanaugh 1983; Jannasch 1984a, b, 1985); (2) predatory or parasitic species affected most directly by the abundance of their prey or host (e.g., Septibranch bivalves and most abyssal gastropods of the families Eulimidae, Aclididae, Epitoniidae and Turridae); and (3) species that can utilize and derive a large proportion of their nutrition from dissolved organic material in the pore-water of deep-sea sediments (Southward and Southward 1982; Wright and Stephens 1982). Hence, forms that may be most affected by changes in flux of organic matter are filter feeders (e.g., Pectinidae) and arguably deposit-feeders, particularly those feeding at the sediment surface (e.g., protobranch bivalves). Omnivores, scavengers, and predators (i.e., most abyssal gastropods) may not be directly affected by changes in organic flux.

For those species that inhabit ephemeral or unstable environments, it may be more adaptive to reproduce, if not continuously, then at least intermittently without relation to cues from the surface. Such ephemeral and temporary environments are treated below.

Fecundity Among Deep-Sea Benthic Molluscs

Measurement of Fecundity

Fecundity is defined as the number of gametes that potentially may be produced during the lifetime of an individual, although fecundity is seldom known except among species that reproduce only once during their lifetime. In such instances fecundity can be determined by counting the total number of eggs present in a fully developed female gonad. Among species which reproduce more than once per lifetime (e.g., seasonally), annual fecundity or fertility (i.e., eggs per unit time) can be estimated by summing the annual fecundity for each year of reproduction over the lifetime of an individual. Such procedure may, however, confound and underestimate lifetime fecundity, because there is an increase in fertility with increasing size and inferred age (figure 3.1). Therefore, the fecundity for each year must be determined separately and then summed. Among

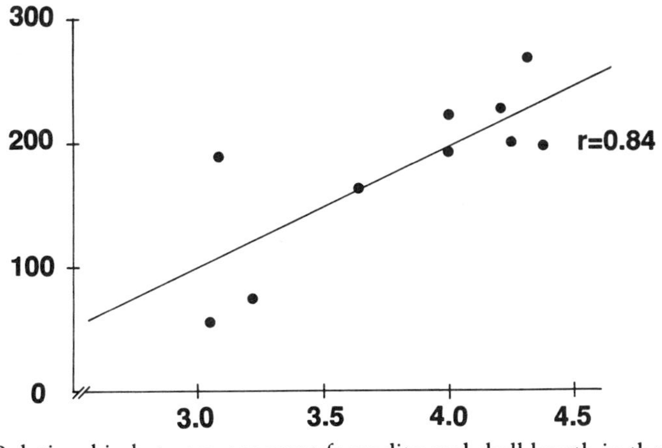

FIGURE 3.1 Relationship between apparent fecundity and shell length in the deep-sea protobranch, *Nucula verrilli*. Ordinate shows number of oocytes in ovary; abscissa indicates length of shell in millimeters. The apparent fecundity increases with increase in shell length.

species that reproduce continuously it is not possible to determine the lifetime fecundity unless the rate of gametogenesis as well as the duration of the sexual maturity is known. If the age-specific rate of gametogenesis is assumed to be roughly similar among all continuously breeding deep-sea bivalve species then a relative fecundity of each species could be derived if their average length of life were known.

However, there are no reliable data on age attained by deep-sea molluscan species. Although the minute structure of protobranch shells has been examined in some detail (Gilkinson et al. 1986) there is no convincing evidence to indicate the time interval represented by each growth interval. Turekian et al. (1975), using a radiometric technique, concluded that 50 years were required for the protobranch bivalve *Tindaria callistima* to reach sexual maturity but the 95 percent confidence interval was 75 years. Gage and Tyler (1991) argue that values obtained by radiometric methods must be regarded with skepticism and that the results obtained with *Tindaria* should not be generalized to all protobranch species.

In practice, most workers report fecundity among long-lived species as the maximum number of gametes at the time when an individual is about to spawn. However, this apparent fecundity is the number of gametes present at a particular instant rather than a rate of gamete production per year or per lifetime. Direct comparisons of such an apparent fecundity among continuously reproducing species tacitly assumes that the number of gametes encountered at any particular time reflects somehow the rate of gametogenesis and consequently is related to the lifetime fecundity. But this need not be so. Notwithstanding, comparisons of apparent fecundity provide the best prospect for guessing the relative reproductive effort of deep-sea bivalves, i.e., the "approximate portion of resources converted to reproduction" (Stearns 1976; Calow 1979:32).

Apparent Fecundity

Summarized in table 3.2 are representative data on the apparent fecundity of some deep-sea molluscan species. The table should not be regarded as comprehensive. There are large differences among deep-sea molluscs, with values known to range between 2 to 30,000 eggs. Even though their lifetime fecundities remain unknown, it seems highly unlikely that a species with an apparent fecundity of only 2 eggs will ever approach 30,000 eggs.

Two infaunal pristiglomid protobranch bivalves restricted to the western Atlantic, viz. *Microgloma yongei* (2000 to 2700 m) and *Microgloma turnerae* (1000 to 2300 m), produce only two eggs at once! These two species may be compared with the siliculid protobranch bivalve *Silicula filatova* (3800–5000 m) which produces ca. 2000 eggs. The known apparent fecundity of other protobranchs falls between those two extreme values. Thyasiridae show a comparable sequence. Table 3.2 further shows that among bivalves there is a rough correlation between the size of a species and its apparent fecundity.

TABLE 3.2 *Apparent Fecundity of Representative Deep-Sea Bivalve Mollusca*

Bivalvia	Shell Length (mm)	Apparent Fecundity (No. oocytes)	Authority
Protobranchia			
Microgloma yongei	0.8–1.1	2	Sanders & Allen 1973
Microgloma turnerae	1	2	Sanders & Allen 1973
Pristogloma alba	1.2	20	Sanders & Allen 1973
Spinula filatovae	2.6	20	Allen & Sanders 1982
Nucula cancellata	3.3	194	Scheltema, R. 1972
Pristogloma nitans	*ca.* 2	*ca.* 200	Sanders & Allen 1973
Nucula granulosa	2.2	217	Scheltema, R. 1972
Tindaria callistiformis	6.5	230	Sanders & Allen 1977
Lametila abyssorum	2.5	260	Allen & Sanders 1973
Nucula verrilli	4.3	260	Scheltema, R. 1972
Tindaria hessleri	2.5	*ca.* 350	Sanders & Allen 1977
Pseudotindaria erebus	6.0	*ca.* 800	Sanders & Allen 1977
Silicula fragilis	*ca.* 4	*ca.* 1000	Allen & Sanders 1973
Silicula mcalesteri	*ca.* 8	*ca.* 1000	Allen & Sanders 1973
Silicula filotova	4	*ca.* 2000	Allen & Sanders 1973
Septibranchia (Verticordiidae)			
Lyonsiella abyssicola	3.5	120	Allen & J. F. Turner 1974
Thyasiridae			
Thyasira incrassata	*ca.* 2	*ca.* 14	Payne & Allen 1991
Thyasira ovata	1.5	44	Payne & Allen 1991
Thyasira transversa	1.6	57	Payne & Allen 1991
Thyasira brevis	*ca.* 3	70	Payne & Allen 1991
Thyasira succisa succisa	1.5	87	Payne & Allen 1991
Thyasira subovata subovata	*ca.* 2	136	Payne & Allen 1991
Thyasira alleni	1.5	160–200	Payne & Allen 1991
Thyasira excavata plicata	15.3	>1000	Payne & Allen 1991
Eulamellebranchia			
Xylophaga 9 spp.		1–10	Knudsen 1961
Xylophaga sp.		30,000	R. D. Turner 1973

Almost no data exist on the fecundity of gastropods. Although prosobranch egg capsules are sometimes found in deep-sea samples (Gustafson et al. 1991), it is difficult to estimate fecundity since there is no way of knowing how many capsules are produced by a single female. Moreover, the egg cases frequently cannot be related to parent species. The gonads of dredged individuals are usually poorly fixed since animals withdraw into their shells when placed in preservative and consequently provide poor material for histologic preparations.

Mode of Development Among Deep-Sea Epifaunal and Infaunal Benthic Molluscs

Thorson (1950:26) observed that "as the phytoplankton can only assimilate, and so multiply, in the uppermost water layers, only these layers seem sufficiently rich in food to support a pelagic larval development." If, therefore, deep-sea species reproduce by means of planktonic larvae they "will have to rise by the aid of their own locomotory organs through a water column often 2000–4000 m . . . . from the [ocean] bottom to the food producing [surface] layer." From such a consideration Thorson (1950:27) speculated that "in the deep sea a nonpelagic development will be totally dominant." However nothing was known about the mode of development among deep-sea molluscs in 1950. Direct field or laboratory observations were then, and remain today, very difficult to make. Consequently most evidence comes from indirect methods.

Method for Determining Mode of Development

Ockelmann was among the first to use indirect methods to determine the mode of development among bivalve molluscs. "Work on material of more than 200 species of bivalves, mainly from the N. Atlantic . . . [showed] that certain quantitative and qualitative relations exists between egg, larval or embryonic shells and type of development in any marine bivalve species (figure 3.2). These relations can be expressed by a few simple rules of wide applicability" (Ockelmann 1965:25). Indirect methods for determining mode of development in bivalves was based upon (1) egg diameter and (2) size of prodissoconch I, i.e., the shell laid down during early development of the larva (see Waller 1981 for details).

The egg diameter and prodissoconch size should be used with caution when making inferences about mode of development, since the size categories sometimes may overlap. Comparison should be made with closely related species where mode of reproduction is already known. In at least one instance prodissoconch morphology of a sublittoral species has been shown to be environmentally modified (O'Foighil 1986).

The use of the protoconch in gastropods as an indicator for mode of development was suggested by Thorson (1950:33), who observed that "as a general rule, a clumsy, large apex points to a nonpelagic development . . . [but] a general rule valid for all apices cannot be given. Only when we know the type of apex derived from pelagic or nonpelagic development within an individual genus [or closely related taxa] can reliable results be obtained" (figure 3.2). The protoconch characters used include (1) size of protoconch I, (2) ornamentation on both protoconch I and II, (3) shape of protoconch II in relation to number of whorls, i.e., relative spire height, (4) presence or absence of sinusigerous lip, and (5) evidence from egg size. Houbrick (1991) warns that in some gastropod species a "secondary protoconch" may replace the true protoconch. The former may be markedly different from the original.

FIGURE 3.2. Schematic illustrations to show characters used to determine differences in mode of development. Upper left: bivalve species with a planktrophic larvae and long planktonic life. Prodissoconch I is small, between 70 to 150 μm in length. A large Prodissoconch II indicates a long period in the plankton. Center left: bivalve species with a lecithotrophic development and short planktonic life. Prodissoconch I is between 135–230 μm. The narrow Prodissoconch II indicates a brief planktonic life. Lower left: bivalve species with direct nonplanktonic development. Prodissoconch I is large, between 230 to 500 μm (or more). Upper right: Schematic figures of the apex of a turrid with a planktotrophic development and long planktonic development. The Turridae are well represented at abyssal depths. The initial whorl, Protoconch I is small, often with pustulate markings. Protoconch II has several whorls with ribs. Lower right: Schematic illustration of the apex of a turrid with direct development as indicated by large bulbous Protoconch I. Differences in shell characters indicating mode of development are not always so distinct. (Left hand figures slightly modified after Ockelmann 1965; Top right: modified after Shuto 1983; Lower right: drawn from scanning electron micrographs Coleman et al. 1986).

EVIDENCE FOR MODE OF DEVELOPMENT AMONG DEEP-SEA BENTHIC SEDIMENT-DWELLING MOLLUSCS

Using the methods of Ockelmann described above, Knudsen (1979) and Schein (1989) have published information on mode of reproduction of 92 species of deep-sea bivalves (table 3.3). The predominant mode of development at both bathyal and abyssal depths appears to be lecithotrophy (i.e., a short, nonfeeding planktonic larval development). Approximately one-half of all bathyal species have lecithotrophic development while at abyssal depths the proportion increases to two-thirds (67 percent). Concomitantly, the percentage of species with direct nonplanktonic development decreases from 41 percent at bathyal depths to 27 percent in the abyss. At both the bathyal and abyss the number of species with long planktonic development is less than 10 percent. Is then lecithotrophy especially adaptive among deep sea bivalves?

Protobranchs are a major faunal element among the deep-sea bivalves; Eulamellibranchia and Septibranchia make only a small contribution. Protobranchs, insofar as known, are entirely restricted to either lecithotrophy, with a nonfeeding, barrel-shaped paricalymma larva remaining in the plankton for only a short time (Drew 1899; Gustafson and Reid 1986) or to a direct nonplanktonic stage of development; no instances of planktotrophy are known. The complete lack of a feeding planktonic stage of development among both sublittoral and deep-sea protobranchs probably can be attributed to phylogenetic constraints and the widespread occurrence of lecithotrophy in deep-sea bivalve molluscs seems to be determined largely by faunal composition namely the predominance of protobranchs.

According to the data of Bouchet (1977) and Rex and Warén (1982) from both the eastern and western Atlantic (table 3.4), almost two-thirds of all bathyal prosobranch gastropods (63 percent) have a direct, nonplanktonic development. At abyssal depths the proportion with nonplanktonic development decreases from 63 percent to 51 percent, whereas a planktonic larval development increases from 37 percent to 49 percent.

TABLE 3.3 *Mode of Development Among Deep-Sea Bivalves of the Eastern Atlantic Inferred from Diameter of Eggs and Length of Prodissoconch*

Depth	Total No. Spp.	Direct Development	"Lecithotrophic" Short Planktonic Development (nonfeeding larva)	"Planktotrophic" Long Planktonic Development (feeding larva)
Bathyal (200–2000 m)	44	18 (41%)	23 (52%)	3 (7%)
Abyssal (2000–5000 m)	48	13 (27%)	32 (67%)	3 (6%)

SOURCE: Data compiled from Knudsen 1979: 216–218; Schein 1989: 101.

TABLE 3.4 *Mode of Development of Deep-Sea Gastropods in the Eastern and Western Atlantic Inferred from Characteristics of the Protoconch*

	Total No. Spp.	No Spp. with Non-Planktonic Development	No. Spp. with Planktonic Development
Bathyal[a] (200–2000 m)			
Archaeogastropoda	16	15	1**
Mesogastropoda	27	15	12
Neogastropoda	27	14	13
Total Prosobranchs	70	44 (63%)	26 (37%)
Abyssal[b] (2000–5000 m)			
Archaeogastropoda	14	14	0
Mesogastropoda	8	5	3
Neogastropoda	29	7	22
Total Prosobranchs	51	26 (51%)	25 (49%)

SOURCE: Data compiled from Bouchet 1977; Rex and Wareń 1982.

[a]Data from Rex and Wareń are based upon 20 stations but only seven are from bathyal depths. A possible bias may be therefore introduced when making comparisons between bathyal and abyssal depths.

[b]*Iphitus tuberatus* a species of uncertain systematic affinity.

From these data Rex and Warén (1982) asserted that "the incidence of planktotrophic development in prosobranchs increases with depth seaward of the Continental Shelf in the western North Atlantic." This statement is, however, misleading since there is also a decrease in the number of individuals and species per sample with increasing depth (see Rex and Warén 1982:176–179, table 1) from an average of 2162 individuals and 68 species at slope depths to 144 individuals and 22 species in the abyss. Although the proportion of species with planktotrophic development was greater in the abyss (i.e., 13 among 22 species), the actual number with planktonic larvae was only little more than one-half that found on the Continental Slope.

Rex and Warén (1982) suggest that increased planktotrophy among prosobranchs "may result from both stronger selections for dispersal ability to track patchy prey resources and relaxed restrictions on dispersal with increasing depth." This hypothesis proposes that the increase in planktotrophy with depth is adaptive in the deep sea. As in bivalves, here again the apparent trend for increased planktotrophy between bathyal and abyssal depth cannot be understood without considering difference in faunal composition. At bathyal depths the proportion of species with planktonic development is about equally divided between Meso- and Neogastropoda. At abyssal depths approximately 80 percent of planktotrophic species are Neogastropoda. Concomitant with the increase in planktotrophy among abyssal Neogastropoda is the reduction in the number of families. Especially conspicuous in the abyss are Turridae and Eulimidae (Bouchet and Warén 1980, 1985, 1986). Indeed turrids are by far the most abundantly represented family at abyssal depths both in numbers of species and individuals. Accordingly the increase in proportion of planktonic species in the abyss must be attributed largely to differences in faunal composition at the family level (viz. an increase in the proportion of Turridae and Eulimidae) rather than to an evolutionary tendency toward planktotrophy among species. Once lost, planktotrophy is rarely reacquired (Strathmann 1978), and it seems unlikely that the increase in planktonic development can be attributed to natural selection at the species level.

"Ontogenetic Vertical Migration" and Phytoplanktotrophy by Larvae

Bouchet (1976a) was among the first to make a detailed study of the protoconchs of deep-sea prosobranch gastropods and demonstrated that planktotrophy occurs among at least some deep-sea prosobranch species. Bouchet proposed (p. 947) that "genetic exchanges could occur through such planktonic stages between populations of deep basins separated by ridges, e.g., North American and West European Basins across the mid-Atlantic ridge." In a subsequent paper Bouchet (1976b:821) put forward a further hypothesis that "planktonic larvae migrate up to the surface to assure their growth which depends upon the phytoplankton, then migrate to the bottom at the time of metamorphosis"* This conclusion was based entirely upon the occurrence of the veliger larva of a single abyssal gastropod species *Benthonella tenella* (range 2000–4900 m [fam. Rissoinidae]) which had been illustrated from surface plankton in a publication by Richter and Thorson (1975). Subsequently the larvae of only one additional bathyal species of gastropod, *Benthomangelia marca* (Watson) (800–1100 m), has been identified in surface waters.

Veliger larvae belonging to the Turridae also occur in surface waters of the open ocean (Scheltema 1986:244, table 2), but at present none of these larvae have yet been identified as deep-sea species. In order to establish whether such larvae belong to sublittoral species or have originated from bathyal or abyssal depths requires that the veliger larvae be identified. Turner et al. (1985) found that larvae taken in a plankton tow 2 m above the bottom near a hydrothermal field included specimens of a single species of turrid that "were remarkably similar to a single, late veliger taken at the surface . . . but subtle differences in the operculum and shell structure suggest that they are probably distinct species."

* Ses larves planctoniques migrent jusqu'a' la surface pour assurer leur croissance auz dépens du phytoplancton, puis migrent vers la fond du moment de la métamorphose.

To illustrate the magnitude of the "ontogenetic vertical migration" proposed, consider the following simple calculation. Gastropod larvae of sublittoral species according to existing laboratory data swim between 0.1 and 1.0 cm sec^{-1} (Chia et al. 1984). Assume a rate of approximately 0.5 cm sec^{-1} (higher than that encountered among most veliger larvae), and suppose that an early-stage veliger can sustain such a rate, swimming continuously upward until reaching the surface. From a depth of 4000 m such an ascent would require approximately 9.25 days; this presumably must be accomplished without any significant source of food. Increasing buoyancy might offer an alternative means for a larvae to rise to the surface but no such mechanism has been demonstrated for any deep-sea mollusc larva. Nor has upward vertical transport by upwelling or eddy-diffusion yet been demonstrated.

Turner et al. (1985), a decade after Bouchet proposed his hypothesis, allowed that "there is still some question as to whether some or all of the deep-water species of bivalves and gastropods with an apparent planktotrophic mode of development . . . might feed at the sea surface."

Bouchet and Fontes (1981) have submitted additional indirect evidence to support the hypothesis of "ontogenetic vertical migration." Differences found in the ratio of the oxygen isotopes ^{16}O and ^{18}O (usually expressed as $\delta\ ^{18}O$) and from carbon ^{12}C and its isotope ^{13}C found between the protoconch and teleconch (post-larval shell) of three bathyal and upper abyssal prosobranch species, namely *Pleurotomella packardi, Gymnobela subaraneosa* and *Amphissa haliaeeti,* seemed to show that protoconchs were deposited at higher temperatures than teleconchs and moreover that the larval shells were formed in water masses different from that of the adult shell. Surprisingly analysis of a fourth abyssal species, *Benthonella tenella,* the only one among the four examined whose larvae heretofore had actually been found at the sea surface showed no differences in isotope ratios between the larval and post-larval shell. Bouchet and Fontès concluded notwithstanding the equivocal results with *Benthonella* that "isotope analysis show that larvae and post-larvae of certain abyssal gastropods live in different water masses and temperature" and that "the results confirm the hypotheses of an upward migration of veliger larvae" (in translation).

Killingley and Rex (1985) expanded on the experiments of Bouchet and Fontes and proposed that "the isotope makeup of larval and adult shells in lecithotrophic [non-planktonic] species will remain the same because both are deposited at bottom temperatures . . . [whereas] planktotrophic [planktonic] species will show a significant difference between larval and adult shells because larval shells are deposited in warmer water." Nine species, five inferred to have a planktonic larva (viz. *Benthonella tenella, Benthomangelia antonia, Theta lyronuclea, Theta chariessa,* and *Pleurotemella sandersoni*) and four with a nonplanktonic development (viz. *Oenopota graphica, Oenopota ovalis, Boretrophon abyssorum,* and *Cingula leptalea*) were selected. Those species inferred to have a planktonic development all showed differences in $\delta\ ^{18}O$ between their protoconch and teleconch whereas no such differences occurred among the nonplanktonic species. Computations on the extent of vertical migration "using paleotemperature equations with temperature and salinity values" and based upon data from *Benthonella tenella* and *Benthomangelia antonia* (the only deep-sea species whose larvae have been actually demonstrated at the surface) give values for vertical migration that account for ascent to within a few hundred meters from the surface. Killingley and Rex concluded that "the pattern of dispersal in deep sea prosobranchs that use planktotrophic development, at least those analyzed here, involves ontogenic migration to the photic zone where larvae can exploit phytoplankton as a food resource." It is not at all clear from these results, however, that the planktonic larvae of deep-sea species are obligate phytoplanktotrophs [= phytoplankton feeders] and must necessarily come to the surface to complete development.

Bouchet and Warén (1979b) make a "very rough estimate" that about 30 percent of eastern North Atlantic bathyal and abyssal gastropods (those found at > 1000 m) have a planktotrophic development, while Rex and Warén (1982, table 1) indicate about 50 percent have planktotrophic planktonic development in the western North Atlantic. If indeed the larvae of all these species are obliged to come to the surface to feed, then it should be possible to find them by sampling at or near the surface of the open ocean.

An alternative to ontogenetic vertical larval migration to the surface is the hypothesis that the veliger larvae of gastropod species can remain at or near the bottom, within the nephlos or bottom boundary layer, feeding upon heterotrophic species including bacteria, flagellates and ciliates or upon particulate or dissolved organic matter. For example, the sublittoral species, *Crassostrea virginica,* heretofore regarded an obligate phytoplanktotroph in laboratory experiments is also able to feed omnivorously, ingesting a wide range of food particles, protozoa and bacteria and concluded that these "play an important supplementary role in larval nutrition" (Baldwin and Newell 1991). Shimek (1986) has observed in laboratory culture that the larvae of the turrid *Oenopota levidensis* spend a week to 10 days as typical free-swimming veligers, whereupon their behavior changes and they become demersal. Larvae remain on the bottom, feeding with their velum either fully extended or sometimes folded around their shell, and complete their development and metamorphose there. Is it not possible that heterotrophs, particulates and dissolved organic matter provide for the nutrition of deep-sea planktonic larvae just as suspension feeders (e.g., pectinid bivalves, serpulids and sponges) on the bottom subsist omnivorously upon suspended particulate matter (Sokolova 1972)? The hypotheses of phytoplanktotrophy and heteroplanktotrophy of course are not mutually exclusive and neither have been satisfactorily nor fully confirmed. Meanwhile the question remains—"Where do all the larvae go?"

Fecundity and Mode of Development in the Changing or Temporary Deep-Sea Benthic Environment

Thorson (1950:4, fig. 1) demonstrated that among benthic invertebrates there exists a relationship between fecundity and mode of development. Among those species with indirect, planktonic development (taking into account differences among major taxa), the number of eggs is consistently greater than among species with a direct, nonplanktonic development. The question of how such differences in fecundity and mode of development are related to different kinds of environment was considered already in 1954 by Cole. Species of molluscs that exclusively occupy unpredictable environments can be expected to have a short life span, to quickly attain sexual maturity, and to reproduce only once during their lifetime, producing large numbers of gametes all released at once. Development includes a passively dispersed life-history stage. Such species are termed semelparous and can be regarded "opportunistic" and adapted for rapidly colonizing ephemeral or temporary habitats. In contrast, in relatively unchanging stable environments iteroparity can be expected, i.e., species that live longer and reproduce more than once, either intermittently or continuously throughout their sexually mature lifetime.

Ephemeral and Temporary Deep-Sea Benthic Environments

Until about fifteen years ago it was generally assumed that ephemeral and temporary habitats were rare or almost nonexistent at great depths in the ocean. It has since become evident that this is not so. Ephemeral habitats are defined as those resulting from irregularly reoccurring events and can be occupied by benthic species for only short periods on a time-scale of one year or less. In the deep sea, plant remains of terrestrial or shallow water origin such as algae, fruits, or rhizomes of coastal marine plants may act as ephemeral habitats providing either a substratum or food. Wolff (1979) summarizes records of 52 species of bathyal and abyssal bivalves known to be associated with plant remains, including 8 mytilids and 35 pholadid wood borers. The precise spatial and temporal distribution of wood and other plant remains in the deep sea is largely unpredictable, but the greatest influx is in tropical regions and seems to be related to the monsoon season, when large amounts of debris are carried down rivers into the sea. Marshall (1986) reported a number of other substrata such as bone, algal holdfasts, cephalopod beaks, and skate eggs, among others to which molluscs become attached. The associations are sometimes only facultative; in other instances they may be obligatory (e.g., wood for boring bivalves).

There are, in addition, short-term disturbances such as "nekton falls," for example, the carcasses of fish. Smith (1985a) found that "fish and invertebrate scavengers [including certain deep-sea gastropods] consume and disperse carcass parcels so efficiently . . . that little carrion remains for direct utilization by the sediment community." On the other hand the partially buried remains of blue or fin whales produce a microhabitat and a fauna supported by chemical-reducing bacteria that is distinct from the surrounding benthic community (Smith et al. 1989).

On a larger time-scale are temporary habitats; an example is areas effected by "deep sea storms" such as observed by Hollister and McCave (1984) and Gross et al. (1988) in the western Atlantic on the Scotian Rise 500 km east of Halifax, Nova Scotia. Here at a depth of 5000 m "the mean deposit thickness of 10 years net accumulation can be removed in a few weeks and redeposited in a few months." Currents may reach maximum velocities as high as 75 cm sec^{-1}. During the period of a storm of several days to a few weeks, currents have often been known to reverse, moving alternately first eastward and then westward at velocities exceeding 30 cm sec^{-1}. Tucholke et al. (1985) noted that "many bedforms, especially of smaller scale, are constructed and destroyed by strong variable current events on a time scale of 3–6 months or less."

Steep submarine slopes and canyons, where slumps and turbidity currents are likely to arise, also are considered temporary and unpredictable habitats. Earth tremors on the continental shelf south of Newfoundland in 1929 (Heezen and Ewing 1952) resulted in a turbidity current that traveled down slope with a velocity up to 750 cm sec^{-1} onto the continental rise and subsequently to the abyssal ocean floor extending well over 450 miles [> 735 km] beyond its point of origin. Such a large-scale turbidity current displaces huge amounts of sediment. Commonly, velocities of turbidity currents are very much lower, between 50–75 cm sec^{-1} (Dengler et al. 1984).

Yet another kind of temporary habitat is the hydrothermal fields found in areas of accretion between adjacent crustal plates. Such fields are characterized by large chimneys of various types from which emerge great plumes of hot water up to 350°C and "smoke" which precipitates quantities of metallic sulfides (J. F. Grassle 1985). Vent fields have been explored at numerous locations in the east Pacific, from the Galapagos Rift and the East Pacific Rise to Juan de Fuca Ridge off the coast of Vancouver, Canada and are found along ridge axes at depths between ca. 2000–4000 m. Similar hydrothermal vents recently have been found in the Mariana Trough in the West Pacific (Hessler and Lonsdale 1991). Explorations along the Mid-Atlantic Ridge have led to the discovery of geothermal fields at 26°N, 45°W located at abyssal depths of almost 4000 m (Rona et al. 1986).

The age of active vents can be measured in years or sometimes decades. Estimates of heat loss at 350°C vents suggest a life span of 1–10 years (McDonald et al. 1980). Assemblage of hydrothermal phenomena is not limited to intermediate-to-fast-spreading ridges (Rona et al. 1986). Slowly spreading oceanic ridges extend "through the Atlantic Ocean and Western Indian Ocean, comprising more than half the 55,000 km global length of seafloor spreading centers" (Rona et al. 1986:33).

Other deep-sea reducing environments recently discovered include (1) cold water-sulfur seeps described by Hecker (1985) at 3270 m in the Gulf of Mexico at the base of the Florida escarpment and (2) subduction zone cold seeps such as found off the Oregon coast at 2036–2380 m (Suess et al. 1985) and at the Japanese subduction zone at 3850–6000 m depth (Juniper and Sibuet 1987).

Reproduction in Ephemeral and Temporary Habitats

How is reproduction of molluscan species adapted for changing deep-sea environments? Turner (1973) describes reproduction of an extraordinary wood-boring bivalve, *Xylophaga* sp., collected from wooden panels submerged in the bathyal at 1830 meters over a period of 140 days. In this short time, large numbers of sexually mature individuals up to a maximum of 20 mm length had largely destroyed their own habitat. From their high apparent fecundity (30,000 eggs) Turner concluded that the species must have a planktonic larva, but was unable to obtain veligers from near the bottom. It was concluded that "high population densities, high reproductive rates, early maturity, rapid growth, apparent ease of dispersal, and ability to utilize a transient habitat make these wood borers 'classic' examples of opportunistic [or semelparous] species" living in an ephemeral habitat. However, Knudsen (1961) found that of the 17 species of *Xylophaga* taken on the Galathea Expedition, nine could be shown to have some form of brood protection. Egg masses were often attached to the outside of the shell, and the number of larvae that were produced (i.e., the apparent fecundity) varied between one and ten. Clearly, the species of *Xylophaga* examined by Knudsen differed in their reproduction from those encountered by Turner by having a low fecundity, a seemingly low capacity for dispersal, and are very limited in geographic distribution, thus being iteroparous notwithstanding their ephemeral habitat. Just how such species find and colonize new pieces of wood remains a puzzle.

Thistle et al. (1985) made a study on the effect of deep-sea storms upon the infaunal populations on the Scotian Rise in the Western North Atlantic. It was found that the abundance of bivalves at 4626 m depth was "conspicuously high compared to those from comparable depths" in regions not affected by such strong and variable currents often exceeding 30 cm sec^{-1} and sometimes even exceeding velocities of 70 cm sec^{-1}. Most of the bivalves were surface deposit feeders such as the protobranch *Malletia abyssorum* or the thyasirid *Thyasira subovata*. The latter made up 25 percent of the molluscan fauna. The large bivalve populations were attributed to "enhanced food availability caused by strong near-bottom currents that flow through the area." There was a "high species dominance" [low species diversity] and populations consisted mostly of juvenile individuals, indicating recent recruitment. Thistle et al. (1985) concluded that "the similarity between our data and those of recolonization experiments suggest that some factor . . . is creating patches that are defaunated . . . and are [subsequently] colonized by larvae from the water column."

Hydrothermal vents have a spectacular and largely unique fauna (J. F. Grassle 1985, 1986; Tunnicliffe 1991:344–353, table 2), with most species living on or near vents being indigenous to a hydrothermal environment. They must grow to maturity and reproduce within a sufficiently short time so that the next generation may colonize new active regions. One might expect that species found on such "island habitats" would have an effective means of long-distance dispersal. The subject of larval dispersal is treated elsewhere in this volume by Gustafson and Lutz and need not be reviewed in detail here (see also Lutz et al. 1984). Turner et al. (1985:175) concluded that "vent species . . . exhibit a variety of developmental patterns, but all are typical and within the range of variation of the systematic group to which they belong." Lutz (1988) suggested that "nonplanktonic development with a free swimming but nonfeeding [lecithotrophic], larval stage is [among molluscs] a rule, rather than an exception, at deep-sea hydrothermal vents."

Many species co-occur between widely separated active hydrothermal vent fields. This suggests that nonfeeding lecithotrophic larvae may have "a dispersal capability sufficient to maintain chains of far-flung populations through step-wise, larval dispersal" (Lutz 1988, see also Leal and Bouchet 1991). Mullineaux et al. (1991) and Mullineaux (this volume) show that hydrothermal vent plumes may act as larval highways and have determined that "traces of plumes and their larval passengers can be found many kilometers away from their original vent source."

Aside from larval transport other means for passive dispersal may be possible. Sigurdsson et al. (1976) have shown in laboratory experiments with sublittoral species that young postlarval bivalves can be transported by "byssus drifting," a mode of dispersal by which early postlarvae may be carried along the bottom owing to increased drag provided by a single byssus thread. Active dispersal, such as the migration of adults, encountered among some sublittoral gastropod species, may also account for dispersal on a relatively small spatial scale during the period of a single lifetime or on a larger scale over the course of many generations.

Some Consequences Arising from Different Modes of Development

Mode of development may affect the life history of a deep-sea molluscan species in at least three ways: (1) it may influence the patterns of colonization and recruitment and consequently the demography of its populations; (2) it may affect the amount of genetic exchange between its disjunct populations; and (3) it may direct the geographical distribution or biogeography of a species.

Colonization

Colonization in the deep sea has been studied mostly in sedimentary ("soft") bottom communities using trays or boxes containing abiotic sediments placed on the bottom and subsequently recovered at various time intervals (Grassle 1977; Desbruyères et al. 1985; Grassle and Morse-Porteus 1987). The general conclusion of these experiments is that the rates of colonization in the deep sea are very slow, although initial colonizers appear within a few months; "trays of azoic sediment resting on the bottom for 5 years did not achieve the density of individuals or species found in the [surrounding] natural community" (Grassle and Morse-Porteus 1987).

Smith and Hessler (1987) in a recent review state that "Trays deployed for months or years are often dominated by juveniles, suggesting that stable age distributions have not been attained and that [therefore] larval recruitment is important in tray colonization" (p. 36). Such a pattern of recruitment differs from sublittoral habitats which show "much higher rates of colonization, with background levels of abundance, species richness and adult size classes attained within days to weeks."

Tray experiments cannot measure the usual course of recruitment and mortality in naturally occurring populations; instead they reveal the consequences of a disturbance on the bottom. Smith (1985b) demonstrated moreover that tray experiments as they are usually conducted lead to heavily biased results "due to altered hydrodynamics, isolation, and arbitrary space and sampling times" and tend to overestimate the time required for colonization of a disturbed area.

Recent studies have also dealt with hard substrata, notably hydrothermal vents (Van Dover et al. 1988) and manganese nodules (Mullineaux 1988). The presence of larval stages of vent organisms (including molluscan veligers) in the water column overlying established vent communities has been observed by Berg and Van Dover (1987), but it has been difficult to directly relate the presence of larvae to populations on the bottom. Van Dover et al. (1988) found that "size distributions of individuals within a taxon on [experimental] panels deployed for 1,216 days [in the region of hydrothermal vents] are broad, suggesting intermittent or continuous recruitment in many of the vent associated species rather than a single episodic recruitment event." The results, however, are confounded by "some combination of competition, migration and predation [which] maintains . . . populations in immature stages."

Smith and Hessler (1987) noted that because most megafaunal hydrothermal vent species are endemic to a vent environment and "because most species are sessile as adults, planktonic dispersal of larvae must be essential for colonization." Paradoxically most hydrothermal vent molluscs do not have long-lived planktotrophic larvae, but rather short lecithotrophic development with a nonfeeding planktonic larva. But there is some evidence that lecithotrophic larvae may have a greater potential for dispersal than previously supposed (Leal and Bouchet 1991).

Mullineaux (1988) has examined the role of current velocity on settlement onto hard substrata and demonstrated that "settlement onto manganese nodules resting on the sediment in a relatively 'low flow' environment was significantly greater than onto nodules suspended in a 'higher flow' regime." The nature of the flow as well as the velocity at the bottom boundary layer is also important for settlement. Mullineaux and Butman (1990) performed a series of experiments on seamounts using circular flat plates of differing thickness and surface characteristics, and concluded that "larval settlement may be a function of very small-scale variations in the boundary-layer flow, reflecting . . . larval supply to the plate, larval retention on the plate surface, and active larval responses to the flow regime over the plate."

GENETIC EXCHANGE

Comparison of genetic structure allows inferences about genetic exchange between populations. Extensive dispersal in the absence of strong selection can be expected to result in genetic similarity between populations. On the other hand, intensive differentiating selection, even with substantial exchange of larvae, may result in genetic differentiation. However, there may be an apparent similarity even with very restricted gene flow if there are only limited genetic variants to serve as markers of genetic divergence. Restriction in genetic variation can result from repeated colonization and extinction, characteristic of species that live in transitory habitats such as hydrothermal vents, and other disturbed areas. Such restriction in genetic variability results from the founder effect repeated over many successive generations (i.e., only a small fraction of the genetic variation of a species is transmitted into newly available habitat by each generation of founding colonists).

Genetic differentiation between deep-sea populations have seldom been studied. J. P. Grassle (1985) compared the genetic differentiation between widely separated populations of the hydrothermal vent mytilid, *Bathymodiolus thermophilus*. Populations from two regions were examined, one at 13°N on the East Pacific Rise, the other on the Galapagos Rift. Grassle's research demonstrated that "the uniformly large mussels from the sparse 13°N population show low genetic variability and [yet] are clearly differentiated from Galapagos populations at three loci . . . . this indicates that there has been no recent exchange of plankotrophic larval immigrants between these distant populations" 2200 km apart. Further study of two populations on the Galapagos Rift 7.8 km apart showed that recent recruits "have high levels of genetic similarity . . . as do large mussels . . . from the two sites." There are, however, significant differences among recent recruits and large mussels; such differences occur at both sites. It was concluded that "over the lifetime of the [two neighboring] vents, recruits have come from the same small number of genetically distinct populations." The geographic distribution of *Bathymodiolus thermophilus* may be accomplished in part by stepwise dispersal over time.

To summarize, existing data are insufficient to make broad generalizations about patterns of genetic variation, but one may speculate that the mode of development as well as the spatial and temporal scale of environmental heterogeneity will prove important in understanding the genetic differentiation or similarity among populations of deep-sea molluscs.

Biogeography

Geographical distribution is determined by the ability of species to disperse and by physical barriers and ecological constraints that limit such dispersal. The origin of the deep-sea benthic molluscan fauna of the Mediterranean, composed almost entirely of eastern Atlantic species, provides an interesting example (Bouchet and Taviani 1992). Among the total invertebrate macrofauna, including about 6300 species, only 0.3 percent live exclusively below 2000 m. The sill depth of only 280 m at the Straits of Gibraltar is a physical barrier for colonization or exchange of deep-sea species between the eastern Atlantic and Mediterranean. Water from the east originating in the Atlantic enters along the surface; denser water leaves the Mediterranean along the bottom. Mediterranean bottom water has a temperature of ca. 13°C, significantly higher than that of the Atlantic. Hence deep-sea benthic organisms entering the Mediterranean must overcome not only the physical barrier of the sill at Gibraltar, but also the ecological constraint imposed by the high temperature of bottom water, at least 10°C greater than that of the eastern Atlantic.

Bouchet and Taviani proposed the hypothesis that the entry of eastern Atlantic deep-sea species into the Mediterranean is accomplished by planktonic larvae making "vertical ontogenetic migrations" to the surface followed by transport of the larvae across the sill and hence into the western Mediterranean. After entering, the larvae must drop to the bottom, metamorphose, and survive to reproduce at a temperature much higher than that of their origin. Bouchet and Taviani overcome this objection by proposing that much of the Mediterranean deep-sea benthic fauna consists of expatriate or nonreproducing "pseudopopulations" and that its existence depends entirely upon the continuous influx of larvae from the eastern Atlantic Ocean. Evidence for this interesting hypothesis comes from the geographic distribution of the prosobranch family Turridae. Forty-seven species are known from the Ibero-Moroccan Gulf west of the sill at the Straits of Gibraltar. Among these, 18 are also found in the western Mediterranean but only 4 have nonplanktonic development. Whether reproduction occurs among indigenous populations is unknown because of a lack of living material occurring in bottom samples. Indeed most deep-sea dredge hauls taken in the Mediterranean consist almost entirely of dead shells. Veliger larvae of none of the 14 species have been reported in surface waters of either the eastern Atlantic, Straits of Gibraltar, or eastern Mediterranean.

The geographic distributions of hydrothermal vent faunas offer an insight into the possible role of larval dispersal for the biogeography and evolution of deep-sea benthos (Van Dover 1990). Tunnicliffe (1988) has compared the faunas of the Northeast Pacific Ridges with those of the equatorial East Pacific Rise. Within each of the two regions there was a high degree of faunal similarity. However between the widely separated regions of the Juan de Fuca Ridge and the East Pacific Rise, a distance of > 3000 km, there was differentiation of molluscs at the species level but considerable similarity in genera. Tunnecliffe, on the basis of faunal evidence, concluded that "a physical disjunction—a vicariance event—. . . may have caused sufficient separation in time and space to establish an endemic northeast fauna, with few species but many genera in common . . . [with the East Pacific Rise]." This "vicariance event" was the southwesterly motion of the North American Plate which "interposed as a barrier between the northeast and equatorial spreading ridges about 35 Ma BP" (see Tunnicliffe 1988:357–358, figs. 2 and 3), thereby separating into two portions a once continuous ridge and interrupting a natural route of "stepping stones" for dispersal and gene flow along its axis. Apparently dispersal allows species similarity and genetic continuity between hydrothermal vents within relatively close range of each other but seldom between the 3000 kilometers between Juan de Fuca and the East Pacific Rise.

The dearth of information on the mode of reproduction and the capacity for dispersal makes biogeographical explanations still largely speculative and emphasizes the need for more information on the reproduction of the deep-sea benthos.

Conclusions and Summary

Adaptations for reproduction among deep-sea benthic molluscs are fully as varied as those encountered among more familiar sublittoral species. Such diversity in reproduction corresponds with the wide diversity of deep-sea environments heretofore unsuspected and only recently discovered. Thus there are large areas of physically monotonous, seemingly stable benthic environments such as ocean basins underlying oligotrophic seas. In contrast are temporary and temporally unpredictable and unstable deep-sea benthic environments, such as (1) regions of deep-sea storms along the western borders of oceans; (2) canyon slopes giving rise to turbidity currents; (3) hydrothermal fields arrayed along a global mid-ocean ridge system; and (4) a number of other ephemeral habitats. At a smaller scale, local topographic features may affect flow along the bottom and serve as microhabitats for benthic species.

Most molluscan species for which data are available reproduce continuously and show no seasonal periodicity. Only regions with periodic and high seasonal productivity at the sea surface have reoccurring cycles of seasonal reproduction, although the patterns do not seem temporally coupled to surface productivity.

Apparent fecundity varies widely among deep-sea bivalves but neither the rate of gametogenesis nor the absolute or lifetime fecundity can be determined for continuously or intermittently reproducing iteroparous species whose longevity is undetermined. Estimates of reproductive effort therefore are not possible for most deep-sea molluscan species.

As in the sublittoral, the mode of development among bathyal and abyssal molluscs differs characteristically for different higher taxa. The predominant mode of development among bivalves is a short lecithotrophic planktonic development (ca. two-thirds of those species for which data exist). Such development is a reflection of the predominance of Protobranchia, none of which are known ever to have a planktotrophic development (i.e., feeding planktonic larvae). Among gastropods, planktotrophic development with feeding planktonic larvae is common among approximately one-third of all species, but the ecology of the veliger larvae is entirely unknown.

Hypotheses about "ontogenetic vertical migration" are based largely on indirect evidence and, until now, the larvae of only two deep-water eurybathic gastropod species have been seen in surface waters. The alternate hypothesis is proposed that many deep-sea molluscan larvae that feed in the plankton may not be obligate phytoplanktotrophs but rather heteroplanktotrophs feeding upon protozoa, flagellate and other heterotrophs or alternately utilizing dissolved organic material on or near the seafloor.

Both semelparity and iteroparity are encountered among deep-sea molluscs. Semelparous species, those with a short life span that quickly attain sexual maturity, produce large numbers of gametes and reproduce only once during their lifetime, are encountered among some deep-sea ephemeral habitats (e.g., some species of *Xylophaga*). Iteroparous species that live longer and reproduce either intermittently or continuously throughout their sexually mature life generally are to be found in more stable deep-sea environments. The latter appear to predominate among deep-sea molluscs.

The mode of development has important consequences for the biogeography and evolution of deep-sea molluscs, but until more data are accumulated it will not be possible to evaluate the role of dispersal[3].

Author's Note: In writing this article I have not attempted to be exhaustive in covering all relevant literature; this would probably require a book-length manuscript and would need to consider reproduction in other invertebrate taxa. Rather, I have selected areas for further research in the discussion and summary above.

Acknowledgments

I am much indebted and thank my colleagues Lauren Mullineaux, Cindy Lee Van Dover, and Robert Hessler for reading an earlier draft of this article. I have incorporated many of their suggestions. The writing of this paper was partially supported by the Woods Hole Oceanographic Institution. Contribution Number 7223 from the Woods Hole Oceanographic Institution.

References

Allen, J. A. and H. L. Sanders. 1973. Studies on deep-sea protobranchia (Bivalvia): The families Siliculidae and Lametilidae. *Bull. Mus. Comp. Zool., Harvard* 145:263–309.

Allen, J. A. and H. L. Sanders. 1982. Studies on the deep-sea protobranchia; the sub-family Spinulinae (family Nuculanidae). *Bull. Mus. Comp. Zool., Harvard* 150:1–30.

Allen, J. A. and J. F. Turner. 1974. On the functional morphology of the family Verticordiidae (Bivalvia) with descriptions of new species from the abyssal Atlantic. *Phil. Trans. Roy. Soc. Lond.* B 268:401–536.

Baldwin, B. S. and R. I. E. Newell. 1991. Omniverous feeding by planktotrophic larvae of the eastern oyster *Crassostrea virginica*. *Mar. Ecol. Prog. Ser.* 78:285–301.

Barnes, H. and M. Barnes. 1958. The rate of development of *Balanus balanoides* L. larvae. *Limnol. Oceanogr.* 3:29–32.

Berg, C. J. 1985. Reproductive strategies of mollusks from abyssal hydrothermal vent communities. *Biol. Soc. Washington Bull.* 6:185–197.

Berg, C. J. and C. L. Van Dover. 1987. Benthopelagic macrozooplankton communities at and near deep-sea hydrothermal vents in the eastern Pacific ocean and the Gulf of California. *Deep-Sea Res.* 34:379–401.

Billett, D. S. M., R. S. Lampitt, and A. L. Rice. 1983. Seasonal sedimentation of phytoplankton to the deep-sea benthos. *Nature* 302:520–522.

Blaume, F., G. Graf, P. Linke, W. Ritzran, J. Rumohr, A. Schelz, L. Thomsen, and W. Ziebis. 1991. Processes in the bottom nepheloid layer at the western Barents Sea Shelf and the Continental Slope. *6th Deep-Sea Biol. Symp. Copenhagen, Abstr.*, pp. 16–17.

Bouchet, P. 1976a. Mise en évidence de stades larvaires, planctoniques chez des Gastéropodes Prosobranchs des étages bathyal et abyssal. *Bull. Mus. natl. d' Histoire, natr.* Ser. 3, No. 400, Zool. 277, pp. 947–972.

Bouchet, P. 1976b. Mise en évidence d 'une migration de larves véligéres entre l' étage abyssal et la surface. *C. R. Acad. Sci. Paris* 283:321–824.

Bouchet, P. 1977. Le developpement larvaire des Gastéropodes de gande profoundeur considérations sur la spéciation these-Diplome de Docteur de 3e cycle à l' universite Pierre et Marie Curie Paris, 38 pp, 13 pl.

Bouchet, P. and J.-C. Fontes. 1981. Migrations Verticales des larves de Gastéropodes abyssaux: Arguments nouveaux dûs analyse isotopique de la cozuille larvaire et postlarvaire. *C. R. Acad. Sci. Paris* 292 (sec. 3):1005–1008.

Bouchet, P. and M. Taviani. 1992. The Mediterranean deep-sea fauna: Pseudopopulations of Atlantic Species? *Deep-Sea Res.* 39:169–184.

Bouchet, P. and A. Warén. 1979a. The abyssal molluscan fauna of the Norwegian Sea and its relation to other faunas. *Sarsia* 64:211–243.

Bouchet, P. and A. Warén. 1979b. Planktotrophic larval development in deep-water gastropods. *Sarsia* 64:37–40.

Bouchet, P. and A. Warén. 1980. Revision of the Northeast Atlantic Bathyal and Abyssal Turridae (Mollusca, Gastropoda). *J. Moll. Std.* Suppl. 8, pp. 1–119.

Bouchet, P. and A. Warén. 1985. Revision of the Northeast Atlantic Bathyal and Abyssal Neo-gastropoda (Mollusca, Gastropoda). *Bollettino Malacologico* (Supplemento 1) [Milano], pp. 121–296.

Bouchet, P. and A. Warén. 1986. Revision of the Northeast Atlantic Bathyal and Abyssal

Aclididae, Eulimidae, Epitoniidce (Mollusca, Gastropoda). *Bollettino Malacologico* (Supplemento 2) [Milano], pp. 297–576.

Butman, C. A., W. D. Grant, and D. Stolzenbach. 1986. Predictions of sediment trap bias in turbulent flows: A theoretical analysis based on observations from the literature. *J. Mor. Res.* 44:601–644.

Calow, P. 1979. The cost of reproduction—a physiological approach. *Biol. Rev.* 54:23–40.

Cavanaugh, C. M. 1983. Symbiotic chemoautotrophic bacteria in marine invertebrates from sulfide-rich habitats. *Nature, Lond.* 302:58–61.

Chia, F.-S., J. Buckland-Nicks, and C. M. Young. 1984. Locomotion of marine invertebrate larvae: A review. *Canadian J. Zool.* 62:1205–1222.

Cole, L. C. 1954. The population consequences of life history phenomena. *Quart. Rev. Biol.* 29:103–137.

Colman, J. G. and P. A. Tyler. 1988. Observations on the reproductive biology of the deep-sea trochid *Calliotropis ottoi* (Philippi). *J. Moll. Stud.* 54:239–242.

Colman, J. G., P. A. Tyler, and J. D. Gage. 1986. Larval development of deep-sea gastropods (Prosobranchia: Neogastropoda) from the Rockall Trough. *J. Mar. Biol. Ass. U. K.* 66:951–965.

Davis, G. 1985. The physical characteristics, population structure, and reproductive strategy of the deep-sea bivalve *Nueula cancellata. 4th Deep-Sea Biol. Symp.*

Dengler, A. T., W. Wilde, E. K. Noda, and W. R. Normark. 1984. Turbidity currents generated by Hurricane Iwa. *Geo-Mar. Letters* 4:5–11.

Desbruyères, D., J. W. Deming, and A. Dinet. 1985. Réactions de l' écosystéme benthique profond aux perturbations: Nouveaux résultats experimentaux. In L. Laubier and C. Monniot, eds., *Peuplements profonds du Golfe de Gascogne Campagnes Biogas,* pp. 193–208. Institut Francais de Recherche Pour l' Eploitation des la Mer (IFREMER), 629 pp.

Deuser, W. G. and E. H. Ross. 1980. Seasonal change in the flux of organic carbon to the deep sea. *Nature* 283:364–365.

Drew, G. A. 1899. Some observations on the habits, anatomy, and embryology of members of the protobranchia. *Anat. Anz.* 15:493–519.

Gage, J. D. 1991. Biological rates in the deep sea: A perspective from studies on processes in the benthic boundary layer. *Rev. Aquatic Sci.* 5:49–100.

Gage, J. D. and P. A. Tyler. 1991. *Deep-Sea Biology: A Natural History of Organisms at the Deep-Sea Floor.* Cambridge: Cambridge University Press.

Gage, J. D., P. A. Tyler, G. Davies, and R. Harvey. 1986. Life-history studies on deep-sea protobranch bivalves from the Rockall Trough (N.E. Atlantic). *Ninth Internal. Malacological Congress, Unit as Malacologica Abstracts,* p. 27.

Giles, L. A., P. A. Tyler, and J. D. Gage. 1991. Diet and reproduction in the deep-sea protobranch *Ledella pustulosa* and *Malletia cuneata. 6th Deep-Sea Biol. Symp. Copenhagen, Abstr.,* p. 84.

Gilkinson, K. D., J. A. Hutchings, P. E. Oshel, and R. L. Haedrich. 1986. Shell microstructure and observations on interval banding patterns in the bivalves *Yoldia thraciaeferins* Storer, 1838, and *Nuculana permula* Miller, 1779 (Nuculanidae), from a deep-sea environment. *Veliger* 29:70–77.

Grant, A. and P. A. Tyler. 1983. The analysis of data in studies of invertebrate reproduction. Part 2: The analysis of oocyte size/frequency data, and comparison of different types of data. *Internat. J. Invert. Reprod.* 6:271–183.

Grassle, J. F. 1977. Slow recolonization of deep-sea sediment. *Nature* 265:618–619.

Grassle, J. F. 1985. Hydrothermal vent animals: Distribution and biology. *Science* 229:713–717.

Grassle, J. F. 1986. The ecology of deep-sea hydrothermal vent communities. *Adv. Mar. Biol.* 23:301–362.

Grassle, J. F. and L. S. Morse-Porteous. 1987. Macrofaunal colonization of disturbed deep-sea environments and the structure of deep-sea benthic communities. *Deep-Sea Res.* 34:1911–1950.

Grassle, J. P. 1985. Genetic differentiation in populations of hydrothermal vent mussels *Bathymodiolus thermophilous* from the Galapagos Rift and 13°N on the East Pacific Rise. *Biol. Soc. Washington Bull.* 6:429–442.

Gross, T. F., A. J. Williams III, and A. R. M. Nowell. 1988. A deep-sea sediment transport storm. *Nature* 331:518–520.

Gustafson, R. G. and R. G. B. Reid. 1986. Development of the pericalymma larva of *Solemeya reidi* (Bivalvia: Cryptodonta: Solemyidae) as revealed by light and electron microscopy. *Mar. Biol.* 93:411–427.

Gustafson, R. G., D. T. J. Littlewood, and R. A. Lutz. 1991. Gastropod egg capsules and their content from deep-sea hydrothermal vent environments. *Biol. Bull.* 180:34–55.

Hecker, B. 1985. Fauna from a cold sulfur-seep in the Gulf of Mexico: Comparison with hydrothermal vent communities and evolutionary implications. *Biol. Soc. Washington Bull.* 6:465–473.

Heezen, B. C. and M. Ewing. 1952. Turbidity currents and submarine slumps and the 1929 Grand Banks earthquake. *Amer. J. Sci.* 250:849–873.

Hessler, R. R. and P. F. Lonsdale. 1991. Biogeography of Mariana Trough hydrothermal vent community. *Deep-Sea Res.* 38:185–199.

Hinga, K. R., J. McN. Sieburth, and G. Ross Heath. 1979. The supply and use of organic material at the deep sea floor. *J. Mar. Res.* 37:557–578.

Hollister, C. D. and I. N. McCave. 1984. Sedimentation under deep-sea storms. *Nature* 309:220–225.

Houbrick, R. S. 1991. Functional inference from gastropod shell morphology. *Lethaia* 24:265–270.

Hutchings, J. A. and R. L. Haedrich. 1984. Growth and population structure of two species of bivalves (Nuculanidae) from the deep-sea. *Mar. Ecol. Prog. Ser.* 17:135–142.

Jannasch, H. W. 1984a. Chemosynthesis: The nutritional basis for life at deep-sea vents. *Oceanus* 27:73–78.

Jannasch, H. W. 1984b. Microbial processes of deep-sea hydrothermal vents. In P. A. Rona, K. Bostrom, L. Laubier, and K. L. Smith, eds., *Hydrothermal Processes at Seafloor Spreading Centers,* pp. 677–709. New York: Plenum.

Jannasch, H. W. 1985. The chemosynthetic support of life and the microbial diversity at deep- sea hydrothermal vents. *Proc. Roy. Soc. Lond.* B 225:277–297.

Juniper, S. K. and M. Sibuet. 1987. Cold seep benthic communities in Japan subduction zones: Spatial organization, trophic strategies, and evidence for temporal evolution. *Mar. Ecol. Prog. Ser.* 40:115–126.

Killingley, J. S. and M. A. Rex. 1985. Mode of larval development in some deep-sea gastropods indicated by oxygen-18 values of their carbonate shells. *Deep-Sea Res.* 32:809–818.

Knudsen, J. 1961. The bathyal and abyssal Xylophaga (Pholadidae, Bivalvia). *Galathea Rept.* 5:163–209.

Knudsen, J. 1979. Deep Sea Bivalves. In S. V. D. Spoel, A. C. V. Bruggen, and J. Lever, eds., *Pathways in Malacology,* pp. 195–224. Utrecht: Scheltema, Holkema.

Lampitt, R. S. 1985. Evidence for the seasonal deposition of detritus to the deep-sea floor and its subsequent resuspension. *Deep-Sea Res.* 32:885–897.

Lampitt, R. S., T. Moji, and B. Von Bodungen. 1990. What happens to zooplankton fecal pellets? Implications for material flux. *Mar. Biol.* 104:15–23.

Leal, J. H. and P. Bouchet. 1991. Distribution patterns and dispersal of prosobranch gastropods along a seamount chain in the Atlantic Ocean. *J. Mar. Biol. Ass. U. K.* 71:11–25.

Lightfoot, R. H., P. A. Tyler, and J. D. Gage. 1979. Seasonal reproduction in deep-sea bivalves and brittlestars. *Deep-Sea Res.* 26A:967–973.

Lutz, R. A. 1988. Dispersal of organisms at deep-sea hydrothermal vents: A review. *Oceanol. Acta. Hydrothermlism, Biol. Ecol. Symp. Paris,* pp. 23–29.

Lutz, R. A., D. Jablonski, D. C. Rhoads, and R. D. Turner. 1980. Larval dispersal of a deep-sea hydrothermal vent bivalve from the Galapagos Rift. *Mar. Biol.* 57:127–133.

Lutz, R. A., D. Jablonski, and R. D. Turner. 1984. Larval development and dispersal at deep-sea hydrothermal vents. *Science* 226:69–72.

Marshall, B. A. 1986. New Zealand deep-sea mollusca from biogenic substrata. *Ninth Internatl. Malacal. Congr.; Unitas Malacologica, Edinburgh, Scotland [Abstr.].*

McCave, I. N. 1975. Vertical flux of particles in the ocean. *Deep-Sea Res.* 22:491–502.

McDonald, K. C., K. Becker, F. N. Spiess, and R. D. Ballard. 1980. Hydrothermal heat flux of the "black smoker" vents on the East Pacific Rise. *Earth Planet. Sci. Lett.* 48:1–7.

Meadows, P. S. and J. Tait. 1985. Bioturbation, geotechnics, and microbiology at the sediment water interface in deep-sea sediments. In P. E. Gibbs, ed., *Proceedings of the 19th European Marine Biology Symposium,* pp. 191–199. Cambridge: Cambridge University Press.

Menzies, R. J. 1965. Conditions for the existence of life on the abyssal sea floor. *Oceanogr. Mar. Biol. Ann. Rev.* 3:195–210.

Mullineaux, L. S. 1988. The role of settlement in structuring a hard-substratum community in the deep sea. *J. Exp. Mar. Biol. Ecol.* 120:247–261.

Mullineaux, L. S. and C. A. Butman. 1990. Recruitment of encrusting benthic invertebrates in boundary-layer flows: A deep-water experiment on cross seamount. *Limnol. Oceanogr.* 35:409–423.

Mullineaux, L. S., P. H. Wiebe, and E. T. Baker. 1991. Hydrothermal vent plumes: Larval highways in the deep sea? *Oceanus* 34:64–68.

Ockelmann, K. W. 1965. Developmental types in marine bivalves and their distribution along the Atlantic Coast of Europe. In L. R. Cox and J. F. Peake, eds., *Proc. 1st Europ. Malac. Congr. London 1962,* pp. 25–35.

O'Foighil, D. 1986. Prodissoconch morphology is experimentally modified in the breeding bivalve *Lasaea subviridis. Mar. Biol.* 92:317–524.

Orton, J. H. 1920. Sea-temperature, breeding and distribution of marine animals. *J. Mar. Biol. Assn. U. K.* 12:339–366.

Payne, C. M. and J. A. Allen. 1991. The morphology of deep-sea Thyasiridae (Mollusca: Bivalvia) from the Atlantic Ocean. *Phil. Trans. Roy. Soc. Lond.* B 334:481–562.

Rex, M. A. and A. Warén. 1982. Planktotrophic development in deep-sea prosobranch snails from the western North Atlantic. *Deep-Sea Res.* 29:171–184.

Rex, M., C. A. Van Ummersen, and R. D. Turner. 1979. Reproductive patterns in the abyssal snail, *Benthonella tenella* (Jeffreys). In S. E. Stancyk, ed., *Reproductive Ecology of Marine Invertebrates,* pp. 173–188. Columbia: University of South Carolina Press.

Richter, G. and G. Thorson. 1975. Pelagische Prosobranchier-larven des Golfes von Neapel. *Ophelia* 13:109–185.

Rokop, F. J. 1974. Reproductive patterns in the deep-sea benthos. *Science* 186:743–745.

Rokop, F. J. 1977. Seasonal reproduction of the brachiopod *Frieleia halli* and the scaphopod *Cadulus californicus* at bathyal depths in the deep sea. *Mar. Biol.* 43:237–246.

Rona, P. A., G. Klinkhammer, T. A. Nelsen, J. H. Trefry, and H. Elderfield. 1986. Black smokers, massive sulfides, and vent biota at the Mid-Atlantic Ridge. *Nature* 321:33–37.

Sanders, H. L. and J. A. Allen. 1973. Studies on deep-sea protobranchia (bivalvia); prologue and pristiglomidae. *Bull. Mus. Comp. Zool., Harvard* 145:237–262.

Sanders, H. L. and J. A. Allen. 1977. Studies on deep sea protobranchia (bivalvia): The family tindariidae and the genus *Pseudotindaria*. *Bull. Mus. Comp. Zool., Harvard* 148:23–59.

Schein, E. 1989. Pectinidae (mollusca, bivalvia) bathyaux et abyssaux des carmpagnes BIOGAS (Golf de gascogne). Systematique et Biogéographique. *Ann. Inst. Oceanogr. (Paris)* 65:59–129.

Scheltema, R. S. 1967. The relationship of temperature to the larval development of *Nassarius obsoletus* (gastropoda). *Biol. Bull.* 132:253–265.

Scheltema, R. S. 1972. Reproduction and dispersal of bottom dwelling deep-sea invertebrates: A speculative summary. In R. W. Brauer, ed., *Barobiology and Experimental Biology of the Deep Sea,* pp. 58–66. University North Carolina, North Carolina Sea Grant Program.

Scheltema, R. S. 1986. Long-distance dispersal by planktonic larvae of shoal-water benthic invertebrates among central Pacific Islands. *Bull. Mar. Sci.* 39:241–256.

Schenck, H. G. 1939. Revised nomenclature for some nuclid pelecypods. *J. Paleontology* 13:21–41.

Shimek, R. L. 1986. The biology of the Northeastern Pacific Turridae. V. Demersal development, synchronous settlement and other aspects of the larval biology of *Oenopota levidensis*. *Internat. J. Invert. Reprod. Develop.* 10:313–333.

Shuto, T. 1983. New turrid taxa from the Australian waters. *Mem. Fac. Sci. Kyushu Univ., Ser. D. Geol.* 25:1–26.

Sigurdsson, J. B., C. W. Titman, and P. A. Davies. 1976. The dispersal of young post-larval bivalve molluscs by byssus threads. *Nature* 262:386–387.

Smith, C. R. 1985a. Food for the deep sea: Utilization, dispersal, and flux of nekton falls at the Santa Catalina Basin floor. *Deep-Sea Res.* 32:417–442.

Smith, C. R. 1985b. Colonization studies in the deep-sea: Are results biased by experimental design? In P. Gibbs, ed., *Proceedings of the 19th European Marine Biology Symposium,* pp. 183–189. Cambridge: Cambridge University Press.

Smith, C. R. and R. R. Hessler. 1987. Colonization and succession in deep-sea ecosystems. *Trends Ecol. Evol.* 2:259–263.

Smith, C. R., H. Kukert, R. A. Wheatchroft, P. A. Jumars, and J. W. Deming. 1989. Vent fauna on whale remains. *Nature* 341:27–28.

Sokolova, M. N. 1972. Trophic structure of deep-sea macrobenthos. *Mar. Biol.* 16:1–12.

Southward, A. J. and E. C. Southward. 1982. The role of dissolved organic matter in the nutrition of deep sea benthos. *Amer. Zool.* 22:647–658.

Stearns, S. C. 1976. Life-history tactics: A review of the ideas. *Quart. Rev. Biol.* 51:3–47.

Strathmann, R. R. 1978. The evolution and loss of feeding larval stages of marine invertebrates. *Evol.* 32:894–906.

Suess, E., B. Carson, S. D. Ritger, J. C. Moore, M. L. Jones, L. D. Kulm, and G. R. Cochrane. 1985. Biological communities at vent sites along the subduction zone off Oregon. *Biol. Soc. Washington Bull.* 6:475–484.

Thistle, D., J. Y. Yingst, and K. Fauchald. 1985. A deep-sea benthic community exposed to strong near-bottom currents on the Scotian rise (Western Atlantic). *Mar. Geol.* 66:91–112.

Thorson, G. 1946. Reproduction and larval development of Danish marine bottom invertebrates. *Medd. Dan. Fisk. Havunders., Ser. Plankton* 4:1–523.

Thorson, G. 1950. Reproductive and larval ecology of marine bottom invertebrates. *Biol. Rev.* 25:1–45.

Tucholke, B. E., C. D. Hollister, P. E. Biscaye, and W. D. Gardner. 1985. Abyssal current character determined from sediment bed forms on the Nova Scotian Continental Rise. *Mar. Geol.* 66:43–57.

Tunnicliffe, V. 1988. Biogeography and evolution of hydrothermal-vent fauna in the eastern Pacific Ocean. *Proc. Roy. Soc. Lond.* B 233:347–366.

Tunnicliffe, V. 1991. The biology of hydrothermal vents: Ecology and evolution. *Oceanogr. Mar. Biol. Ann. Rev.* 29:319–407.

Turekian, K. K., J. K. Cochran, D. P. Kharkar, R. M. Cerrato, J. R. Visnys, H. L. Sanders, J. F. Grassle, and J. A. Allen. 1975. Slow growth rate of a deep-sea clam determined by ^{228}Ra chronology. *Proc. Nat. Acad. Sci. USA* 72:2829–2832.

Turner, R. D. 1973. Wood-boring bivalves, opportunistic species in the deep sea. *Science* 180:1377–1379.

Turner, R. D., R. A. Lutz, and D. Jablonski. 1985. Modes of molluscan larval development at deep-sea hydrothermal vents. *Biol. Soc. Washington Bull.* 6:167–184.

Tyler, P. A. 1988. Seasonality in the deep sea. *Oceanogr. Mar. Biol. Annu. Rev.* 22:227–250.

Van Dover, C. L. 1990. Biogeography of hydrothermal vent communities along seafloor spreading centers. *Trends Ecol. Evol.* 5:242–246.

Van Dover, C. L., C. J. Berg, and R. D. Turner. 1988. Recruitment of marine invertebrates to hard substrates at deep-sea hydrothermal vents on the East Pacific Rise and Galapagos spreading center. *Deep-Sea Res.* 35:1833–1849.

Vangriesheim, A. and A. Khripounoff. 1990. Near-bottom particle concentration and flux: Temporal variations observed with sediment traps and nephelometer on the Meriadzek terrace, Bay of Biscay. *Prog. Oceanogr.* 24:103–116.

Waller, T. R. 1981. Functional morphology and development of veliger larvae of the European oyster, *Ostrea edulis* Linné. *Smithsonian Contr. Zool.,* no. 328, 70 pp.

Wiebe, P. H., L. P. Madin, L. R. Haury, G. R. Harbison, and L. M. Philbin. 1979. Diel vertical migration by *Salpa aspera* and its potential for large-scale particulate organic matter transport to the deep sea. *Mar. Biol.* 53:249–255.

Wolff, T. 1979. Macrofaunal utilization of plant remains in the deep sea. *Sarsia* 64:117–136.

Wright, S. H. and G. C. Stephens. 1982. Transepidermal transport of amino acids in the nutrition of marine invertebrates. In W. G. Ernst and J. G. Morin, eds., *The Environment of the Deep Sea.* Ruby Vol. 2., pp. 302–323. Englewood, Cliffs, N.J.: Prentice Hall.

CHAPTER FOUR

Molluscan Life History Traits at Deep-Sea Hydrothermal Vents and Cold Methane/Sulfide Seeps

Richard G. Gustafson and Richard A. Lutz

Abstract

Despite their apparent geographic isolation, ephemeral nature, and generation of hot, acidic water, deep-sea hydrothermal vents are characterized by a remarkably rich molluscan fauna. Similar isolation and production of toxic effluents characterize deep-sea cold-water methane/sulfide seeps where a dense assemblage of molluscs is also present. These high molluscan population densities in conjunction with extreme environmental conditions suggests the evolution of adaptive life history strategies. We review the available information on mode of larval development (inferred from the size and morphology of larval shells), spawning periodicity, sexuality, fecundity, egg size, sperm morphology, accessory sex glands, and copulatory organs in molluscs at deep-sea vents and seeps. The larval shell dimensions and morphology of two deep-sea bivalves associated with methane/sulfide seeps, and two bivalves and three gastropods from deep-sea hydrothermal vents are described for the first time. Many species demonstrate life history traits that appear to be adaptive in these severe environments. These include rapid growth, continuous reproduction, and internal fertilization. Although most vent and seep molluscs appear to develop without planktotrophic larvae, it is equally true that the mode of development in deep-sea vent and seep molluscs reflects the developmental pattern common to the systematic group to which each species belongs.

Deep-sea hydrothermal vent communities are characterized by long-term ecological instability, geographic isolation, and localized nutrient enrichment (for review see Tunnicliffe 1991). Little is known concerning the longevity of deep-sea methane/sulfide seeps, although they are also patchily distributed. Both seep and vent environments may be thought of as nutritional oases in a deep-sea nutritional desert. Coupled with the constant supply of nutrients present in the highly acidic vent waters are compounds such as sulfides and heavy metals that act as metabolic poisons. Toxic hydrocarbon compounds and sulfides are also present in the mud but not the water column at methane/sulfide seep sites. Despite their apparent ephemeral nature and geographic isolation, deep-sea hydrothermal vent fields located along contiguous ridge systems have remarkably similar faunal components (Tunnicliffe 1991). Likewise, widely separated sites of methane/sulfide seepage in the deep sea share many morphospecies (RAL, RGG unpublished results). How invertebrate populations are established and maintained in habitats that are scattered through time and space, and that appear to us to be basically hostile to life, has intrigued scientists since the discovery of deep-sea hydrothermal vents in the late 1970s.

It has been postulated that invertebrates living in geographically disjunct and temporally unstable hydrothermal vent environments would have an *r-type* life history strategy that includes fast growth, early reproduction, and widespread dispersal of larvae (Desbruyères and Laubier 1984: Grassle 1985). Because bivalve and gastropod molluscs preserve in their shells a relatively complete ontogenetic record of growth, including the larval shell from which the early developmental strategies may be inferred (see below), molluscs have provided much of the available information on both growth rates and early life histories of hydrothermal vent fauna (Lutz 1988).

Because laboratory culture of deep-sea organisms has proved difficult (Turner et al. 1985), our perceptions about reproduction and mode of larval development in vent and seep molluscs have, by necessity, been inferred from indirect evidence. Anatomical studies have yielded information on spawning periodicity, sexuality (gonochorism or hermaphroditism), fecundity, egg size, sperm morphology, and secondary sexual structures such as copulatory organs, prostate glands, sperm receptacles, and capsule glands (Fretter et al. 1981; Berg 1985; McLean and Haszprunar 1987; Fretter 1988, 1989, 1990; Warén and Bouchet 1986, 1989; Haszprunar 1989a, b; Okutani et al. 1989; McLean 1990b; Healy 1990; Scheltema and Kuzirian 1991; Warén and Ponder 1991).

The mode of larval development in prosobranch gastropods and bivalves from seeps and vents has been classified as either planktotrophic or nonplanktotrophic based on larval shell morphology, using criteria formulated by many researchers for these taxa in shallow water (see review of topic in Jablonski and Lutz 1980, 1983). The amount of nutrient in the egg is reflected in the size of the initial larval shell, Prodissoconch I (PI) in bivalves and Protoconch I (PI) in gastropods. The degree to which feeding in the plankton occurs is reflected in the size of the second part of the larval shell, which includes the PI and is termed Prodissoconch II (PII) in bivalves and Protoconch II (PII) in gastropods. The most reliable method for determining developmental mode from prodissoconch and protoconch morphologies is to make comparisons with confamilial or congeneric species whose developmental history is known (Jablonski and Lutz 1983). A caution is in order: criteria used to infer mode of larval development are based on observations of taxonomically related shallow-water species and the extent to which these same criteria reflect the early life history of deep-sea bivalves and gastropods is debatable.

The terms lecithotrophic or nonplanktotrophic imply only that a larva obtains its entire energy supply in the form of yolk derived from the parent, whereas planktotrophic implies that larval energy is obtained from feeding in the plankton, in addition to a certain amount of yolk also supplied by the parent. By definition, planktotrophic larvae are planktonic and as a consequence may become widely dispersed. Likewise, nothing concerning the larva's dispersal ability is implied by the terms lecithotrophy or nonplanktotrophy, although if the larvae are planktonic significant dispersal may occur (Crisp 1974; Hines 1986; Strathmann 1987).

In this contribution we review published information on adult reproductive anatomy, morphology of larval shells, and the contents of egg capsules of hydrothermal vent and methane/sulfide seep molluscs and present some new data that address the question: Have hydrothermal vent and methane/sulfide seep molluscs modified their life history traits in response to life in these unique environments?

Materials and Methods

Specimens were collected on various DSV Alvin and DSRV Johnson-Sea-Link I dives. Maximum dimensions of prodissoconchs and protoconchs and the number of volutions of gastropod larval shells were measured from scanning electron micrographs. Scanning electron microscopy methods are described elsewhere (Gustafson et al. 1991). Maximum diameter of the entire protoconch was defined as the straight-line distance from the protoconch-teleoconch boundary to the opposite side of the protoconch in the region of greatest width. Maximum diameter of the PI was defined as the greatest straight-line distance from the PI-PII boundary in species with planktotrophic development, or from the PI-teleoconch boundary to the opposite side of the PI in the region of greatest width in nonplanktotrophic species lacking a PII (Lima and Lutz 1990). Maximum diameter measurements of the protoconch should not be confused with the "first whorl diameter," which is a measurement of the straight-line distance tangent to the straight beginning of the suture and extended in both directions to where it intersects the nearest suture (Lima and Lutz 1990, their fig. 1A).

Results

Approximately 144 described and putative new species of molluscs (8 Aplacophora, 1 Polyplacophora, 65 Bivalvia, 68 Gastropoda, and 1 Cephalopoda) are known to occur in hydrothermal vent and cold-water seep environments worldwide. No information is available concerning life history strategies of Polyplacophora or Cephalopoda from vents and seeps.

Reproduction of only one aplacophoran has been studied. As in other aplacophorans, *Helicoradomenia juani* Scheltema and Kuzirian from Northeast Pacific hydrothermal vents is hermaphroditic and practices internal fertilization (Scheltema and Kuzirian 1991). Aplacophorans exhibit development ranging from brooding to planktonic nonplanktotrophy (Hyman 1967). The type of development in *H. juani* cannot be deduced from the available evidence.

Information pertaining to life history strategies is available for only 3 of the 66 bivalves known from deep-sea vents and seeps. In this contribution, the prodissoconchs of 4 more species of vent and seep bivalves are illustrated for the first time (figures 4.1–4.6).

The hydrothermal vent mytilid *Bathymodiolus thermophilus* Kenk and Wilson is native to several sites on the East Pacific Rise (EPR) and has a PI ranging from 95 to 110 μm in length, of undetermined sculpture due to erosion, and a PII of over 400 μm in length, with a sculpture consisting of concentric growth lines alone. These larval shell characters are suggestive of a planktotrophic larval stage (Lutz et al. 1980; Turner and Lutz 1984). Likewise, Berg (1985) reported a juvenile *B. thermophilus* collected at the Galapagos Rift with a PI length of 108 μm and a PII length of 470 μm. Kenk and Wilson (1985) postulated that the small size of the PI in this species suggests high fecundity. Analysis of gonad sections indicates that *B. thermophilus* individuals from the Galapagos Rift delay reproduction, not becoming fully mature until about 44 percent of the maximum size, are protandrous hermaphrodites, spawn continuously, and have eggs of approximately 50 μm in diameter (Berg 1985). On the other hand, Le Pennec et al. (1983) report that *Bathymodiolus* sp. from 11°N–13°N on the EPR (presumably the same species that occurs at the Galapagos Rift: *B. thermophilus*) are gonochoristic in contrast to the protandrous hermaphroditism reported for *B. thermophilus* from the Galapagos Rift (Berg 1985). However, *Bathymodiolus* sp. from 11°N–13°N are reported to spawn continuously and have egg sizes of 40–50 μm (Le Pennec et al. 1983; Le Pennec 1988), characters shared with Galapagos Rift *B. thermophilus* (Berg 1985). The small egg size postulated large fecundity, and size and morphology of the PI and PII suggest that *B. thermophilus* develops planktotrophically.

Species descriptions of a number of new deep-sea vent and seep mussels in the bivalve family Mytilidae are in preparation in collaboration with Dr. Ruth Turner. In the interim, several of these species are referred to here by vernacular names established to avoid the creation of manuscript names and to supply a standard of reference so that all researchers can recognize the species we are dealing with. These specimens are referred to as either "Seep Mytilid" or "Vent Mytilid," followed by an uppercase roman numeral designation and in some cases by a lowercase letter designation where subspecific categories are suspected. Turner and Lutz (1984) illustrated a post-larval specimen of the common mussel from the West Florida Escarpment that has subsequently been assigned the name "Seep Mytilid Va" in preliminary systematic studies (RAL, RGG, and Dr. Ruth Turner unpublished results). This specimen has a PI of approximately 137 μm in length and a PII of about 468 μm in length, as measured from a published micrograph (Turner and Lutz 1984, their fig. 8). Prodissoconchs of two separate mytilid mussel species from the Louisiana Slope hydrocarbon seeps that have been assigned the names "Seep Mytilid Ia" and "Seep Mytilid III" in preliminary systematic studies (RAL, RGG, and Dr. Ruth Turner unpublished results), are illustrated for the first time in figures 4.1–4.2 and figure 4.3, respectively. The PI and PII of "Seep Mytilid Ia" range in length from 100 to 110 μm and from 385 to 404 μm, respectively (figures 4.1, 4.2). Although the PI sculpture is eroded in all "Seep Mytilid Ia" specimens available, the sculpture of concentric growth lines alone on the PII is consistent with planktotrophic development (figures 4.1, 4.2). The PII, on a single specimen of the rare "Seep Mytilid III," measured approximately 460 μm in length (figure 4.3) and was devoid of sculpture with the exception of concentric growth lines. The PI was hidden from view due to the prosogyrous nature of shell growth in this species. The PII of an unidentified mytilid mussel from the hydrothermal vents on the southern Juan de Fuca Ridge that has been assigned the name "Vent Mytilid III" in preliminary systematic studies (RAL, RGG, and Dr. Ruth Turner unpublished results), has a length of 400 μm (figure 4.4). The PI of this specimen is eroded. The larval shell size and morphology of Seep Mytilids Va, Ia, III, and Vent Mytilid III are all indicative of planktotrophic development.

Based on measurement of the vacant eroded space once occupied by the prodissoconch, Schein-Fatton (1985) estimated the prodissoconch length of the hydrothermal vent scallop *Bathypecten vulcani* Schein-Fatton, from 13°N on the EPR, at 80 to 100 μm. Actual PI and PII lengths of *B. vulcani* from the Galapagos Rift range from 147 to 159 μm and from 156 to 175 μm, respectively (figures 4.5, 4.6), suggesting nonplanktotrophic development. Other closely related nonvent deep-sea pectinids *B. eucymatus* (Dall), *Cyclopecten ambiannulatus* Schein, and *Hyalopecten undatus* (Verrill and Smith) have similar prodissoconch sizes and morphologies (Schein 1989), indicating that nonplanktotrophy in *B. vulcani* is not a modification attributable to life in a vent habitat.

Although the prodissoconch morphology of the hydrothermal vent vesicomyid bivalve *Calyptogena magnifica* Boss and Turner is unknown (Turner et al. 1985), estimates of a maximum oocyte diameter of 309 μm (Berg and Turner 1980) and a mean ovarian oocyte diameter of 150 μm (Berg 1985) suggest this species develops nonplanktotrophically. Analysis of gonad sections reveals that

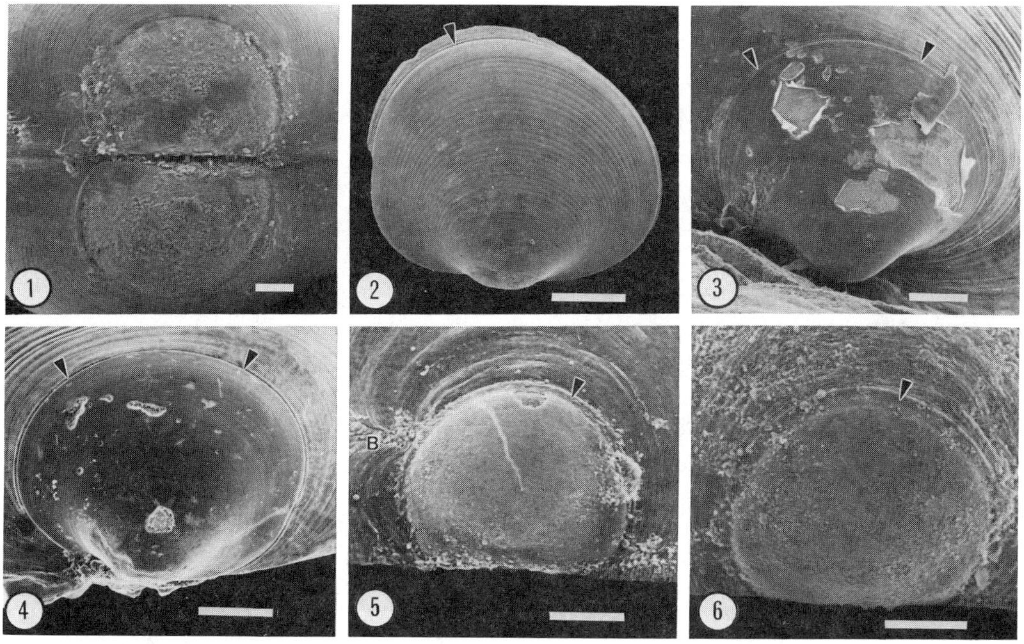

FIGURE 4.1. Enlarged view of corroded Prodissoconch I retained on post-larval specimen of Seep Mytilid Ia from Louisiana Slope hydrocarbon seeps showing hinge line and remnants of punctate sculpture, Johnson-Sea-Link I Dive 3137 (Green Canyon - 272).

Scale bar = 20 μm.

FIGURE 4.2. Prodissoconch II of Seep Mytilid Ia from Louisiana Slope hydrocarbon seeps showing conspicuous concentric sculpture, Johnson-Sea-Link I Dive 3137 (Green Canyon - 272). Arrowhead, Prodissoconch II - dissoconch boundary.

Scale bar = 100 μm.

FIGURE 4.3. Larval shell, Prodissoconch II, retained on adult specimen of Seep Mytilid III from Louisiana Slope hydrocarbon seeps showing conspicuous concentric sculpture (near Garden Banks - 386). Arrowheads, Prodissoconch II - dissoconch boundary.

Scale bar = 100 μm.

FIGURE 4.4. Larval shell, Prodissoconch II, retained on adult specimen of Vent Mytilid III from hydrothermal vents at South Cleft on the Juan de Fuca Ridge showing conspicuous concentric sculpture, Alvin Dive 2427. Arrowheads, Prodissoconch II - dissoconch boundary.

Scale bar = 100 μm.

FIGURE 4.5. Larval shell, Prodissoconch I and II, retained on lower (right) valve of adult specimen of *Bathypecten vulcani* from hydrothermal vents at Galapagos Rift, Alvin Dive 1529. Arrowhead, extremely narrow Prodissoconch II; B, byssal notch of adult shell.

Scale bar = 50 μm.

FIGURE 4.6. Larval shell, Prodissoconch I and II, retained on upper (left) valve of adult specimen of *Bathypecten vulcani* from hydrothermal vents at Galapagos Rift, Alvin Dive 1529. Arrowhead, extremely narrow Prodissoconch II.

Scale bar = 50 μm.

Calyptogena magnifica is gonochoristic, produces "large numbers" of eggs, spawns continuously, and delays reproduction until individuals reach approximately 40 percent of their maximum shell length (Berg 1985). Berg and Van Dover (1987) and Wiebe et al. (1988) report the collection in plankton tows of larvae thought to belong to *Calyptogena* spp.; however, identification of larval *Calyptogena* will remain tentative until either an unambiguous juvenile specimen is found with an intact prodissoconch or members of the family Vesicomyidae are successfully cultured.

Of the 68 gastropods known from deep-sea vents and seeps, significant information pertaining to life history strategies is available for 46 known species, protoconch dimensions and/or sculpture are known for 35 species, and aspects of the reproductive anatomy and/or secondary sexual characters have been described for 40 species. The protoconch dimensions and morphologies of an additional 3 gastropod species are presented here. The protoconch dimensions of the following species have been either measured directly from published micrographs or represent values presented in the cited authors' text. The use of nonstandard protoconch terminology, lack of agreement as to reference points for measurement, natural variation between specimens, and the inherent errors of measurement in the SEM (as great as 10–15 percent) may explain some of the discrepancies among values reported by different authors.

The scissurellid slit limpets *Temnocinclis euripes* McLean and *Temnozaga parilis* McLean, from East Pacific hydrothermal vents, are gonochoristic with sperm that lack tails or have very short tails. The lack of spermatozoa with long flagellae and the fact that female *Temnocinclis euripes* have a "low number" of "large yolk-rich eggs" suggests that fertilization in these two limpets occurs in the female mantle cavity, despite the lack of a male penis (Haszprunar 1989a). The protoconch sculpture of *Temnocinclis euripes* is unknown, although it has a reported "diameter" of 250 μm (McLean 1989b, his fig. 4A); whereas the unsculptured protoconch of *Temnozaga parilis* has a reported "diameter" of 170 μm (McLean 1989b, his fig. 4C). The scissurellid slit limpet *Sutilizona theca* McLean is also gonochoristic, lacks flagellated sperm, produces "large (maximum diameter 200 μm) yolky eggs," and likely fertilizes the eggs in the female mantle cavity. Eggs in all stages of development occur concurrently (Haszprunar 1989b), which suggests that spawning is continuous. The protoconch of *S. theca* has a reported "diameter" of 250 μm and a sculpture consisting of a series of spirally arranged pits (McLean 1989b, his figs. 4E, F). The large egg size, low fecundity, and protoconch morphology suggests that these three slit limpets have nonplanktotrophic development.

In the large neomphalid limpet *Neomphalus fretterae* McLean, from hydrothermal vents on the Eastern Pacific Rise, the sexes are separate. The males have a large prostate gland and the left tentacle is modified as a penis. The females have a receptaculum seminis and the oviduct has two different glandular areas suggesting that the eggs are coated with two extra-embryonic layers before leaving the oviduct. Fertilization most likely occurs internally within the female oviduct before the addition of nutritive albumen and a protective coat of a jellylike material (Fretter et al. 1981). The ovary contains "many" eggs with a diameter of 100–150 μm (Fretter et al. 1981) and reproduction is continuous (McLean 1981). Similarly, the sexes are separate and the left tentacle is modified to form a penis in the only other known neomphalid limpet, *Symmetromphalus regularis* McLean, from the Marianas Back–Arc Basin hydrothermal vents (McLean 1990b). McLean (1981) reported the protoconch of *N. fretterae* to have a sculpture of irregular low ridges, 1.2 rounded whorls, and a "maximum protoconch diameter" of 200 μm (his figs. 10A, B). McLean's (1981) protoconch dimensions agree well with one other reported protoconch diameter of 207 μm for *N. fretterae* (Warén and Bouchet 1989, their fig. 79) but not with a third value for the maximum PI diameter of 287 μm for *N. fretterae,* as estimated from published micrographs (Turner et al. 1985, their fig. 3c). The protoconch of *S. regularis* (McLean 1990b, his figs. 13, 14) is reported to have a "length" of 220 μm and a sculpture of irregular netlike ridges (McLean 1990b). The protoconch dimensions and sculpture of these two neomphalid limpets suggests that they both develop nonplanktotrophically. Batten (1984) speculated that multiple shell layers in the protoconch of *N. fretterae* suggests a larval stage with a lengthy planktonic period.

The left tentacle of male specimens of the cyathermid gastropods *Cyathermia naticoides* Warén and Bouchet and *Lacunoides exquisitus* Warén and Bouchet, from hydrothermal vents on the EPR, is modified to form a penis and a large prostate gland suggesting copulation and internal fertilization (Warén and Bouchet 1989). *C. naticoides* has a protoconch "diameter" of 210 μm as reported in Warén and Bouchet (1989, their fig. 80), that contrasts with a maximum PI diameter of 260 μm, as measured from a published micrograph (Turner et al. 1985, their fig. 14c). *L. exquisitus* has a reported PI "diameter" of 160 μm (Warén and Bouchet 1989, their fig. 81). The sculpture of the protoconchs of both *C. naticoides* and *L. exquisitus* consists of an initial irregular net-like pattern that grades into a smooth distal portion. These two cyathermids most likely develop non-planktotrophically based on their protoconch dimensions and morphologies.

Although males of the peltospirid limpet *Rhynchopelta concentrica* McLean, from 13°N and 21°N on the EPR, lack a penis or a well-developed prostate gland, the presence of sperm in the female receptacular and ovarian ducts indicates oocytes are fertilized internally within the ovary. Sperm, which have long tails, presumably must be liberated in the immediate vicinity of the female inhalent current and then undergo a brief swimming phase before entering the female mantle cavity and receptacular duct (Fretter 1989). Berg (1985) reported egg sizes in this species to average 152 by 132 μm (n = 142), whereas Fretter (1989) found ova within the ovarian duct to measure "10% greater" (n = 20) than those measured by Berg (1985). Egg sizes of other peltospirid limpets, from vents on the Eastern Pacific Rise, are of a similar size and it has been assumed that aspects of the reproductive anatomy of *Peltospira operculata* McLean, *P. delicata* McLean, *Nodopelta heminoda* McLean, *N. subnoda* McLean and *Echinopelta fistulosa* McLean are also similar to *R. concentrica* (Fretter 1989). The brown protoconch of *R. concentrica* has strong parallel raised ridges over the first three quarters of the shell. The reported "diameter" of 225 μm (McLean 1989a; his figs. 37, 38) contrasts sharply with a maximum PI diameter of 323 μm, as measured from a published micrograph (Turner et al. 1985, their fig. 5c). The protoconch of *P. operculata* is dark brown, has a sculpture of longitudinal spiral ridges running the length of the shell and has a reported "length" of 180 μm (McLean 1989a, his fig. 5), in contrast to a maximum PI diameter of 218 μm as estimated from a published micrograph (Lutz et al. 1986, their fig. 1B). The protoconch "diameter" of *P. delicata* is also reported to be 180 μm, although an intact protoconch has not been illustrated to date (McLean 1989a; 53). The protoconch of *N. subnoda* is also dark brown and has a sculpture of parallel raised ridges and a reported "diameter" of 160 μm (McLean 1989a, his fig. 29). The protoconchs of *N. heminoda* and *Hirtopelta hirta* McLean are to date unknown, however, characters of the protoconch of *E. fistulosa* are reported and illustrated here for the first time (figures 4.7–4.9). The PI of *E. fistulosa* has a maximum diameter of 212 μm (as measured from an electron micrograph) and has a protoconch sculpture of longitudinal ridges, as was hypothesized by McLean (1989a:64). Non-planktotrophic development is common to all the peltospirid limpets based on their protoconch dimensions and morphologies.

The oocytes of the coiled peltospirid gastropod *Pachydermia laevis* Warén and Bouchet, from 13°N and 21°N on the EPR, were reported to measure on average 105 by 130 μm (n = 100) (Berg 1985; "Species K"). The protoconch of *P. laevis* consists of half a whorl, has an initial fine net-like sculpture that gradually becomes smooth, and has a reported "diameter" of 210 μm (Warén and Bouchet 1989, their figs. 85–86). The latter measurement contrasts with that of the maximum PI diameter of 255 μm, as estimated from a published micrograph (Turner et al. 1985, their fig. 15c). These protoconch characters are consistent with nonplanktotrophic development.

The gonochoristic coiled peltospirid gastropod *Depressigyra planispira* Warén and Bouchet, from 13°N and 21°N on the EPR, has a reported protoconch "diameter" of "about" 215 μm (Warén and Bouchet 1989, their fig. 84) in contrast with a maximum PI diameter of 264 μm as measured from a published micrograph (Turner et al. 1985, their fig. 6c). The initial protoconch sculpture consists of a network of low ridges that gradually becomes smooth distally. Although a protoconch of *D. globulus* Warén and Bouchet, from northeast Pacific hydrothermal vents, with well-preserved sculpture has not been found, the maximum PI diameter, as measured from a published micrograph, is approximately 190 μm (Warén and Bouchet 1989, their fig. 83).

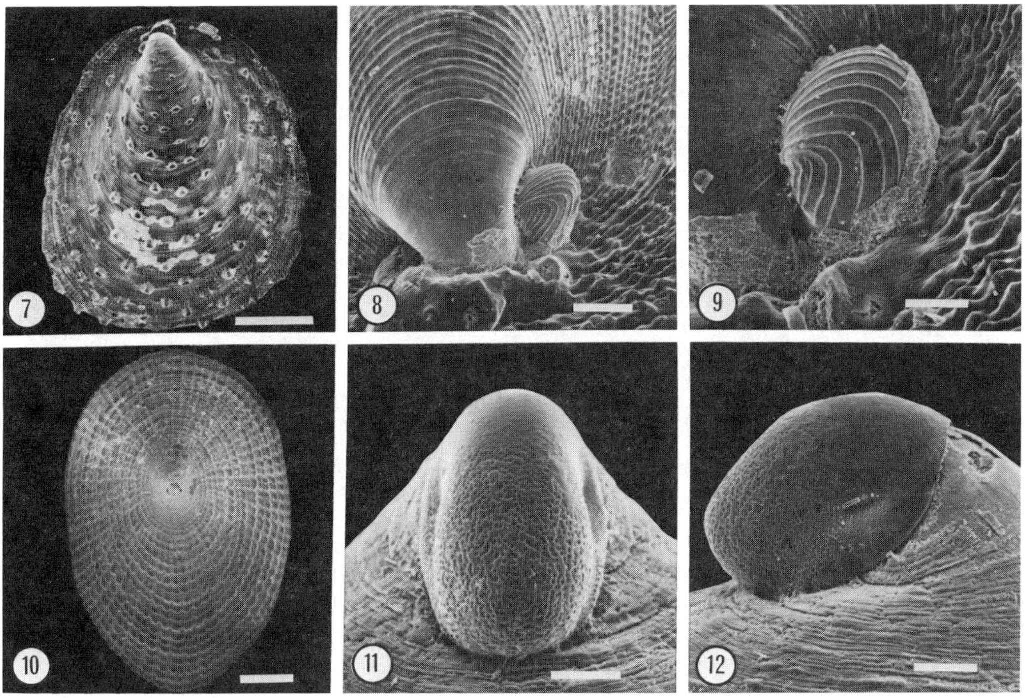

FIGURE 4.7. *Echinopelta fistulosa* from 21°N, Alvin Dive 2233.

Scale bar = 1 mm.

FIGURE 4.8. Posterior view of same specimen of *Echinopelta fistulosa* as depicted in figure 4.7, showing protoconch sculpture of longitudinal ridges.

Scale bar = 100 μm.

FIGURE 4.9. Enlarged view of protoconch I of same specimen of *Echinopelta fistulosa* as depicted in figures 4.7 and 4.8, partially obscured by overlying mineral precipitate.

Scale bar = 50 μm.

FIGURE 4.10. *Neolepetopsis densata* from inactive sulfide mounds on the Galapagos Rift, Alvin Dive 1650.

Scale bar = 500 μm.

FIGURE 4.11. Apical view of Protoconch I of *Neolepetopsis densata* from inactive sulfide mounds on the Galapagos Rift showing punctate sculpture and lateral pouches common to the order Patellogastropoda, Alvin Dive 1651.

Scale bar = 50 μm.

FIGURE 4.12. Lateral view of Protoconch I of *Neolepetopsis densata* from inactive sulfide mounds on the Galapagos Rift showing initial punctate sculpture and lateral pouch common to the order Patellogastropoda, Alvin Dive 1651.

Scale bar = 50 μm.

The gonochoristic coiled peltospirid gastropod *Solutigyra reticulata* Warén and Bouchet, from 13°N and 21°N on the EPR, has a large male prostate gland and a glandular oviduct and displays a protoconch of half a whorl with an irregular net-like sculpture that becomes smooth distally and a reported "diameter" of 200–240 μm (Warén and Bouchet 1989, their figs. 89, 90). Protoconch dimensions and morphologies of the two species of *Depressigyra* and of *S. reticulata* suggest that these peltospirods develop nonplanktotrophically.

The gonochoristic coiled peltospirid vent gastropod *Lirapex humata* Warén and Bouchet, known only from 21°N on the EPR, has a protoconch consisting of half a whorl that is sculptured by 5–6 strong spiral ridges over the initial portion and has a reported "maximum diameter" of 180 μm (Warén and Bouchet 1989, their fig. 87). *L. granularis* Warén and Bouchet, known only from 21°N on the EPR, has a sculpture similar to that of the protoconch of *L. humata* and a reported protoconch "diameter" of 220 μm (Warén and Bouchet 1989, their fig. 88) that essentially equates with a maximum PI diameter of 248 μm as measured from a published micrograph (Turner et al. 1985, their fig. 13c). Both species of *Lirapex* develop nonplanktotrophically based on the morphology and dimensions of their protoconchs.

The coiled vent archaeogastropod *Melanodrymia aurantiaca* Hickman, native to several sites on the EPR, is a gonochoristic species. Both male cephalic tentacles are modified for copulation and the females possess a receptaculum seminis. Fertilization is thus inferred to be internal (Warén and Bouchet 1989, Haszprunar 1989b). The presence of a glandular oviduct suggests that a spawn mass or egg capsule is produced and that fertilization probably occurs in the female gonoduct. Gametogenesis is likely continuous in this species as all stages of sperm and egg development occur concurrently (Haszprunar 1989b). The report that *M. aurantiaca* has "large, few eggs" is suggestive of nonplanktotrophic development (Haszprunar 1989b). The protoconch of *M. aurantiaca* is consistent with conditions of nonplanktotrophy; it consists of slightly more than half a whorl and has a reported "diameter" of 260 μm (Warén and Bouchet 1989, their fig. 82) that equates well with a maximum PI dimension of 273 μm, as estimated from a published micrograph (Turner et al. 1985, their fig. 12c).

Internal fertilization appears to be the rule in the gonochoristic lepetodrilid vent limpets, native to several sites on the Eastern Pacific Rise, *Lepetodrilus pustulosus* McLean, *L. elevatus elevatus* McLean, *L. elevatus galriftensis* McLean, *L. ovalis* McLean, *L. cristatus* McLean, *L. guaymasensis* McLean, and *L. fucencis* McLean. Males of the species have a prominent penis of epipodial origin arising just beneath the right cephalic tentacle and a prostate gland. Females possess a receptaculum seminis (Fretter 1988). Berg (1985) reported average egg measurements of 104 by 82 μm (n = 100) for *L. pustulosus* and 95 by 74 μm (n = 100) for *L. elevatus elevatus*. Oocytes of *L. pustulosus, L. cristatus,* and *L. fucencis* had respective diameters of 120 μm,

140–150 μm, and 100–140 μm (Fretter 1988). Fretter (1988) observed four to six heavily yolked eggs, without extra-embryonic membranes, free within the mantle cavity of three separate specimens of *L. pustulosus.* Observations on *L. pustulosus* suggest that sperm are stored in the female receptaculum subsequent to copulation and that fertilization and copulation are disjunct events. Gonad condition in *L. pustulosus* suggests that spawning may be continuous and that sexual maturity is reached at an early stage since ripe eggs and sperm were observed in specimens one third of their maximal size (Fretter 1988). The finely pitted protoconchs of *L. pustulosussis, L. elevatus elevatus, L. elevatus galriftensis,* and *L. fucencis* have a reported "length" of 120 μm; whereas those of *L. ovalis, L. cristatus,* and *L. guaymasensis* have a reported "length" of 130 μm (McLean 1988). Turner et al. (1985) illustrated protoconchs of *L. cristatus* (their fig. 2c), *L. ovalis* (their fig. 6c), *L. elevatus elevatus* (their fig. 7c), and *L. elevatus galriftensis* (their fig. 8c) that have maximum PI diameters as estimated from the above micrographs of 163 μm, 187 μm, 180 μm, and 200 μm, respectively.

Males of the lepetodrilid limpets *Gorgoleptis emarginatus* McLean and *G. patulus* McLean from East Pacific hydrothermal vents possess an unusual, putative copulatory organ that arises from the left wall of the snout. A prominent prostate gland is present in *G. emarginatus* but not in *G. patulus,* whereas no sperm or receptaculum seminis has been found in females of either species of *Gorgoleptis* (Fretter 1988). The protoconchs of *G. emarginatus* and *G. patulus* are reported to be finely pitted with respective maximum protoconch "lengths" of 130 μm and 120 μm (McLean 1988). Turner et al. (1985) illustrated a protoconch of *G. emarginatus* (their fig. 1c) that has a maximum PI diameter, as estimated from this micrograph, of 190 μm. The protoconch morphologies and dimensions of both lepetodrilid genera, *Lepetodrilus* and *Gorgoleptis,* are consistent with non-planktotrophic development.

The clypeosectid slit limpets *Clypeosectus delectus* McLean, *C. curvus* McLean (from East Pacific hydrothermal vents), and *Pseudorimula marianae* McLean (from hydrothermal vents at Marianas Back-Arc Basin) are gonochoristic, have a large prostate gland, and large (maximum diameter 200 μm) yolky eggs. Eggs in all stages of development are present concurrently, suggesting that spawning occurs continuously or over an extended reproductive season. Both species of *Clypeosectus* display a penis near the base of the right cephalic tentacle and a receptaculum seminis is present in the posterior female mantle cavity suggesting that fertilization occurs in the female mantle cavity. Although the latter two structures are absent in *P. marianae,* the presence of a large prostate gland and aberrant sperm with a short tail in *Pseudorimula* suggests internal fertilization also occurs in this genus (Haszprunar 1989a). The protoconch of this species is sculptured with a series of weak spiral ridges and randomly spaced fine pits and has a reported "diameter" of 200 μm (McLean

1989b). The protoconch of *P. marianae* has a finely reticulated surface sculpture and a "length" of 200 μm (McLean 1989b, his fig. 13B). All three species of clypeosectid slit limpets appear to have nonplanktotrophic development.

The pseudococculinid limpet *Amphiplica (Gordabyssia) gordensis* McLean (from Gorda Ridge) has a protoconch with a reported "length" of 200 μm, a sculpture of reticulated raised ridges (McLean 1991, his figs. 29, 30) and most likely develops nonplanktotrophically.

The pyropeltid limpets *Pyropelta musaica* McLean from the Juan de Fuca Ridge and *P. corymba* McLean from Guaymas Basin are hermaphroditic with separate testis and ovary. The protoconch morphology of these species is unknown. The right cephalic tentacle is modified to form a copulatory organ indicating internal fertilization, and the presence of "large and yolk-rich" eggs suggests nonplanktotrophic development (McLean and Haszprunar 1987).

The neolepetopsid limpets *Paralepetopsis floridensis* McLean (from methane/sulfide seeps at the West Florida Escarpment) and *Eulepetopsis vitrea* McLean, *Neolepetopsis occulta* McLean, *N. densata* McLean, and *N. gordensis* McLean (from East Pacific hydrothermal vents) are gonochoristic and lack a copulatory organ, a prostate gland, and a receptaculum seminis. They apparently liberate unprotected eggs and sperm directly through the mantle cavity to the external environment. Eggs are heavily yolked (300 μm in diameter in both *E. vitrea* and *N. occulta*), suggesting that development is lecithotrophic (Fretter 1990). Protoconchs of neolepetopsids are rarely encountered as they are shed at an early age, a common patellogastropod trait (McLean 1990a). However, the protoconchs of *E. vitrea* and *P. floridensis* have been recently illustrated (McLean 1990a; his plate VIIh and plate Xf-h, respectively) and are both reported to have a "length" of 400 μm. The protoconch of *N. densata* also has a reported "length" of 400 μm, although an intact protoconch has not previously been illustrated (McLean 1990a; his plate IIIg). A representative adult specimen and two views of the protoconch of *N. densata* collected from an inactive sulfide chimney at 9–10°N on the EPR are shown in figures 4.10–4.12. Comparison of the radula of these specimens to the radula of *N. densata* illustrated in McLean (1990a) verifies the specific identification. Initially the protoconch is finely pitted, becoming gradually smooth distally. The maximum PI diameter of *N. densata* of two specimens from 9–10°N averaged 202 μm, slightly greater than half the size reported above for this species by McLean (1990a). As in *E. vitrea* and *P. floridensis* (McLean 1990a), the protoconch of *N. densata* (figures 4.11, 4.12) has lateral folds and its anterior-posterior axis is offset approximately 10% from the axis of the teleoconch, characters present in several other limpet groups (McLean 1990a:515). Protoconch characters of three neolepetopsid limpets support the suggestion that these limpets develop nonplanktotrophically.

Male specimens of the large white hydrothermal vent trochid gastropod *Bathymargarites symplector* Warén and Bouchet, from several hydrothermal vent sites on the Eastern Pacific Rise, have a highly modified left neck lobe that is hypothesized to function as a penis for internal fertilization (Warén and Bouchet 1989). Berg (1985) reported ovarian oocytes of *B. symplector* (designated "Species M") to measure on average 125 by 161 μm. This large egg size is suggestive of nonplanktotrophic development. The trochid gastropod *Cataegis toreuta* McLean and Quinn, first described from a nonseep environment, is common at the Louisiana Slope hydrocarbon seeps and has a reported protoconch "diameter" of 400 μm (McLean and Quinn 1987), suggestive of nonplanktotrophic development.

The presence of a copulatory organ near the base of the right tentacle in specimens of the neritacean limpet *Shinkailepas kaikatensis* Okutani, Saito and Hashimoto from a hydrothermal vent on Kaikata Seamount off Japan (Okutani et al. 1989) suggests that fertilization is likely internal in this species.

Sexes are separate and fertilization occurs internally within vent and seep gastropods of the family Provannidae. Although males lack a penis, female members of the genus *Provanna* have a series of from 1 to 7 seminal receptacles that store the male dimorphic sperm consisting of euspermatozoa and paraspermatozoa, the later acting as simple nurse cells (Healy 1990; Warén and Ponder 1991). Male *Provanna sculpta* Warén and Ponder and *P. variabilis* Warén and Bouchet have a well-developed prostate gland, whereas the prostate glands of *P. ios* Warén and Bouchet and *P. muricata* Warén and Bouchet are not as elaborate. Eggs from a species of *Provanna* from 21°N, designated "Species F" (either *P. ios* or *P. muricata*) measured, on average, 112 by 146 μm (Berg 1985). Ovarian oocytes of *P. sculpta* measured up to 180 μm by 80 μm by 55 μm (Warén and Ponder 1991). Warén and Bouchet (1986) reported immature oocytes of an unspecified provannid to measure 100 μm by 150 μm. An albumen gland and an extensive capsule gland are present in female *P. sculpta* and *P. variabilis* (Warén and Ponder 1991). The PI of an unspecified species of the genus *Provanna* illustrated by Warén and Ponder (1991; their figs. 1D, F) was reported to have 1.5 whorls with an initial sculpture of fine granules that gradually changes to a sculpture of weak and then strong axial ribs with intervening wavy spiral lines. The first whorl diameter and the maximum diameter of the PI of this illustrated specimen (Warén and Ponder 1991, their fig. 1F) are 380 μm and 411 μm, respectively, as measured from the published micrograph. The lack of a PII (Warén and Ponder 1991) and the size of the PI suggests that a planktotrophic stage does not occur in *Provanna* spp., although questions concerning the presence or absence of a planktonic stage cannot be resolved based on the available evidence. An adult specimen and the PI of *P. variabilis,* from hydrothermal vents on Gorda Ridge, are illustrated in figures 4.13–4.15. The larval shell of *P. variabilis* (figures 4.14, 4.15), which matches that of *Provanna* sp. described above (Warén and Ponder 1991), consists of 1.5 whorls, has a maximum PI diameter of 390 μm, lacks a PII, and has a terminal sculpture of rather strong axial ribs.

FIGURE 4.13. *Provanna variabilis* from hydrothermal vents at Escanaba Trough on Gorda Ridge, Alvin Dive 2040.

Scale bar = 1 mm.

FIGURE 4.14. Lateral view of Protoconch I of *Provanna variabilis* retained on juvenile specimen from hydrothermal vents at Escanaba Trough on Gorda Ridge showing eroded apical portion and weak axial growth lines, Alvin Dive 2035. Arrowhead, distinct Protoconch I - teleoconch boundary.

Scale bar = 100 μm.

FIGURE 4.15. Apical view of Protoconch I of *Provanna variabilis* retained on juvenile specimen from hydrothermal vents at Escanaba Trough on Gorda Ridge showing eroded apical portion, Alvin Dive 2035. Arrowhead, distinct Protoconch I - teleoconch boundary.

Scale bar = 100 μm.

FIGURE 4.16. *Phymorhynchus* sp. from hydrothermal vents at Galapagos Rift, Alvin Dive 1528.

Scale bar = 1 mm.

FIGURE 4.17. Lateral view of portion of the Protoconch II of the same specimen of *Phymorhynchus* sp. as depicted in figure 4.16, showing the complex reticulate sculpture of spiral raised ridges and axial riblets. The first three whorls including the Protoconch I are missing.

Scale bar = 200 μm.

FIGURE 4.18. Late Protoconch I larva removed from turrid egg capsule retrieved from hydrothermal vents on the Galapagos Rift showing complex pattern of reticulate sculpture, Alvin Dive 2031.

Scale bar = 50 μm.

The turrid gastropod *Phymorhynchus* sp. most likely lays the large lenticular egg capsules that have been retrieved from the Galapagos Rift hydrothermal vents (Berg 1985, Gustafson et al. 1991). These capsules contain eggs measuring on average 192 by 136 μm (n = 100) (Berg 1985) that give rise to over 1000 veliger larvae per capsule (Gustafson et al. 1991). Ovarian oocytes with dimensions that average 101 by 121 μm (n = 20) have been reported for this species (Berg 1985). The Protoconch II morphology of juvenile specimens of this undescribed species of turrid gastropod (figures 4.16, 4.17) resembles the morphology of specimens from within the above mentioned capsules (figure 4.18; Gustafson et al. 1991, their figs. 12–14) and specimens with intact protoconchs illustrated by Turner et al. (1985, their figs. 27a–e). The protoconch morphology and large number of eggs per capsule of *Phymorhynchus* sp. suggests that, after encapsulation, this species develops planktotrophically (Gustafson et al. 1991). Juvenile specimens of another vent turrid, from 21°N on the EPR, have similar PI and PII morphologies suggesting that this second species of vent turrid also develops planktotrophically (Turner et al. 1985, their figs. 26a–e). An undetermined species that spawns inflated, triangular egg capsules at the Galapagos Rift develops planktotrophically after hatching, whereas larvae of another species from the Juan de Fuca Ridge probably develop nonplanktotrophically in pouchlike egg capsules (Gustafson et al. 1991).

Discussion

One of the characteristics of an *r-type* life history strategy, fast growth, has been documented for *Bathymodiolus thermophilus* and *Calyptogena magnifica,* using several techniques (Lutz et al. 1985). Early reproduction, also a character of an *r-type* life history strategy, has not been documented in vent and seep molluscs. In fact, both *Bathymodiolus thermophilus* and *Calyptogena magnifica* delay reproduction until a large size is reached (Berg 1985).

However at least 2 vent bivalves and 6 vent gastropods are thought to reproduce continuously in these habitats. Continuous reproduction may be thought of as a reproductive adaptation to the relatively constant supply of nutrition at hydrothermal vent and cold methane/sulfide seeps. Although several nonvent deep-sea molluscs also spawn throughout the year, perhaps in response to a lack of seasonal stimuli, this is by no means a universal strategy among deep-sea molluscs (Gage and Tyler 1991:307–308).

The majority of archaeogastropods from hydrothermal vents appear to have internal fertilization, which is rarely encountered among archaeogastropods (Warén and Bouchet 1989). In fact, *Bathymargarites symplector* is the only member of the family Trochidae known to have a penis (Warén and Bouchet 1989). On the other hand, some nonvent members of the archaeogastropod suborder Cocculiniformia and the families Acmeidae (Warén and Bouchet

1989) and Skeneidae (Höisaeter 1968) have a copulatory organ. It has been suggested that internal fertilization is a strategy adopted by vent molluscs to ensure that sperm and unfertilized eggs are not directly exposed to the high temperature, acidic vent waters that may be laden with potentially toxic sulfides and heavy metals (Fretter 1988, 1989; Warén and Bouchet 1989). Besides development of a copulatory organ, several other anatomical modifications such as aflagellate sperm, large male prostate glands, and female sperm receptacles are produced in species that undergo internal fertilization (see discussion above). Incorporation of sperm within a viscous prostate secretion is also thought to insulate sperm from the vent water (Warén and Bouchet 1989). Internal fertilization in the aplacophoran *Helicoradomenia juani* and the neritoidean limpet *Shinkailepas kaikatensis* is not unexpected in these taxa and is not likely a strategy adopted specifically to deal with the vent environment. Neolepetopsid limpets are the only group of vent and seep archaeogastropods not thought to fertilize the eggs internally (see discussion above).

The two pyropeltid limpets (McLean and Haszprunar 1987), the aplacophoran *Helicoradomenia juani* (Scheltema and Kuzirian 1991), and the reportedly protandrous hermaphrodite *Bathymodiolus themophilus* (Berg 1985) are the only known hermaphrodites among approximately 59 vent molluscs thus far examined for this trait. This apparent scarcity of hermaphroditism in vent molluscs was viewed as unexpected by McLean (1985) in light of ecological theory, which predicts a high degree of hermaphroditism in animals inhabiting isolated, ephemeral habitats (Ghiselin 1969). Hoagland (1978) notes that molluscs inhabiting environments "patchy in time and space" have more often than not evolved protandry. However, ecological theory also predicts hermaphroditism in conditions of low population density (Ghiselin 1969), a condition not often encountered in the nutrient-rich vent and seep environments.

Based on the above review (larval shell morphology, egg size, fecundity), a total of 5 vent or seep bivalves (all mytilids) and 3 vent gastropods appear to develop planktotrophically, whereas 2 known vent bivalves and 39 vent and/or seep gastropods appear to develop nonplanktotrophically. In most cases, the inferred larval developmental patterns of vent and seep molluscs are also typical of shallow-water members of the respective systematic groups. Likewise, the hydrothermal vent scallop *Bathypecten vulcani* has prodissoconch characters similar to closely related deep-sea scallops from nonvent habitats (Schein 1989). The high incidence of nonplanktotrophy evident amongst vent and seep gastropods is parsimoniously explained by the fact that most are archaeogastropods (Tunnicliffe 1991) and archaeogastropodes appear to be evolutionarily constrained to nonplanktotrophy (Gustafson et al. 1991). It is likely that phylogenetic constraints and limitations imposed by small adult size have restricted the range of potential larval life history traits available to vent and seep molluscs.

Despite the ephemeral nature of vent habitats and the patchy distribution of both vents and methane/sulfide seeps, many molluscan morphospecies are present at more than one eastern Pacific vent field (6 are shared by Juan de Fuca and Explorer Ridges; 5 are shared by Juan de Fuca and Gorda Ridges; 4 are shared by Juan de Fuca, Gorda, and Explorer Ridges; 9 are shared by Galapagos Rift and 13°N; 11 are shared by Galapagos Rift and 21°N; 20 are shared by 13°N and 21°N; and 8 are shared by Galapagos Rift, 13°N, and 21°N) (see references in Gustafson et al. 1991; RAL, RGG unpublished results). In addition, many species from Galapagos Rift and the EPR sites are shared with newly explored sites at 11°25'N and 9°45'–52'N on the EPR (RAL, RGG unpublished results), whereas 5 molluscs are common to both the hydrothermal vent fields at Guaymas Basin and a recently investigated cold-water methane/sulfide seep on the Guaymas Transform Fault (RAL, RGG unpublished results). The presence of a series of far-flung populations of vent and seep molluscan morphospecies implies a dispersal capability sufficient to maintain this distribution. Life history theory predicts that pelagic larval strategies would be advantageous in sessile organisms occupying habitats that are variable in time and space (Crisp 1974), such as deep-sea hydrothermal vent and cold methane/sulfide seeps. In fact, sampling with larval traps and plankton nets above hydrothermal vents reveals the presence of significant numbers of larval and post-larval molluscs (Berg and Van Dover 1987; Wiebe et al. 1988; Mullineaux 1991).

It is becoming evident that traditional generalizations, derived from studies of shallow-water species, concerning the larval dispersal capability of planktonic larvae are not applicable in many cases to the situation in the deep sea. The effect of low temperature on durations of both competent and precompetent larval periods may be dramatic. Larvae of the bathyal echinoid *Aspidodiadema jabobyi* Agassiz develop from eggs 98 μm in diameter but grow for three weeks at 16°C entirely on their own yolk supply, prior to feeding (Young et al. 1989). Larvae should be able to live on yolk reserves even longer at abyssal depths where water temperature a short distance from a hydrothermal vent may be only 1.5–2.5°C. Several researchers have speculated that molluscan larvae from hydrothermal vents may remain in the plankton for several weeks (Turner et al. 1985) or months (Warén and Bouchet 1989) relying entirely on yolk reserves. Nutrient acquisition from dissolved organic matter (DOM) in the deep-sea may also extend the planktonic period in both planktotrophic and non-planktotrophic molluscan larvae (Manahan 1990; Gage and Tyler 1991). Even without invocation of reduced temperature, delayed metamorphosis, or DOM many nonplanktotrophic archaeogastropods have planktonic stages of a week or more (Heslinga 1981; Hadfield and Strathmann 1990).

Thorson (1950) suggested high mortality should be offset by high fecundity, implying that predation pressure and other hazards of planktonic life would require animals with long planktonic larval periods to have high fecundity, whereas animals that stay in the plankton for shorter periods would likely produce fewer gametes. Again, this generalization may not hold in the deep sea. Demersal larvae which may be able to utilize detrital material and/or DOM in the rich Benthic Boundary Layer probably experience less predation than planktonic larvae in the euphotic zone. With increasing depth the importance of suspension feeders declines and most abyssal taxa become deposit feeders (Gage and Tyler 1991). Generalist planktonic predators are relatively rare in the Benthic Boundary Layer; at 4 km depth the biomass of pelagic organisms is only 1 percent of that in the euphotic zone (Gage and Tyler 1991). Therefore the low fecundity in vent gastropods (scissurellid slit limpets, *Melanodrymia aurantiaca*, etc.) does not necessarily preclude a lengthy demersal planktonic larval stage.

The morphology of larval protoconchs (Bouchet and Warén 1979) and ^{18}O and ^{13}C isotope analyses of larval shells (Bouchet and Fontes 1981; Killingley and Rex 1985) suggest that veligers of deep-sea prosobranchs may migrate to surface waters to feed and develop. The few vent and seep species with planktotrophic larvae may also have this ability to ascend to the surface and feed, or they may remain near the bottom and feed on detrital material in the Benthic Boundary Layer. In either case their dispersal capability is likely to be considerable.

Many molluscs have developed unique life history strategies to cope with life in the nutrient rich, highly toxic hydrothermal vent and methane/sulfide seep environments. Many vent molluscs grow much more rapidly than other deep-sea organisms. Many vent molluscs reproduce continuously, probably because their source of nutrients is relatively constant. Numerous anatomical modifications have arisen in vent and seep archaeogastropods to accommodate internal fertilization, which is thought to protect gametes and early embryonic stages from toxic sulfides, heavy metals, and low pH conditions in the water column. The adaptation of molluscan larval development patterns to life in hydrothermal vent and methane/sulfide seep environments is less clear. The validity of using larval shell criteria derived from observations of shallow-water organisms to infer modes of larval development, and particularly dispersal capability, in deep-sea organisms is questionable. Laboratory culture of a number of vent and seep molluscs will be necessary to confirm the link between larval shell characteristics, mode of development, and dispersal. In the meantime, ongoing genetic analyses of widely separated populations of vent and seep molluscs are expected to provide estimates of gene flow between and among these deep-sea environments, that may give insights into the actual mechanisms of dispersal.

ACKNOWLEDGMENTS

We thank the pilots and crew of R/V *Atlantis II/Alvin* and R/V *Seward Johnson/Johnson-Sea-Link I* for assistance with retrieval of specimens; Fred Grassle and Rose Petrecca of the Institute of Marine and Coastal Sciences, Rutgers University, for specimens from Gorda Ridge; Bob Carney of the Coastal Ecology Institute, Louisiana State University and Ian

MacDonald of the Geochemical and Environmental Research Group, Texas A&M University, for specimens from the Louisiana Slope; and John Grazul of the Electron Microscopy Facility, Nelson Biological Laboratories, Rutgers University, for assistance with SEM. This is Contribution #92–17 of the Institute of Marine and Coastal Sciences, Rutgers University and New Jersey Agricultural Experiment Station Publication No. F-32402–1–92, supported by State funds and by NSF Grants OCE-87–16591 to R. A. L. and OCE-89–17311 to R. A. L. and Robert Vrijenhoek.

References

Batten, R. L. 1984. Shell structure of the Galapagos Rift limpet *Neomphalus fretterae* McLean, 1981, with notes on muscle scars and insertions. *Am. Mus. Novitates* No. 2776:1–13.

Berg, C. J., Jr. 1985. Reproductive strategies of mollusks from abyssal hydrothermal vent communities. *Biol. Soc. Wash. Bull.* 6:185–197.

Berg, C. J., Jr. and R. D. Turner. 1980. Description of living specimens of *Calyptogena magnifica* Boss and Turner with notes on their distribution and ecology. *Malacologia* 20:183–185.

Berg, C. J., Jr. and C. L. Van Dover. 1987. Benthopelagic macrozooplankton communities at and near deep-sea hydrothermal vents in the eastern Pacific Ocean and the Gulf of California. *Deep-Sea Res.* 34:379–401.

Bouchet, P. and J.-C. Fontes. 1981. Migrations verticales des larves de Gastéropodes abyssaux: arguments nouveaux dûs a l'analyse isotopique de la coquille larvaire et postlarvaire. *C. R. Hebd. Seances Acad. Sci.* Ser. 3, 292:1005–1008.

Bouchet, P. and A. Warén. 1979. Planktotrophic larval development in deep-water gastropods. *Sarsia* 64:37–40.

Crisp, D. J. 1974. Energy relations of marine invertebrate larvae. *Thalass. Jugoslav.* 10:103–120.

Desbruyères, D. and L. Laubier. 1984. Primary consumers from hydrothermal vent animal communities. In P. A. Rona, K. Bostrom, L. Laubier, and K. L. Smith, Jr., eds., *Hydrothermal Processes at Seafloor Spreading Centers,* pp. 711–734. New York: Plenum Press.

Fretter, V. 1988. New archaeogastropod limpets from hydrothermal vents; Superfamily Lepetodrilacea. II. Anatomy. *Philos. Trans. R. Soc. Lond. B Biol. Sci.* 318:33–82.

Fretter, V. 1989. The anatomy of some new archaeogastropod limpets (Superfamily Peltospiracea) from hydrothermal vents. *J. Zool.* 218:123–169.

Fretter, V. 1990. The anatomy of some new archaeogastropod limpets (Order Patellogastropoda, Suborder Leptopsina) from hydrothermal vents. *J. Zool.* 222:529–555.

Fretter, V., A. Graham, and J. H. McLean. 1981. The anatomy of the Galapagos Rift limpet, *Neomphalus fretterae. Malacologia* 21:337–361.

Gage, J. D. and P. A. Tyler. 1991. *Deep-Sea Biology: A Natural History of Organisms at the Deep-Sea Floor.* Cambridge: Cambridge University Press.

Ghiselin, M. T. 1969. The evolution of hermaphroditism among animals. *Quart. Rev. Biol.* 44:189–208.

Grassle, J. F. 1985. Hydrothermal vent animals: Distribution and biology. *Science* 229:713–725.

Gustafson, R. G., D. T. J. Littlewood, and R. A. Lutz. 1991. Gastropod egg capsules and their contents from deep-sea hydrothermal vent environments. *Biol. Bull.* 180:34–55.

Hadfield, M. G. and M. F. Strathmann. 1990. Heterostrophic shells and pelagic development in trochoideans: Implications for classification, phylogeny, and palaeoecology. *J. Moll. Stud.* 56:239–256.

Haszprunar, G. 1989a. New slit-limpets (Scissurellacea and Fissurellacea) from hydrothermal vents. Part 2: Anatomy and relationships. *Contrib. Sci. (Los Angel.)* 408:1–17.

Haszprunar, G. 1989b. The anatomy of *Melanodrymia aurantiaca* Hickman, a coiled archaeogastropod from the East Pacific hydrothermal vents (Mollusca, Gastropoda). *Acta Zool.* 70:175–186.

Healy, J. M. 1990. Taxonomic affinities of the deep-sea genus *Provanna* (Caenogastropoda): New evidence from sperm ultrastructure. *J. Moll. Stud.* 56:119–122.

Heslinga, G. A. 1981. Larval development, settlement, and metamorphosis of the tropical gastropod *Trochus niloticus*. *Malacologia* 20:349–357.

Hines, A. H. 1986. Larval problems and perspectives in life histories of marine invertebrates. *Bull. Mar. Sci.* 39:506–525.

Hoagland, K. E. 1978. The evolution of protandry and labile sex determination in the Mollusca. *Bull. Am. Malacol. Union* 1977:97.

Höisaeter, T. 1968. Taxonomic notes on the north-European species of *Cyclostrema* sensu Jeffreys 1883 (Prosobranchia, Diotocardia). *Sarsia* 33:43–58.

Hyman, L. H. 1967. *The Invertebrates.* Vol. 6: *Mollusca 1*. New York: McGraw Hill.

Jablonski, D. and R. A. Lutz. 1980. Molluscan larval shell morphology: Ecological and paleontological applications. In D. C. Rhoads and R. A. Lutz, eds., *Skeletal Growth of Aquatic Organisms,* pp. 323–377. New York: Plenum Press.

Jablonski, D. and R. A. Lutz. 1983. Larval ecology of marine benthic invertebrates: paleobiological implications. *Biol. Rev.* 58:21–89.

Kenk, V. C. and B. R. Wilson. 1985. A new mussel (Bivalvia, Mytilidae) from hydrothermal vents in the Galapagos Rift zone. *Malacologia* 26:23–271.

Killingley, J. S. and M. A. Rex. 1985. Mode of larval development in some deep-sea gastropods indicated by oxygen-18 values of their carbonate shells. *Deep-Sea Res.* 32:809–818.

Le Pennec, M. 1988. Alimentation et reproduction d'un Mytilidae des sources hydrothermales profondes du Pacifique oriental. In *Proceedings of the Hydrothermalism, Biology, and Ecology Symposium,* Paris, November 4–7, 1985. *Oceanol. Acta* Special vol., no. 8:181–190.

Le Pennec, M., A. Lucas, and H. Petit. 1983. Études preliminaires sur un Mytilidae des sources hydrothermales du Pacifique. *Haliotis* 13:69–82.

Lima, G. M. and R. A. Lutz. 1990. The relationship of larval shell morphology to mode of development in marine prosobranch gastropods. *J. Mar. Biol. Assoc. U. K.* 70:611–637.

Lutz, R. A. 1988. Dispersal of organisms at deep-sea hydrothermal vents. In *Proceedings of the Hydrothermalism, Biology, and Ecology Symposium,* Paris, November 4–7, 1985. *Oceanol. Acta* Special vol., no. 8:23–29.

Lutz, R. A., D. Jablonski, D. C. Rhoads, and R. D. Turner. 1980. Larval dispersal of a deep-sea hydrothermal vent bivalve from the Galapagos Rift. *Mar. Biol.* 57:127–133.

Lutz, R. A., L. W. Fritz, and D. C. Rhoads. 1985. Molluscan growth at deep-sea hydrothermal vents. *Biol. Soc. Wash. Bull.* 6:199–210.

Lutz, R. A., P. Bouchet, D. Jablonski, R. D. Turner, and A. Warén. 1986. Larval ecology of mollusks at deep-sea hydrothermal vents. *Am. Malacol. Bull.* 4:49–54.

Manahan, D. T. 1990. Adaptations by invertebrate larvae for nutrient acquisition from seawater. *Am. Zool.* 30:147–160.

McLean, J. H. 1981. The Galapagos rift limpet *Neomphalus:* Relevance to understanding the evolution of a major Paleozoic-Mesozoic radiation. *Malacologia* 21:291–336.

McLean, J. H. 1985. Preliminary report on the limpets at hydrothermal vents. *Biol. Soc. Wash. Bull.* 6:159–166.

McLean, J. H. 1988. New archaeogastropod limpets from hydrothermal vents. Superfamily Lepetodrilacea. Part 1: Systematic descriptions. *Philos. Trans. R. Soc. Lond. B Biol. Sci.* 319:1–32.

McLean, J. H. 1989a. New archaeogastropod limpets from hydrothermal vents: New family Peltospiridae, new superfamily Peltospiracea. *Zool. Scr.* 18:49–66.

McLean, J. H. 1989b. New slit limpets (Scissurellacea and Fissurellacea) from hydrothermal vents. Part 1. Systematic descriptions and comparisons based on shell and radular characters. *Contrib. Sci. (Los Angel.)* 407:1–29.

McLean, J. H. 1990a. Neolepetopsidae, a new docoglossate limpet family from hydrothermal vents and its relevance to patellogastropod evolution. *J. Zool.* 222:485–528.

McLean, J. H. 1990b. A new genus and species of neomphalid limpet from the Mariana vents with a review of current understanding of relationships among Neomphalacea and Peltospiracea. *Nautilus* 104:77–86.

McLean, J. H. 1991. Four new pseudococculinid limpets collected by the deep-submersible Alvin in the Eastern Pacific. *Veliger* 34:38–47.

McLean, J. H. and G. Haszprunar. 1987. Pyropeltidae, a new family of cocculiniform limpets from hydrothermal vents. *Veliger* 30:196–205.

McLean, J. H. and J. F. Quinn, Jr. 1987. *Cataegis,* new genus of three new species from the continental slope (Trochidae: Cataeginae New Subfamily). *Nautilus* 101:111–116.

Mullineaux, L. S., P. H. Wiebe, and E. T. Baker. 1991. Hydrothermal vent plumes: Larval highways in the deep-sea? *Oceanus* 34:64–68.

Okutani, T., H. Saito and J. Hashimoto. 1989. A new neritacean limpet from a hydrothermal vent site near Ogasawara Islands, Japan. *Venus* 48:223–230.

Schein, E. 1989. Pectinidae (Mollusca, Bivalvia) bathyaux et abyssaux des campagnes BIOGAS (Golfe de Gascogne) systématique et biogéographique. *Annales de l'Institut Océanographique, Paris* 65:59–125.

Schein-Fatton, E. 1985. Découverte sur la ride du Pacifique oriental à 13°N d'un Pectinidae (Bivalvia, Pteromorpha) d'affinites paléozoïques. *C. R. Hebd. Seances Acad. Sci.* Sér. III 301:491–496.

Scheltema, A. H. and A. M. Kuzirian. 1991. *Helicoradomenia juani* gen. et sp. nov., a Pacific hydrothermal vent Aplacophora (Mollusca: Neomeniomorpha). *Veliger* 34:195–203.

Strathmann, R. R. 1987. Larval feeding. In A. C. Giese, J. S. Pearse, and V. B. Pearse, eds., *Reproduction of Marine Invertebrates,* vol. 9: *General Aspects: Seeking Unity in Diversity,* pp. 465–550. Palo Alto, California: Blackwell Scientific Publ.

Thorson, G. 1950. Reproductive and larval ecology of marine bottom invertebrates. *Biol. Rev.* 25:1–45.

Tunnicliffe, V. 1991. The biology of hydrothermal vents: Ecology and evolution. *Oceanogr. Mar. Biol. Ann. Rev.* 29:319–407.

Turner, R. D. and R. A. Lutz. 1984. Growth and distribution of mollusks at deep-sea vents and seeps. *Oceanus* 27:55–62.

Turner, R. D., R. A. Lutz, and D. Jablonski. 1985. Modes of molluscan larval development at deep-sea hydrothermal vents. *Biol. Soc. Wash. Bull.* 6:167–184.

Warén, A. and P. Bouchet. 1986. Four new species of *Provanna* Dall (Prosobranchia, Cerithiacea?) from East Pacific hydrothermal sites. *Zool. Scr.* 15:157–164.

Warén, A. and P. Bouchet. 1989. New gastropods from East Pacific hydrothermal vents. *Zool. Scr.* 18:67–102.

Warén, A. and W. F. Ponder. 1991. New species, anatomy, and systematic position of the hydrothermal vent and hydrocarbon seep gastropod family Provannidae fam. n. (Caenogastropoda). *Zool. Scr.* 20:27–56.

Wiebe, P. H., N. Copley, C. Van Dover, A. Tamse, and F. Manrique. 1988. Deep-water zooplankton of the Guaymas Basin hydrothermal vent field. *Deep-Sea Res.* 35:985–1013.

Young, C. M., J. L. Cameron, and K. J. Eckelbarger. 1989. Extended pre-feeding period in planktotrophic larvae of the bathyal echinoid *Aspidodiadema jacobyi. J. Mar. Biol. Assoc. U. K.* 69:695–702.

CHAPTER FIVE

Ontogenetic Migration and Dispersal of Deep-Sea Gastropod Larvae

Phillipe Bouchet and Anders Warén

Abstract

Mode of development has been inferred from protoconch morphology in more than thirty families of shelled gastropods from the deep sea. Planktotrophic development is taxonomically widespread in the deep sea, having been recorded down to at least 5400 m depth, and in some species from hydrothermal vents. Evidence that veligers undergo ontogenetic migrations to epipelagic layers comes from several independent observations: occurrence of larvae in surface plankton, isotopic analysis of shell carbonates, presence/absence of eyes, and biogeographic patterns. Planktotrophic development is a plesiomorphic character that is associated with good dispersal and gene flow. Nonplanktotrophic development is a heterogeneous category in terms of dispersal ability. In caenogastropods, nonplanktotrophic development restricts larval dispersal, but in archaeogastropods, nonplanktotrophic development can involve a demersal dispersal phase. Modes of development in deep-sea snails strongly reflect phylogenetic constraints and should not necessarily be viewed in terms of adaptation to life in the deep sea.

Information on reproductive biology and developmental patterns is necessary to understand the ecology, evolution, and distribution of benthic marine invertebrates. Both fields are usually difficult and time consuming to study and much of our present knowledge is anecdotal. This is particularly the case for animals inhabiting bathyal, abyssal, and hadal environments. Although the present review certainly exemplifies this paucity of data on larvae of deep-sea animals, we nevertheless felt a need to summarize what is known, to define some problems, and to encourage future work.

Fifteen years ago, we pointed out for the first time that many deep-sea snails exhibit protoconch morphologies that would indicate planktotrophic development in shallow water species. Some of our early observations and speculations (Bouchet 1976, 1977) have been widely accepted, while others remain controversial. Since those observations, a considerable amount of new information has been added, and our whole perception of life in the deep sea has changed. In the present paper, we review recent evidence and speculate on the evolutionary importance of the different modes of development that occur in deep-water gastropods.

Mode of Development and Protoconch Morphology

There are different ways of classifying developmental patterns of marine invertebrates. From an ecological perspective, it is generally considered that the planktotrophy/nonplanktotrophy dichotomy best describes the larval life histories of marine snails (Jablonski and Lutz 1983).

In planktotrophic development, the veliger larva of a snail hatches from an egg-capsule and spends up to several months in the plankton, where it feeds on phytoplankton. After this obligatory planktonic existence, larvae settle to the bottom, metamorphose, and assume a benthic lifestyle.

In nonplanktotrophic development, the embryo gets its nourishment from the yolk-rich egg or from intracapsular reserves such as albumen or nurse eggs. A short nonfeeding planktonic stage occurs in some species, whereas others undergo metamorphosis in the egg capsule and hatch as crawling juveniles.

The occurrence of both modes of development in the same species has been termed poecilogony. However, no species of marine snail has been proven to exhibit both kinds of development. Reputed cases of poecilogony appear to involve two closely related species, each with its own mode of development (Hoagland and Robertson 1988; Bouchet 1989). In gastropods, mode of development is therefore a species-specific character.

It is important to stress that, among the gastropods, there are different forms of lecithotrophy (nonplanktotropy). In the Archaeogastropoda, except for the Neritimorpha, the eggs are small, free-spawned, and unprotected by an egg-case. These eggs develop into nonfeeding free-swimming larvae. In Neritimorpha, Caenogastropoda and Heterobranchia, the eggs are deposited in an egg-case; development of the large, yolk-rich eggs, or eggs developing with nurse eggs or other resources (Fioroni 1982), is mostly intracapsular, bypassing the planktonic larval stage. At the completion of development, a benthic juvenile emerges from the capsule. These two kinds of lecithotrophy have very different dispersal consequences.

The mode of development of shelled gastropods and bivalves can be inferred from the morphology of the protoconch. The protoconch of a gastropod is the aragonitic shell formed by the veliger or the embryo before metamorphosis. It is retained at the apex of the adult postmetamorphic shell (teleoconch), unless it is broken or eroded. The relationship between protoconch morphology and mode of development has been reviewed in several recent papers (e.g., Jablonski and Lutz 1980). The shell of species with planktotrophic development is formed in three successive stages: (i) protoconch I is secreted by the shell gland of the embryo inside the egg-capsule; (ii) after hatching, protoconch II is formed by the mantle of the planktonic veliger; (iii) after metamorphosis, the teleoconch is the "adult" shell formed by the mantle of the crawling snail (figures 5.1B, 5.2B, 5.3B). The shell of species with lecithotrophic development consists of only two successive stages: protoconch and teleoconch (figs 5.3D, 5.5). The single transition between these two shells marks the stage of hatching and/or metamorphosis in the life cycle of the snail.

Planktotrophy in Deep-Sea Snails

When Thorson (1946, 1950) first proposed the "shell apex theory," the planktotrophy/lecithotrophy dichotomy discussed above seemed typical of most marine phyla and of most habitats except for polar waters and the deep sea, where lecithotrophic development appeared to predominate. Available information on the development of Arctic invertebrates supported his hypothesis. However, it has now been recognized that "Thorson's Rule," although valid for molluscs, does not apply to all Antarctic invertebrates (Pearse et al. 1991; Pearse, this volume). With respect to developmental patterns of deep-sea animals, remarkably few data were available at Thorson's time, and it is likely that Thorson assumed the cold stenothermic deep-sea environment to be analogous to high-latitude environments.

When protoconch characters were used to infer mode of development in deep-sea snails, a different pattern emerged (Bouchet 1977; Bouchet and Warén 1979a; Rex and Warén 1982; Colman et al. 1986). In Archaeogastropoda (except the Neritimorpha), the paucispiral protoconchs do not differ from those in shallow water, suggesting lecithotrophic development. Protoconchs that indicate a planktotrophic development are present in Neritimorpha (Archaeogastropoda), Caenogastropoda, and Heterostropha. Such protoconchs are present in the majority of gastropod families occurring in the deep sea (table 5.1). Indeed, planktotrophy may even be the dominant mode of development in species-rich families such as Turridae and Epitoniidae. It also occurs in taxa restricted to the deep sea, such as Laubierinidae, Haloceratidae, and Pisanianurinae.

Rex and Warén (1982) plotted the percentage of inferred planktotrophy against depth along the Gay Head-Bermuda transect and showed that the percentage of species with planktotrophic development increases from the upper bathyal zone to the abyss. The same trend was observed in turrid assemblages from the eastern and western North Atlantic (Potter and Rex 1992). Table 5.1 gives the deepest record of species with inferred planktotrophic mode of development in each family. Planktotrophy is not known from depths greater

TABLE 5.1 *Occurrences of Gastropod Families (Neritimorpha, Caenogastropoda and Heterostropha only) in the Deep Sea (Deeper than 1000 m) and Frequency of Planktotrophy*

Taxon	Occurrence in the Deep Sea[a]	Planktotrophy[b]	Deepest Occurrence[c]
Phenacolepadidae	+	+ +	2600
Provannidae	+	+	5000
Abyssochrysidae	+	0	—
Cerithidae	+ +	+	1000
Cerithiellidae	+ +	+	1200
Rissoidae	+ +	+	5400
Epitoniidae	+ +	+ +	3300
Eulimidae	+ + +	+	5200
Aclididae	+ +	+	1000
Hipponicidae	+	0	—
Vanikoridae	+ +	+	2000
Naticidae	+ +	+	5000
Ovulidae	+	+ +	1000
Lamellariidae	+	+	4500
Capulidae	+ +	+	2500
Haloceratidae	+ +	+ +	3700
Laubierinidae	+	+ +	3700
Cassidae	+	+	5200
Ficidae	+	?	—
Nassariidae	+ +	+	1000
Fusinidae	+ +	+	1700
Buccinidae	+ +	0	—
Muricidae	+ +	+	2700
Cancellariidae	+ +	+	2900
Turbinellidae	+ +	+	2200
Marginellidae	+ +	0	—
Volutidae	+ +	0	—
Turridae	+ + +	+ +	5400
Pyramidellidae	+ +	+	2500
Architectonicidae	+ +	+ +	2100
Mathildidae	+ +	+ +	1800
Hyalogyrinidae	+	+	2600
Xylodisculidae	+	+	2600

SOURCES: Bouchet and Wareń (1980, 1985, 1986, 1992) and unpublished observations on material from deep-sea cruises in MNHN.

[a]Occurrence in the deep sea: + indicates that the family is known deeper than 1000 m from fewer than 10 species worldwide; + +, *ditto,* with 10–100 species worldwide; + + +, *ditto,* with several hundred species worldwide.

[b]Planktotrophy: o indicates that planktotrophy is not inferred in any deep-sea representative of the family; + indicates planktotrophy occurs, but is not dominating; + + indicates that planktotrophy is the dominating mode of development in deep-sea representatives of the family.

[c]Deepest: deepest records, in meters, of species with inferred planktotrophy, rounded up to the nearest hundred.

than 5400 m (*Benthonella tenella*), but it is uncertain whether this reflects actual absence of planktotrophy at hadal depths, or lack of adequate material. Specimens from very deep water are often severely corroded (Bouchet and Warén 1979a), with the apical whorls dissolved and replaced by an internal plug. This phenomenon is especially common in specimens taken at depths exceeding 4000 m. Most shells collected below 6000 m have the original sculpture remaining only on the last and penultimate whorls, and no protoconch of a snail from hadal depths has ever been seen.

The majority of gastropods from hydrothermal vents and hydrocarbon seeps belong to the Archaeogastropoda, and their protoconchs are paucispiral, indicating lecithotrophic development (Turner et al. 1985). However, planktotrophic larval development has been inferred for the bivalve *Bathymodiolus thermophilus* (Lutz et al. 1980), for an unnamed *Bathymodiolus* from the Louisiana seeps (pers. obs.), and for several vent Neritimorpha, Caenogastropoda and Heterobranchia (Warén and Bouchet 1989, 1992; Gustafson et al. 1991). Although gastropod larvae have been collected in plankton near east Pacific hydrothermal vents (Turner et al. 1985), these larvae appear to belong to east Pacific shallow-water gastropods (Warén and Bouchet 1989).

It is now clear that planktotrophic development is widespread in deep-sea snails, both in terms of taxonomic representation and numbers of individuals. It is prevalent in deep soft-sediment communities, and not uncommon in chemosynthesis-dependent vent and seep communities.

Ontogenetic Migrations in Deep-Sea Snails

Where in the water column do planktotrophic larvae of deep-sea gastropods live?

Meroplanktonic veligers of shelf species normally occur in the photic zone (upper 200 meters) of the water column, either above the shelf, or in the open ocean (Richter and Thorson 1975). These veligers feed on phytoplankton and grow until they reach competence, then metamorphose when the appropriate settlement cues are encountered (Burke 1983).

Planktonic larvae of deep-sea gastropods could: (1) either live a bathypelagic or demersal benthic life, feeding on bacteria or other organic matter; or (2) migrate to the photic zone to feed on phytoplankton, and return to the deep sea prior to settlement. Evidence from four independent sources suggests that the planktonic larvae of at least some deep-sea gastropods undergo ontogenetic migrations to the photic zone.

Evidence from Plankton Samples

Larvae of seven deep-sea taxa have been recognized in surface waters (table 5.2). It should be noted that absolute adult depth ranges may give a somewhat

TABLE 5.2 *Records of Veligers of Deep-Sea Gastropods Captured in Surface Waters*

Species	Adult Bathymetric Range	Number of Larvae Found	Locality
(Provannidae ?)			
Undescribed species[a]	abyssal	1	Mid-Atlantic
(Rissoidae)		161	Mediterranean
Benthonella tenella[b]	bathyal-abyssal	?	N Atlantic
(Elachisinidae)			
Laeviphitus verduini[c]	upper bathyal	12	Mediterranean
(Tonnoidea)			
Oocorys-type[d]	bathyal	2	Mid-Atlantic
(Columbellidae)			
Amphissa acutecostata[e]	bathyal	7	Mediterranean
		?	N Atlantic
(Turridae)			
Drilliola loprestiana[f]	deep-shelf-bathyal	8	Mediterranean
Benthomangelia macra[f]	bathyal	1	Mediterranean
Famelica-type[a]	deep-sea	1	SW Pacific

SOURCES: [a]this paper; [b]Bouchet 1976; [c]Warén et al. 1990; [d]Warén and Bouchet 1990 (Laursen's 1981 "unidentified larva A"); [e]Bouchet and Warén 1985; [f]Bouchet and Warén 1980.

distorted representation of distribution patterns. For example, although scattered individuals of *Benthonella tenella* have been recorded as shallow as 120 m in the Mediterranean, fewer than 2 percent of all known specimens of that species come from depths above 1900 m. We therefore rank *B. tenella* among the bathyal-abyssal fauna.

We report here two new cases of veligers of deep-sea species taken in surface plankton. Figure 5.1A shows a larva taken by Dr. G. Richter during Cruise 51 of R.V. Meteor in February 1979. It was taken 50–100 m below surface, at 01°39.7′ S, 22°00.1′ W, over 4570 m of water. Although this larva cannot be identified with certainty at species level, there are no shallow-water gastropods known to have a similar larva. An undescribed gastropod from 14°50′ N, 67°20′ W, 4956–5005 m has one remaining protoconch whorl with sculpture identical to that of the Meteor larva (figure 5.1B). The taxonomic position of this snail is not certain, but it is nearest to the Provannidae and Abyssochrysidae, families known only from the deep sea.

The second species, illustrated in figure 5.1C, was taken during Cruise HYDROTHON in February 1979 from 0–100 m, at 24°30' S, 173°57′ E (off Fiji, SW Pacific), above 4030 m of water. Again, this larva cannot be precisely identified at species or even genus level. There are, however, only three turrid genera known to have a similar protoconch: *Famelica, Neopleurotomoides,* and *Aliceia.* These genera are restricted to the deep sea, and it is reasonable to assume that the larva from the Fiji plankton also belongs to a deep-sea species.

There are two reasons why so few records exist of deep-sea gastropod larvae in surface plankton. The first is that in surface waters the veligers of shallow-water species far outnumber the veligers of deep-sea species. Richter and Thorson (1975) identified 30,000 gastropod larvae from surface plankton in the Tyrrhenian Sea off Naples. Of these, 29,800 (99.33%) were identified to known shallow-water species, and slightly fewer than 200 (0.67 percent) have later been recognized as larvae of deep-sea species. Of these, 161 were larvae of *Benthonella tenella.* In the Mediterranean, *B. tenella* is the most common deep-sea gastropod,

FIGURE 5.1. Veligers and protoconchs of deep-sea gastropods with planktotrophic development. The sculpture patterns shown here are unknown in larvae of shallow-water gastropods. A: Larva from surface plankton, Mid-Atlantic at 01°39.7′ S, 22°00.1′ W, water depth 4570 m; height 2045 μm. B: Benthic gastropod from 14°50′ N, 67°20′ W, 4956–5005 m; the protoconch is corroded, but its last whorl has a sculpture identical to that of the larva above; height 238O μm. C: Turrid larva from surface plankton, SW Pacific at 24°30′ S, 173°57′ E; water depth 4030 m; height 1100 μm. D: Protoconch of *Famelica mirmidina,* from the bathyal zone of the Azores, height 1.12 mm; note the similarity of the sculpture pattern with the larva above. E: Young postlarva of *Haloceras tricarinata,* a bathyal species of the NE Atlantic; diameter 2780 μm. F: Protoconch of *Iphitus tuberatus,* a member of the deep-sea subfamily Nystiellinae (Epitoniidae), from bathyal depths in the Bay of Biscay; height 440 μm.

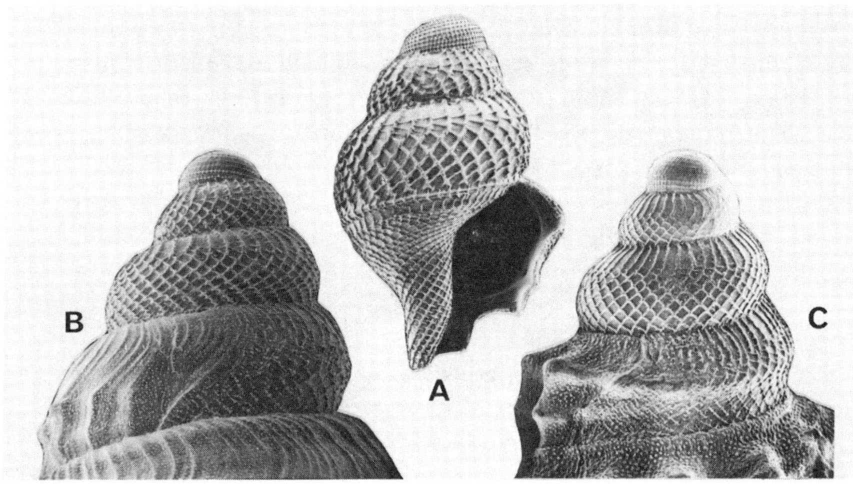

FIGURE 5.2. Daphnelline veliger and protoconchs. This type of protoconch sculpture occurs in hundreds of shallow- and deep-water turrids. A: Larva from surface plankton, SW Pacific, 24°46′ S, 177°13′ E; water depth 4150 m;. height 910 μm. B: Protoconch of *Pleurotomella bureaui,* a bathyal species from the Azores; height 745 μm. C: Protoconch of *Philbertia linearis,* Swedish west coast, shallow water; height 650 μm.

and 86 percent of all gastropods taken alive below 1200 m during the cruises BIOMEDE 1 and 2 in the Western Basin belonged to this species. In the North Atlantic as well, this rissoid is the most abundant deep-sea snail, and its larvae have occasionally been taken in the water column (Roe et al. 1990). The chances of finding the larvae of an average deep-sea gastropod that is 500 times less common than *B. tenella* also decrease 500 times. It would be necessary to sort and identify 100,000 veligers on average before encountering just one of its larvae: a daunting prospect. Plankton samples from the open ocean should, however, have fewer larvae of shallow-water species, and thus an increased proportion of larvae of deep-sea species. Such samples nevertheless are dominated by the teleplanic planktotrophic veligers of shallow-water species from coastal areas. The deep-sea snail larva from the Meteor sample was the only one of its kind in a catch that included 74 larvae of *Bursa corrugata* and 102 other common meroplanktonic gastropod larvae (i.e., a proportion of 0.56 percent). A systematic search for such unusual veligers would probably reveal additional records of migratory veligers of deep-sea species.

The second factor that contributes to an underestimation of veligers of deep-sea species in surface plankton is that not all larvae can be identified with certainty to species level. For instance, the daphnellid type of protoconch is equally frequent in shallow- and deep-sea turrids (figure 5.2). There are at least a thousand turrid species with this type of protoconch, and such larvae can be identified at the specific level only in areas of low species diversity. Veligers of *Phymorhynchus* from hydrothermal vents belong to this category and would not be specifically, or even generically, recognizable if/when taken in plankton hauls. A similar situation is found in the Epitoniinae and the Eulimidae, where there are shallow- and deep-sea species, all having similar protoconchs.

Several taxa of deep-sea gastropods, however, do have remarkable protoconchs with no equivalent in shallow-water faunas (Haloceratidae, Nystiellinae, *Aliceia/ Famelica/ Neopleurotomoides,* Laubierinidae: figures 5.1C–F). We suggest that these should become the primary targets of future attempts to discover additional examples of migratory larvae in surface plankton.

Evidence from Isotopic Analyses

The isotopic analysis of the carbonate shells of deep-sea snails circumvents the difficulties encountered when trying to identify veligers of deep-sea species.

Briefly the technique consists in analyzing $^{16}O/^{18}O$ and $^{13}C/^{12}C$ ratios in the calcium carbonate, separately in the protoconch (formed during the planktonic phase) and the teleoconch (formed during the benthic phase). The $^{16}O/^{18}O$ and $^{13}C/^{12}C$ ratios reflect the composition of the seawater at the time of deposition of the calcium carbonate. The protoconch and teleoconch of a single shell can be analyzed separately; if not enough material is obtained from a single shell, the protoconch and teleoconch fractions of 2 to 5 shells from the same population can be pooled together. The measured isotopic ratios can be compared between the two parts of the shell and used to estimate absolute temperatures. If the planktotrophic larva has migrated to surface layers, the higher temperatures of the water there should be reflected in the isotopic composition of the larval carbonate.

Results of such isotopic analyses were first published by Bouchet and Fontes (1981). They showed a difference in isotopic composition for three planktotrophic species, with a calculated temperature difference of 1.1–3.7°C. However no difference was observed in isotopic composition between the protoconch and teleoconch of *Benthonella tenella,* a species for which larvae had actually been collected in surface plankton. Killingley and Rex (1985) performed an isotopic analysis on both species with inferred lecithotrophic development, and species with inferred planktotrophic development. They obtained unambiguous results. Larval and postlarval carbonates of species with inferred lecithotrophic development were not different. For species with inferred planktotrophic larval development, adult-larval differences in $\delta\ ^{18}O$ corresponded to vertical migrations on the order of 1800–3200 m. They obtained positive results for *Benthonella tenella* as well, and attributed the negative result in Bouchet and Fontes (1981) to secondary deposition of calcium carbonate by the postlarva inside the larval shell. We agree with this interpretation. Bouchet and Warén (1979a) had noted that the larval shell of deep-sea gastropods is frequently much more corroded than the teleoconch, and commonly only an internal mold of the apical whorls remains.

Anatomical Evidence

During our examination of deep-sea gastropods for taxonomic purposes we have often noticed that species with inferred planktotrophy have eyes, whereas species with lecithotrophy frequently are blind. The examples below illustrate this observation.

The eulimid *Rectilabrum lanceolatum,* an amphi-Atlantic species with inferred planktotrophic development, has been recorded from depths of 1770–4825 m. The adult has no eyes, but a juvenile from 2090 m, with 0.8 postlarval whorls, has a pigmented body and pigment spots situated centrally in the bases of the tentacles (Bouchet and Warén 1986:347).

Alvania testae (Aradas and Maggiore 1844) has a bathymetric distribution from 100 to 800 m, with occasional records to 2400 m. Off Western Europe and in the Mediterranean, it is sympatric with *A. subsoluta* (Aradas 1847), which has a bathymetric distribution from 200 to 1200 m, with occasional records from 2000 m. The two species resemble each other closely, and have at times been confused by experienced conchologists. *Alvania testae* (figure 5.3A, B) has eyes and planktotrophic development, while *A. subsoluta* does not possess eyes and has lecithotrophic development (figure 5.3C, D).

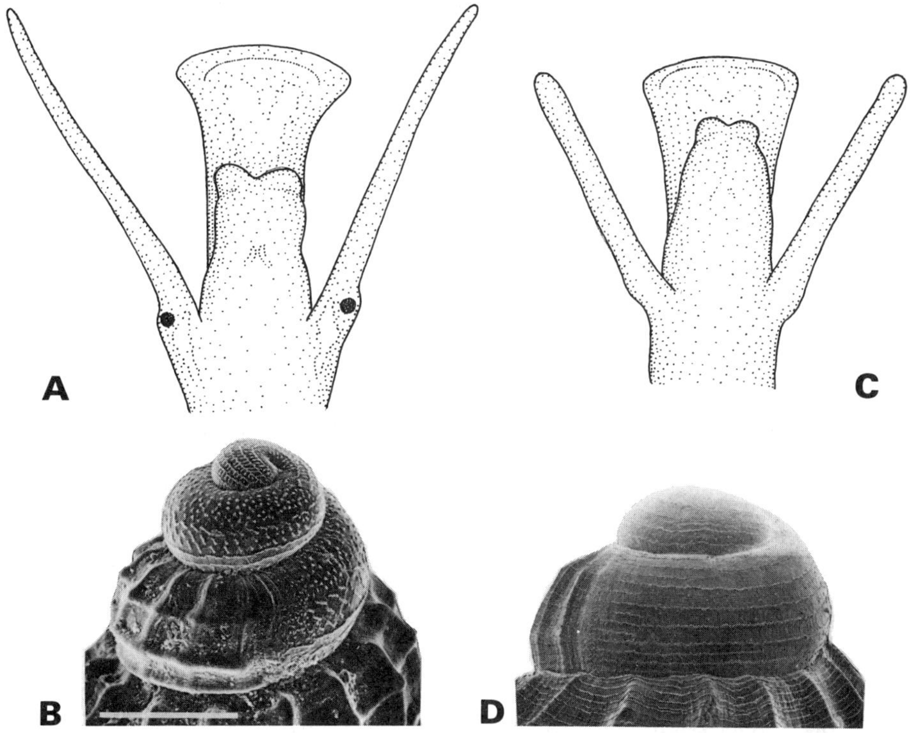

FIGURE 5.3. *Alvania* species pair with different modes of development. The base of the tentacles of *Alvania testae* bear normal eyes (A), and its protoconch (B) is of planktotrophic type. *Alvania subsoluta* is blind (C), and its protoconch (D) is of lecithotrophic type.

Scale line (B, D) 200 μm.

Finally, among gastropods from hydrothermal vents in the SW Pacific, only the young of phenacolepadids, *Alviniconcha* and *Desbruyeresia* have eyes, and all have inferred planktotrophic development. The abyssal provannid corresponding to the Meteor larva (figure 5.1B) also has eyes.

We interpret the presence of eyes in deep-sea species with inferred planktotrophy as evidence that their veliger larvae migrate to the photic zone, whereas the absence of eyes in species with inferred nonplanktotrophy is connected to their development taking place in darkness. However, there is not a perfect association between these characters, and several deep-sea archaeogastropods have well developed eyes (pers. obs.).

Biogeographical Evidence

Biogeography offers a fourth, independent, line of evidence on ontogenic migration of deep-sea planktotrophic veligers.

Bouchet and Warén (1979b) compared the composition of the deep-sea mollusc fauna in the Norwegian Sea and the northern North Atlantic. The two basins are separated by the Wyville-Thomson Ridge (maximum depth 800 m) and the Denmark Strait. Only seven abyssal mollusc species live both north and south of the ridge. Of these, *Anatoma crispata* and *Oenopota ovalis* have a continuous distribution across the sill. The other five are found in deep basins of the North Atlantic and Norwegian Sea, but not on the ridge, and are inferred to have planktotrophic development.

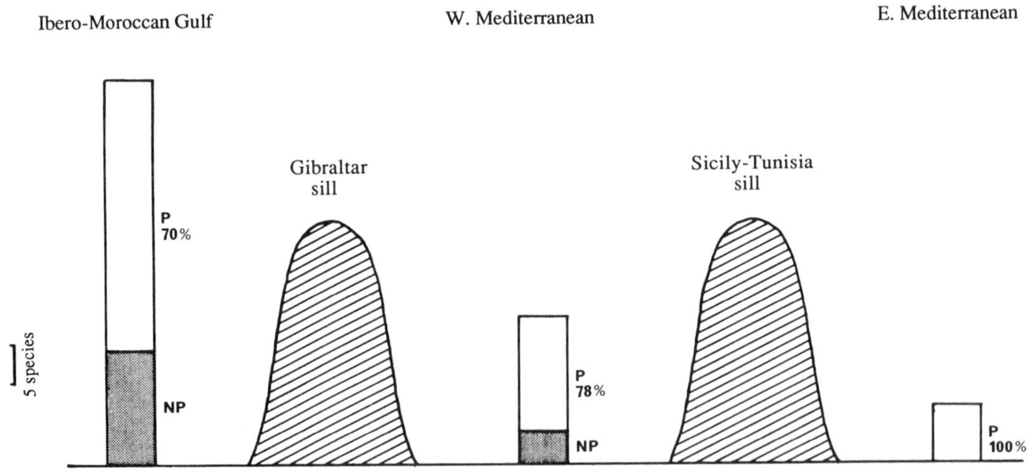

Figure 5.4. The Ibero-Moroccan Gulf has a rich deep-sea turrid fauna, and species with planktotrophic development account for 70 percent of the total. The Gulf is separated by the shallow Gibraltar sill from the western Mediterranean, where deep-sea turrid species with planktotrophic development account for 78 percent of the total. Finally, the eastern Mediterranean is separated from the western basin by the Sicily-Tunisia sill. The impoverished deep-sea turrid fauna consists only of species with planktotrophic development.

The distribution of the deep-sea Turridae east and west of the Strait of Gibraltar offers another example (figure 5.4, after data in Bouchet and Taviani 1992). Many turrids and other gastropods with inferred planktotrophic development have a disjunct Mediterranean population. The most parsimonious explanation is that planktotrophic larvae migrate to the inward (eastern) current, by which they can be brought into the Mediterranean above the sill. Conversely, in species with lecithotrophic development, larvae are confined to the deep benthic environment and excluded from the Mediterranean by the shallow sill.

Ontogenetic Migrations in Snails from Hydrothermal Vents?

Information on the life history of species from hydrothermal vents and seeps is derived from protoconch morphology and contents of egg capsules. Lutz et al. (1980) first pointed out that the protoconch of *Bathymodiolus thermophilus* indicated planktotrophic development. Several gastropod species from vents have later been inferred to have planktotrophic development, namely the turrid *Phymorhynchus* (Warén and Bouchet 1989; Gustafson et al. 1991), the limpet *Shinkailepas,* the "hairy snail" *Alviniconcha* and *Desbruyeresia* spp. (Warén and Bouchet 1992).

Lutz et al. (1980) suggested that ontogenetic migration of *Bathymodiolus* larvae to surface layers was unlikely because vertical migration and dispersal in surface currents should decrease the probability of larvae encountering isolated vent sites. Similarly, Gustafson et al. (1991) refrained from speculating on where in the water column the planktotrophic larvae of *Phymorhynchus* develop. Indeed, it seems unlikely that the larvae of *Bathymodiolus* or *Phymorhynchus* should spend weeks or months in surface waters, become dispersed over vast expanses of ocean, and then become wasted when they cannot find vents for metamorphosis.

Absence of ontogenetic migration in *Bathymodiolus* and *Phymorhynchus* would make them exceptions among deep-sea mollusks with inferred planktotrophic development. We prefer to hypothesize that the veligers of vent species with inferred planktotrophic development do migrate to surface layers. Testing this hypothesis will be difficult. The characteristics of the veligers of most shallow-water Indo-Pacific mytilids and daphnelline turrids are poorly known, and specific recognition of neritimorphs or *Phymorhynchus* larvae in plankton samples from Pacific waters will be difficult, if indeed it is possible. Standard isotopic analysis is also less useful, since surface waters and diluted hydrothermal fluids may have similar temperatures. Elements such as Mn, As, and Se are present at very different concentrations in surface waters and hydrothermal fluids, and may become incorporated in the carbonates or organic matrix of the shell. An analysis of larval and postlarval shell composition may provide unambiguous results.

The Evolutionary Consequences of Developmental Mode

There is little information on the consequences of larval development for the evolution of deep-sea snails. Rex et al. (1988), Rex and Etter (1990), and Etter and Rex (1990) have shown that there is little morphological variation over vast geographical areas in *Benthonella tenella,* a species with planktotrophic development. In contrast, morphological variation is significant over small bathymetric and geographical distances in *Frigidoalvania brychia,* a species with lecithotrophic development. A single paper (Siebenaller 1978) deals with allozyme variation in a deep-sea trochid from five populations off California, and research is in progress on the population genetics of gastropods from hydrothermal sites (Vrijenhoek et al. 1991; Jollivet pers. comm.). However, there is a rich literature on the associations between mode of development, dispersal of larvae and exchange of genetic material between populations of shallow-water gastropods. Generally this work (e.g., Scheltema 1971; Hansen 1983; Janson 1987; Behrens-Yamada 1989) supports the prediction (e.g., Crisp 1978; Grassle and Grassle 1978) that species with good dispersal potential have a higher degree of genetic homogeneity and broader geographic and stratigraphic ranges than species with poor dispersal potential.

However, we want to emphasize that, although planktotrophy confers good dispersal ability, lecithotrophy does not always restrict dispersal.

Lecithotrophy: Different Modes of Dispersal

We have mentioned earlier that lecithotrophy in archaeogastropods (exclusive of Neritimorpha) involves small larvae which are capable of passive dispersal as plankton (figure 5.5A, B). The importance of pelagic lecithotrophy has recently been recognized in Antarctic waters, which share with the abyss poor food conditions and slow rates of development (Pearse et al. 1991). Marshall (1987, and literature therein) has described a spectacular array of limpets from deep-water biogenic substrates off New Zealand. Habitat specializations include old cephalopod beaks (Bathysciadiidae), old whale bones (Osteopeltidae and Pyropeltidae), or sunken wood (Cocculinidae). These are ephemeral, unpredictable and scattered substrates. When the substrate has been exhausted, colonization of a new island must involve drifting of small lecithotrophic larvae. In shallow water, larvae of archaeogastropods are free-swimming for periods of 2–4 days (Bandel 1982). In deep water, however, with much lower temperatures and thus lower metabolic rates, it is likely that these larvae are capable of spending longer periods as demersal, drifting, nonfeeding larvae (see also the discussion in Stuart and Rex, this volume). If that is the case, the dispersal of deep-water archaeogastropods may be much wider than previously suspected, despite their nonplanktotrophic larval development.

Lecithotrophic development in many gastropods from hydrothermal vents (Turner et al. 1985) presents a similar dilemma because vents are temporary

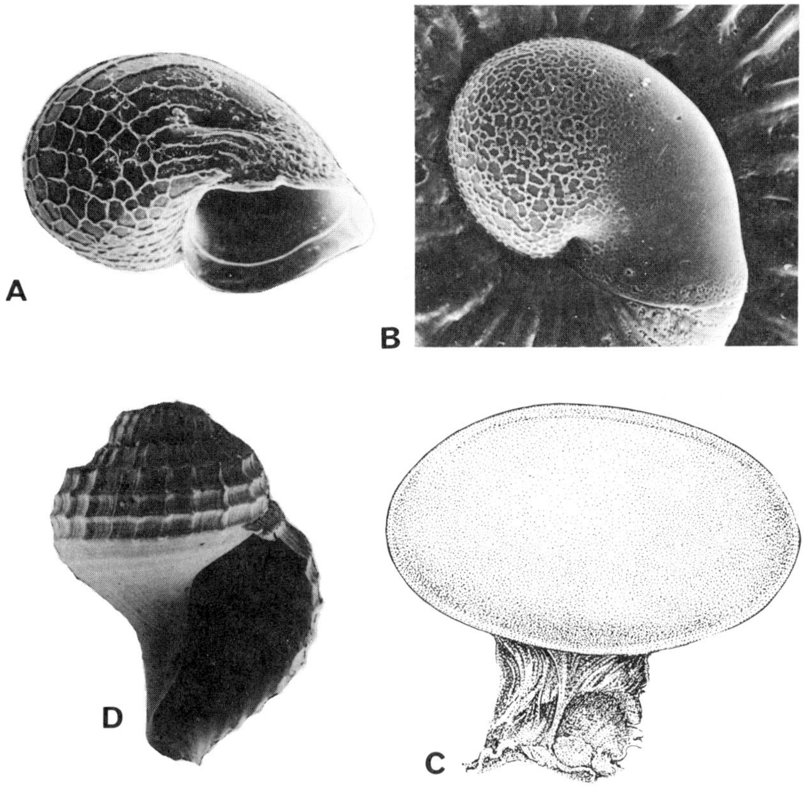

FIGURE 5.5. Different sizes of lecithotrophic larvae mean different dispersal capacities. The larvae of archaeogastropods (A, B) are small, with a free-swimming stage. Those of the more advanced caenogastropods metamorphose in the egg-capsule (C) and hatch as a considerably larger crawl-away juvenile (D). A: *Lepetella laterocompressa,* bathyal of the eastern Atlantic, diameter 178 μm. B: *Solutigyra reticulata,* eastern Pacific hydrothermal vents, diameter 232 μm. C, D: *Mohnia abyssorum,* abyssal of the northeast Atlantic, height 5.1 mm.

and patchy island habitats. We suggest that lecithotrophy in gastropods from this environment merely represents a phylogenetic constraint (Berg 1985; see also Gustafson and Lutz, this volume). Most vent gastropods are archaeogastropods and their nonplanktotrophic mode of development is neither an adaptation nor an obstacle to dispersal. This contrasts with the situation in higher gastropods, where lecithotrophic larvae often emerge in the crawling state from the egg capsule at sizes ranging from 0.25 to 10 mm (figure 5.5C, D). Here, the lack of dispersal ability permits population differentiation and makes populations more vulnerable to local extinction.

We conclude that the mode of development and dispersal are linked, but not in the simple way they are often associated in analyses of the planktotrophy vs. lecithotrophy dichotomy. There exist many variations, from planktotrophy with teleplanic larvae spending months in the plankton, to lecithotrophy where a crawl-away juvenile hatches from the egg-capsule. It is particularly important to recognize that some archaeogastropods may disperse near the bottom and that this may be related to their widespread distribution both at vent sites and in deep soft-sediment environments.

Phylogenetic Constraints on the Mode of Development

In Caenogastropoda and Heterostropha, as in all other marine invertebrates, planktotrophy is the ancestral condition, and lecithotrophy is a derived character state. (It is not known, and it is beyond the scope of this paper to discuss whether the planktonic, nonfeeding larvae of the archeogastropods are primitive or derived.) Strathmann (1978, 1985) discussed the loss of planktotrophy and concluded that reversal was possible, in principle, when the ciliary bands of the velum have not been lost. Despite this theoretical possibility, we know of no fossil record where planktotrophy has been reacquired after it has been lost (but see Reid 1989). We do know fossil cases where planktotrophic development has been lost (Bouchet 1981). Therefore we consider reversion to planktotrophy within lineages to be exceptional, and occurrence of planktotrophy in deep-sea taxa to be a plesiomorphy that primarily reflects phylogeny. Gastropods often have highly specialized diets at the family level (Kohn 1983) and we believe that occurrence of gastropod taxa in the deep sea is related to feeding biology. For example, epitoniids (that feed on Anthozoa) occur in the deep sea because Anthozoa occur in the deep sea, and vermetids (filter-feeders) do not occur in the deep sea because there is too little suspended material in the water. We do not regard the mode of development of epitoniids and vermetids to be directly related to their presence or relative success in the deep ocean.

Reproductive biology of the Laubierinidae offers a remarkable example of deep-sea species where development is phylogenetically constrained. Laubierinid veligers reach 5 mm and rank among the largest, not only of deep-sea gastropods, but of gastropods in general. Laubierinidae belong to the superfamily Tonnoidea, which is characterized by the long-distance planktotrophic larvae for which Scheltema (1971) created the term "teleplanic." Although laubierinids are now restricted to the deep sea, it is likely that their ancestors were shallow-water species that retained their teleplanic planktotrophic mode of development when they invaded the deep sea. Representatives of Laubierinidae are very rare, and the sex has been determined in only six adults collected at different seasons over a period of ten years at abyssal depths in the Indian and Atlantic oceans (Warén and Bouchet 1990). All six adults were females. One very late pediveliger or very young postlarva of a laubierinid was collected at a depth of 3716 m. It still had the remnants of a velum, but had no trace of postlarval shell, and had therefore probably settled very recently (figure 5.6). This very large (shell ca. 5 mm) larva was a sexually mature male, with fully developed penis, male gonad and vesicula seminalis filled with sperm. Warén and Bouchet (1990) interpreted laubierinids to be protandrous hermaphrodites with neotenic males reaching sexual maturity at the end of their planktonic life.

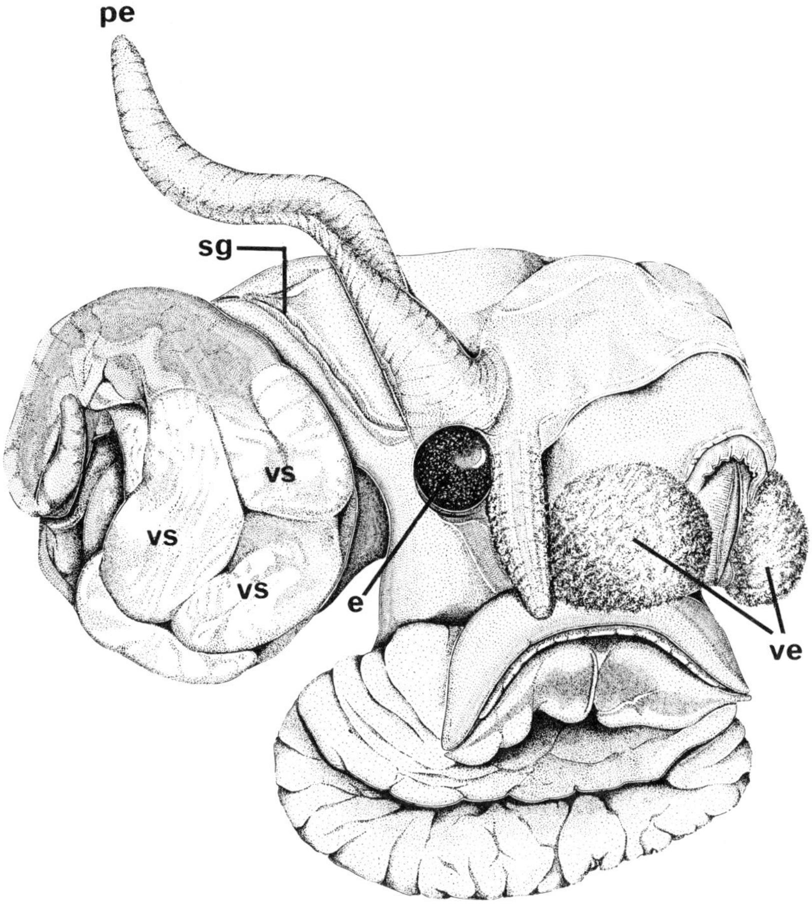

FIGURE 5.6. Very young postlarva of *Laubierina* sp., Mozambique Channel, 3716 m. Remains of velum are still visible, indicating that the animal has only recently metamorphosed. Despite this, the specimen is a functionally mature male. Maximum dimension about 5 mm.

Key: e = eye; pe = penis; sg = seminal groove; ve = remains of velum; vs = vesicula seminalis.

Biological oceanographers were initially skeptical about the existence of planktotrophy in the deep sea because planktotrophy is usually associated with an r-type life history, and the deep sea was thought to be an environment characterized by K-selection (Sanders 1979). Lecithotrophy was seen as an adaptation to life in a stable, low-energy environment, such as the deep sea. In fact, the case of the Laubierinidae suggests that planktotrophy is not a handicap to life in the deep sea. Laubierinid larvae can feed on phytoplankton in surface layers, and convert this rich energy source directly into production of gametes; access to epipelagic energy balances the cost of larval mortality during the very long planktonic phase. Laubierinids could even be considered planktonic species that lay their eggs on the deep-sea bottom, rather than deep-sea species with teleplanic planktotrophic larvae. This example also sheds light on a number of deep-sea turrids such as *Lusitanops,* with protoconchs of 3–4 whorls and teleoconchs of 2–3 whorls (Bouchet and Warén 1980). These could represent species where neoteny is made possible by the large size attained by their planktotrophic veligers.

Concluding Remarks

Several problems remain to be solved by future research. Some of the questions are specific to deep-sea biology, others are more general.

1. The duration of the free-swimming phase in lecithotrophic larvae is an important issue that controls the extent of dispersal of these nonplanktotrophic larvae. We do not know if archaeogastropod larvae go through successive pre-competent and competent phases, and to what extent they are capable of delayed metamorphosis.
2. More work on surface plankton is needed to find identifiable larvae, and to quantify the occurrence of vertical migration in deep-sea species.
3. The energetics of ontogenetic migrations need to be understood. Richter (1973) showed that during their diurnal vertical migrations, veligers are capable of ascending at 30–50 m/hr, and sink at 10 m/hr. If such values are valid for deep-sea snails, a 3000 m ontogenetic migration would last 60–100 hours (2.5–4 days). An hypothesis that needs to be tested is that young planktotrophic veligers have yolk reserves that give a positive buoyancy to the larvae, thereby allowing them to float to the surface.
4. The larval ecology of vent species with inferred planktotrophy needs further study.

Because the mode of development is reflected by protoconch morphology, gastropods will continue to offer good material for the study of evolutionary correlations of the planktotrophy/lecithotrophy dichotomy.

Acknowledgments

This review has benefited from discussions with a number of colleagues, notably M. Rex (Boston), R. Scheltema (Woods Hole), and S. Gofas (Paris). We are also grateful to G. Richter (Frankfurt) for placing some of his unpublished material at our disposal.

References

Bandel, K. 1982. Morphologie und Bildung der frühontogenet-ischen Gehäuse bei conchiferen Mollusken. *Facies* 7:1–198.

Behrens-Yamada, S. 1989. Are direct developers more locally adapted than planktonic developers? *Marine Biology* 103:403–411.

Berg, C. J. 1985. Reproductive strategies of mollusks from abyssal hydrothermal vent communities. *Biol. Soc. Wash. Bull.* 6:185–197.

Bouchet, P. 1976. Mise en évidence d'une migration de larves véligères entre l'étage abyssal et la surface. *C. R. Acad. Sci., Paris* (3)283:821–824.

Bouchet, P. 1977. Mise en évidence de stades larvaires planctoniques chez les Gastéropodes Prosobranches des étages bathyal et abyssal. *Bull. Mus. nat. Hist. nat., Paris* (3)400:947–972.

Bouchet, P. 1981. Evolution of larval development in Eastern Atlantic Terebridae, Neogene to Recent. *Malacologia* 21:363–369.

Bouchet, P. 1989. A review of poecilogony in gastropods. *J. Moll. Stud.* 55:67–78.

Bouchet, P. and J. C. Fontes. 1981. Migrations verticales des larves de Gastéropodes abyssaux: arguments nouveaux dûs à l'analyse isotopique de la coquille larvaire et postlarvaire. *C. R. Acad. Sci., Paris* (3)292:1005–1008.

Bouchet, P. and M. Taviani. 1992. The Mediterranean deep-sea fauna: pseudopopulations of Atlantic species? *Deep-Sea Research* 39:169–184.

Bouchet, P. and A. Warén. 1979a. Planktotrophic larval development in deep-water gastropods. *Sarsia* 64:37–40.

Bouchet, P. and A. Warén. 1979b. The abyssal molluscan fauna of the Norwegian sea and its relation to other faunas. *Sarsia* 64:211–243.

Bouchet, P. and A. Warén. 1980. Revision of the Northeast Atlantic bathyal and abyssal Turridae. *J. Moll. Stud.* suppl.8:1–120.

Bouchet, P. and A. Warén. 1985. Revision of the Northeast Atlantic bathyal and abyssal Neogastropoda, excluding Turridae. *Boll. Malac.* suppl.1:121–296.

Bouchet, P. and A. Warén. 1986. Revision of the Northeast Atlantic bathyal and abyssal Aclididae, Eulimidae, Epitoniidae. *Boll. Malac.* suppl.2:297–575.

Bouchet, P. and A. Warén. 1992. Revision of the Northeast Atlantic bathyal and abyssal Mesogastropoda. *Boll. Malac.* suppl.3. In press.

Burke, R. D. 1983. The induction of metamorphosis of marine invertebrate larvae: Stimulus and response. *Canadian J. Zool.* 61:1701–1709.

Coleman, J. G., P. A. Tyler, and J. D. Gage. 1986. Larval development of deep-sea gastropods from the Rockall Trough. *J. Mar. Biol. Ass. U. K.* 66:951–965.

Crisp, D. J. 1978. Genetic consequences of different reproductive strategies in marine invertebrates. In B. Battaglia and J. A. Beardmore, eds., *Genetics, Ecology, and Evolution,* pp. 257–273. New York: Plenum Press.

Etter, R. J. and M. A. Rex. 1990. Population differentiation decreases with depth in deep-sea gastropods. *Deep-Sea Research* 37:1251–1261.

Fioroni, P. 1982. Entwicklungstypen und Schlüpfstadien dei Mollusken. Einige allgemeine Befunde. *Malacologia* 22:601–609.

Grassle, J. F. and J. P. Grassle. 1978. Life histories and genetic variation in marine invertebrates. In B. Battaglia and J. A. Beardmore, eds., *Genetics, Ecology, and Evolution,* pp. 437–364. New York: Plenum Press.

Gustafson, R. G., D. T. J. Littlewood, and R. A. Lutz. 1991. Gastropod egg capsules and their contents from deep-sea hydrothermal vent environments. *Biological Bulletin* 180:34–55.

Hansen, T. 1983. Modes of larval development and rates of speciation in early Tertiary neogastropods. *Science* 204:501–502.

Hoagland, K. E. and R. Robertson. 1988. An assessment of poecilogony in marine invertebrates: phenomenon or fantasy? *Biological Bulletin* 174:109–125.

Jablonski, D. and R. A. Lutz. 1980. Molluscan larval shell morphology. Ecological and paleontological applications. In D. Rhoads and R. Lutz, eds., *Skeletal Growth of Aquatic Organisms,* pp. 323–377. New York: Plenum.

Jablonski, D. and R. A. Lutz. 1983. Larval ecology of marine benthic invertebrates: paleobiological implications. *Biol. Rev.* 58:21–89.

Janson, K. 1987. Allozyme and shell variation in two marine snails (*Littorina,* Prosobranchia) with different dispersal abilities. *Biol. J. Linn. Soc.* 30:245–256.

Killingley, J. and M. Rex. 1985. Mode of larval development in some deep-sea gastropods indicated by oxygen-18 values of their carbonate shells. *Deep-Sea Res.* 32:809–818.

Kohn, A. J. 1983. Feeding biology of gastropods. In A. S. M. Saleuddin and K. M. Wilbur, eds., *The Mollusca,* 5:1–63. New York: Academic Press.

Laursen, D. 1981. Taxonomy and distribution of teleplanic prosobranch larvae in the North Atlantic. *Dana Report* 89:1–44, pls 1–3.

Lutz, R. A., D. Jablonski, D. C. Rhoads, and R. D. Turner. 1980. Larval dispersal of a deep-sea hydrothermal vent bivalve from the Galapagos Rift. *Mar. Biol.* 57:127–133.

Marshall, B. 1987. Osteopeltidae (Mollusca: Gastropoda): A new family of limpets associated with whale bone in the deep-sea. *J. Moll. Stud.* 53:121–127.

Pearse, J. S., J. B. McClintock, and I. Bosch. 1991. Reproduction of Antarctic benthic marine invertebrates: Tempos, modes, and timing. *Amer. Zool.* 31:65–80.

Potter, E. and M. A. Rex. 1992. Parallel development-depth trends in deep-sea turrid snails from the Eastern and Western North Atlantic. *Nautilus* 106:72–75.

Reid, D. G. 1989. The comparative morphology, phylogeny, and evolution of the gastropod family Littorinidae. *Phil. Trans. Roy. Soc. London* (B)324:1–110.

Rex, M. A. and R. J. Etter. 1990. Geographic variation in two deep-sea gastropods, *Benthonella tenella* (Jeffreys) and *Benthomangelia antonia* (Dall). *Deep-Sea Research* 37:1229–1249.

Rex, M. A. and A. Warén. 1982. Planktotrophic development in deep-sea prosobranch snails from the western North Atlantic. *Deep-Sea Research* 29:171–184.

Rex, M. A., M. C. Watts, R. J. Etter, and S. O'Neill. 1988. Character variation in a complex of rissoid gastropods from the upper continental slope of the western North Atlantic. *Malacologia* 29:325–339.

Richter, G. 1973. Field and laboratory observations on the diurnal vertical migration of marine gastropod larvae. *Neth. J. Sea Res.* 7:126–134.

Richter, G. and G. Thorson. 1975. Pelagische Prosobranchier-Larven des Golfes von Neapel. *Ophelia* 13:109–185.

Roe, H. S. J., D. S. M. Billett, and R. S. Lampitt. 1990. Benthic/midwater interactions on the Madeira abyssal plain; evidence for biological transport pathways. *Prog. Oceanog.* 24:127–140.

Sanders, H. L. 1979. Evolutionary ecology and life-history patterns in the deep sea. *Sarsia* 64:1–7.

Scheltema, R. 1971. Larval dispersal as a means of genetic exchange between geographically separated populations of shallow-water benthic marine gastropods. *Biological Bulletin* 140:284–322.

Siebenaller, J. F. 1978. Genetic variation in deep-sea invertebrate populations: The bathyal gastropod *Bathybembix bairdii. Marine Biology* 47:262–275.

Strathmann, R. 1978. The evolution and loss of feeding larval stages of marine invertebrates. *Evolution* 32:894–906.

Strathmann, R. 1985. Feeding and non-feeding larval development and life-history evolution in marine invertebrates. *Ann. Rev. Ecol. Syst.* 16:339–361.

Thorson, G. 1946. Reproduction and larval development of Danish marine bottom invertebrates, with special reference to the planktonic larvae in the Sound. *Medd. Dan. Fisk. Havunders. (Plankton)* 4(1):1–523.

Thorson, G. 1950. Reproductive and larval ecology of marine bottom invertebrates. *Biol. Reviews* 25:1–45.

Turner, R. D., R. A. Lutz, and D. Jablonski. 1985. Modes of molluscan larval development at deep-sea hydrothermal vents. *Biol. Soc. Wash. Bull.* 6:167–184.

Vrijenhoek, R. C., C. Craddock, S. Schutz, and R. A. Lutz. 1991. Genetic studies at deep-sea hydrothermal vents. *Abstracts 6th Deep-Sea Biology Symposium,* Copenhagen: 69–70.

Warén, A. and P. Bouchet. 1989. New gastropods from East Pacific hydrothermal vents. *Zool. Scripta* 18:67–102.

Warén, A. and P. Bouchet. 1990. Laubierinidae and Pisanianurinae (Ranellidae), two new deep-sea taxa of the Tonnoidea. *Veliger* 33:56–102.

Warén, A. and P. Bouchet. 1992. New species and new records of gastropods from hydrothermal vents and hydrocarbon seeps. *Zool. Scripta.* In press.

Warén, A., F. Carrozza, and R. Rocchini. 1990. *Elachisina versiliensis,* a new Mediterranean species of the family Elachisinidae (Prosobranchia, Truncatelloidea). *Bollettino Malacologico* 25:335–339.

CHAPTER SIX

The Relationship Between Developmental Pattern and Species Diversity in Deep-Sea Prosobranch Snails

Carol T. Stuart and Michael A. Rex

Abstract

In this paper we relate mode of development to large-scale patterns of species diversity in bathyal (500–4000 m) prosobranch assemblages collected from the North Atlantic and the Norwegian Sea. Prosobranchs display a clear latitudinal gradient in species diversity from the equator to 77°N. A multiple regression analysis revealed that local (sample) diversity is a positive and significant function of both regional diversity and the proportion of species with planktotrophic development in the regional species pool. Regional diversity, the most important explanatory variable, decreases with increasing latitude and has a simple monotonic relationship to local diversity. A residual analysis showed that the proportion of species with planktotrophic development has a consistent relationship to diversity, but its effects are regionally specific rather than conforming to a latitudinal gradient. Results provide comparative support for the regional-enrichment model of community structure, and suggest that larval dispersal contributes to maintaining global patterns of species diversity in the deep sea. We also compared latitudinal patterns of development between deep-sea and coastal prosobranch assemblages. The proportion of species with planktotrophic development is higher and decreases more rapidly toward higher latitudes in the coastal fauna than in the deep-sea fauna.

Explanations of deep-sea species diversity have centered primarily on ecological interactions within local communities, including competition (Sanders 1968; Rex et al. 1988) and predation (Dayton and Hessler 1972), physical and biogenic disturbance (Smith 1986; Grassle 1989), or combinations of these mediated by the pattern and rate of nutrient input (Rex 1983). Recently, ecologists have become more aware that local diversity in many biotas can be influenced by regional-scale diversity and dispersal (Cornell 1985a,b; Ricklefs 1987, 1989). According to this regional-enrichment perspective, local diversity represents a balance between extinction, which results from disturbance and biological interaction, and colonization from the regional species pool which is ultimately generated through speciation and adaptive radiation. Communities are viewed as being open (sensu Roughgarden et al. 1988) and loosely organized, so that local diversity develops to reflect the available regional species pool of potential colonists. If local diversity is maintained by enrichment from the regional species pool, then local and regional diversity should be positively correlated over a broad range of values (Cox and Ricklefs 1977; Cornell 1985b). If, on the other hand, local ecological factors alone determine diversity, then local diversity should remain statistically independent of regional diversity. Local communities would reach some saturated level of diversity that is determined by their local environment, regardless of the number of available colonists in the regional species pool (Terborgh and Faaborg 1980). Local ecological determinism should cause communities to converge in their levels of diversity under similar environmental circumstances. However, as Ricklefs (1987, 1989) has pointed out, ecologically similar communities that are geographically separated (such as mangroves, coral reefs and tropical forest systems) often differ markedly in species diversity, suggesting that the history of evolutionary diversification and regional dispersal have made important contributions to community structure.

The mechanism of regional enrichment requires dispersal among local sites. Recent theoretical and empirical studies have shown that dispersal and recruitment dynamics can be important causes of community structure (Menge and Sutherland 1987; Auerbach and Shmida 1987; Roughgarden et al. 1988; Underwood and Fairweather 1989), but it has been exceedingly difficult to make direct measurements of these processes in the deep ocean (Smith et al. 1986; Smith and Hessler 1987; Grassle and Morse-Porteous 1987; Grassle and Maciolek 1992). Josefson (1985) has proposed that the actual proportion of species with planktotrophic (swimming-feeding) larvae within a community is an indication of the importance of larval dispersal and recruitment to benthic community structure. In the North Sea, communities containing a roughly equal mix of species with planktotrophic and nonplanktotrophic development are more diverse than those dominated by species with nonplanktotrophic development. Diversity may be augmented in local communities situated within regions that support a higher incidence of planktotrophic development because local extinction is balanced by rapid recolonization for many constituents (Josefson 1985). The recruitment variability inherent with a mixture of developmental modes may also promote diversity in open nonequilibrial marine systems (Chesson and Warner 1981). In deep-sea snails the mode of development can be ascertained from larval shell morphology in most taxa (Bouchet and Warén 1979a, Gustafson et al. 1991), making it possible to determine the relationship between the ratio of developmental modes within regions and local species diversity.

In this paper, we show that deep-sea prosobranch gastropods display a latitudinal gradient of species diversity in the North Atlantic. We examine this pattern in the context of the regional-enrichment model for taxa in which the mode of development can be established. Elsewhere, we present latitudinal gradients for gastropods as a whole class in both the North and South Atlantic, and consider the ecological and historical causes of global diversity patterns in more detail. Here, we focus on the potential role of larval dispersal in shaping these trends. Documenting the regional-level incidence of planktotrophic development in deep-sea prosobranchs also enabled us to compare latitudinal variation in developmental type between deep-sea and coastal (Thorson 1950) prosobranch faunas of the North Atlantic.

Materials and Methods

We estimated species diversity for prosobranch gastropods from 31 samples collected from soft sediments at bathyal depths (500–4000 m) in major basins of the North Atlantic and the Norwegian Sea (figure 6.1). The gastropod material from the North Atlantic was collected by the Woods Hole Oceanographic Institution's Benthic Sampling Program (Sanders 1977) and included two stations from the Porcupine Seabight in the West European Basin, nine from the Gay Head–Bermuda Transect in the North American Basin, nine in the Guiana Basin off Surinam, three in the Gambia Basin off Senegal, and two in the equatorial region west of the Mid-Atlantic Ridge. Gastropod diversity in the Norwegian Sea was estimated by using the relative abundance data from six epibenthic sled samples reported by Bouchet and Warén (1979b). All samples used in our analysis were taken with an epibenthic sled (Hessler and Sanders 1967). Station data, sample size, and estimates of diversity are given in table 6.1. To obtain comparable estimates of species diversity so that species richness is not an artifact of sample size, the samples were normalized to a common number of individuals ($n = 50$) by using Hurlbert's (1971) expected number of species $E(S_n)$. Normalized estimates of species diversity $E(S_{50})$ were plotted against latitude from the equator to 77°N to explore latitudinal trends in diversity.

A potentially confounding influence on the relationship between diversity and latitude is that diversity can vary with depth (Vinogradova 1962; Sanders

1968; Rex 1981). Since the regions analyzed here have somewhat different bathymetric coverage (table 6.1), depth may introduce a bias in the relationship between diversity and latitude. To test for depth-related effects, a partial correlation was performed between diversity and latitude with depth removed.

We regressed local diversity measured at individual sampling stations against the regional diversity of the deep-sea basins from which they were taken. A positive relationship between local and regional diversity would provide comparative support for regional enrichment (Ricklefs 1987, 1989). To estimate regional diversity, we combined the raw species abundance data for each basin, and again normalized (n = 250) to correct for differences in sample size among basins. We were concerned that the two relatively small equatorial stations did not adequately characterize regional diversity; therefore, the analysis was performed with and without the equatorial stations. When the equatorial stations were included in the analysis, estimates of regional diversity were

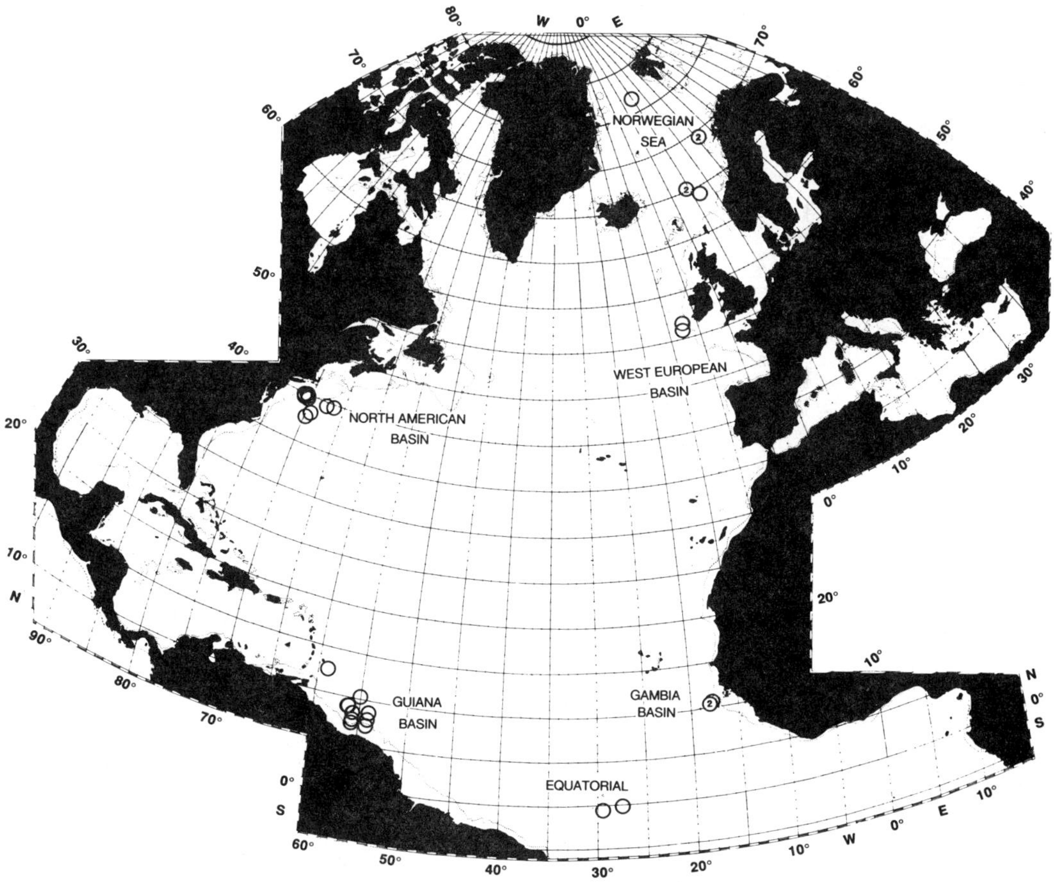

FIGURE 6.1. Map of the North Atlantic and Norwegian Sea showing the localities of the 31 samples of deep-sea prosobranchs analyzed in this study. Station data are provided in table 6.1.

TABLE 6.1 *Station Data, Species Diversity and Frequency of Planktotrophic Larval Development for Deep-Sea Prosobranch Assemblages Used in This Study*

Locality	Latitude	Longitude	Depth (m)	S	N	Local Diversity $E(S_{50})$	Regional Diversity $E(S_{250})$	Regional % Planktotrophic Species
Norwegian Sea				12	2018		8.05	16.67
Station 1								
sample 3	64°23′N	01°40′E	2600	7	123	6.23		
Station 2								
sample 4	65°21′N	00°03′W	2950	5	721	3.98		
sample 5	65°21′N	00°03′W	2950	6	307	4.98		
Station 5								
sample 10	69°32′N	10°27′E	2945	4	410	2.65		
sample 11	69°32′N	10°27′E	2945	7	286	4.67		
Station 6								
sample 13	76°57′N	00°30′E	3197	3	89	2.98		
West European Basin				33	2867		10.87	21.21
Station 313	51°32.2′N	12°35.9′W	1496	14	2613	3.27		
Station 318	50°27.1′N	13°20.4′W	2506	16	188	9.21		
North American Basin				80	14645		19.99	41.25
Station 105	39°56.6′N	71°03.6′W	530	10	570	6.28		
Station 207	39°51.2′N	70°55.4′W	808	21	11590	4.92		
Station 87	39°48.7′N	70°40.8′W	1102	22	1677	8.65		
Station 73	39°46.5′N	70°43.3′W	1400	14	142	10.03		
Station 103	39°43.6′N	70°37.4′W	2022	17	99	13.48		
Station 76	39°38.3′N	67°57.8′W	2862	15	67	12.57		
Station 126	39°37.3′N	66°45.5′W	3806	18	69	15.20		
Station 77	38°00.7′N	69°16.0′W	3806	14	118	11.10		
Station 85	37°59.2′N	69°26.2′W	3834	26	201	15.18		
Gambia Basin				43	268		41.75	27.91
Station 142	10°30.0′N	17°51.5′W	1710	25	97	18.50		
Station 145	10°36.0′N	17°49.0′W	2185	12	69	10.31		
Station 144	10°36.0′N	17°49.0′W	2204	17	63	15.65		

Guiana Basin				101	1893		42.19	29.70
Station 297	7°45.3′N	54°24.0′W	516	21	493	7.05		
Station 295	8°04.2′N	54°21.3′W	1011	22	400	9.78		
Station 293	8°58.0′N	54°04.3′W	1487	30	199	16.69		
Station 299	7°55.1′N	55°42.0′W	2009	27	328	10.53		
Station 301	8°12.4′N	55°50.2′W	2494	30	132	18.21		
Station 303	8°28.8′N	56°04.5′W	2848	16	76	13.90		
Station 306	9°31.1′N	56°20.6′W	3411	28	130	17.27		
Station 307	12°37.6′N	59°04.2′W	3849	13	65	11.31		
Station 291	10°06.4′N	55°14.7′W	3864	17	70	13.97		
Equatorial				39	158			30.77
Station 156	00°46.3′S	29°26.0′W	3459	32	92	22.96		
Station 155	00°03.0′S	27°48.0′W	3759	23	66	20.42		

NOTE: Diversity is estimated by using Hurlbert's (1971) expected number of species in samples normalized to 50 (local) and 250 (regional) individuals. See figure 6.1 for a map of sampling localities. S and N refer to total number of species and individuals respectively.

normalized to n = 150 for all basins, a value that includes the number of individuals present in the two equatorial stations. We assume here that our measure of regional diversity reflects in a relative way the level of faunal diversity of the basins in which the samples were collected. We recognize that while deep-sea basins are delimited by major topographic features, they do not contain geographically isolated benthic faunas, particularly for taxa with larval dispersal. We merely assert that the regional faunas of the widely separated basins shown in figure 6.1 are the primary sources of the species that inhabit local within-basin sites.

Though it is not generally recognized, comparisons of local and regional diversity involve a potential problem of statistical redundancy when the same samples are used to estimate both levels of diversity. In order to make local and regional diversity completely independent, we randomly selected one half of the samples in each basin to calculate local diversities and used the remaining half for regional diversity. It is conceivable that a significant relationship between regional and local diversity could emerge because of sampling error in the process of randomly choosing the local and regional subsets. Spread over 29 (nonequatorial) samples and five separate regions, the probability of this is very small. Nonetheless, to test for the effects of random sampling error, we repeated the procedure of sample selection six times and regressed the different subsets of local diversities against the recalculated regional diversity of the individual basins.

In order to assess the relationship of dispersal ability to species diversity, we used protoconch morphology to classify species as having either planktotrophic or nonplanktotrophic development. The use of morphological criteria to infer mode of development in deep-sea snails has been reviewed several times (e.g., Rex and Warén 1982; Colman et al. 1986; Gustafson et al. 1991). In marine snails, planktotrophic larvae hatch early and develop as swimming feeding veligers which have the potential for large-scale dispersal in ocean currents (see Jablonski and Lutz 1983 for a review). In nonplanktotrophic development, embryos are provisioned with sufficient stored food to undergo development in an egg capsule or gelatinous egg mass, and either hatch as a crawling benthic snail or have only a brief dispersal stage. Studies comparing the oxygen isotope composition of larval and adult shells (Bouchet and Fontes 1981; Killingley and Rex 1985), and actual recovery of larvae in near-surface plankton samples (Bouchet 1976a) have indicated that the planktotrophic larvae of some deep-sea snails migrate vertically in the water column and disperse in surface currents.

We restricted our analysis of development and its implications for dispersal to the Prosobranchia, the largest group of deep-sea snails (Rex 1976) and the one for which the morphological criteria are well established. Lemche (1948) showed that larval shell morphology is associated with mode of development in some opisthobranchs, but this is less well-known for the group as a whole and larval shells are obscured by shell growth in many deep-sea opisthobranchs. We also eliminated the ptenoglossate prosobranchs (Families Eulimidae, Aclididae, and Epitoniidae) from our analysis because protoconchs of species belonging to these groups were too corroded in some regions to determine the mode of development, and because morphological criteria can be less easily recognized in some members of this group (Bouchet and Warén 1986). The remaining nonptenoglossate prosobranchs include 66–95 percent (median 74 percent) of the species and 76–99 percent (median 85 percent) of the individuals for the whole gastropod fauna collected from the 6 regions. For the rest of this paper, the term prosobranch refers to the nonptenoglossate group. Some potential problems of assessing dispersal ability in the archaeogastropods are taken up in the discussion.

We plotted local (sample) diversity against latitude to explore large-scale biogeographical patterns, and against regional diversity and the proportion of species with planktotrophic development as a comparative test of the regional-enrichment model. Then we performed a multiple regression analysis with local diversity as the response variable, and both regional diversity and the regional incidence of planktotrophic development as explanatory variables. We used regional incidence of planktotrophic development (i.e., for the combined species makeup of local samples) as a measure of dispersal potential in the regional pool of available colonists. An alternative is to use the incidence of planktotrophic development in individual samples. However, this reflects local community makeup as it had developed at the time of sampling rather than in the overall potential pool of colonists. Using the local (sample) incidence of planktotrophic development in the analysis yields nearly identical results to those presented below using regional estimates of regional planktotrophy. Depth was also included in the analysis as an independent variable to correct for any differences in bathymetric sampling coverage as discussed earlier. We present a residual analysis to show the precise effect of planktotrophic development in the overall analysis.

Thorson (1950) documented latitudinal gradients in developmental mode for coastal prosobranchs in the North Atlantic. We compare the incidence of planktotrophic development in deep-sea species to Thorson's data.

Results and Discussion

Deep-Sea Patterns

Deep-sea prosobranchs show a clear and highly significant latitudinal gradient in diversity from the equator to 77°N ($r = -0.768$, $p < 0.001$, figure 6.2; see captions for equations and associated F values). The trend appears to be a continuous monotonic decline in diversity within this range. Two basins at the same latitude but on opposite sides of the Atlantic, Guiana Basin in the west and Gambia Basin in the east, show broadly overlapping ranges of diversity, indicating that diversity levels may remain similar across a broad longitudinal band.

As mentioned earlier, because diversity can vary with depth, it is important to remove this potential influence statistically when examining large-scale biogeographic patterns in the deep-sea benthos. Among all of the regions sampled, there is no significant relationship between diversity and depth (figure 6.3). Within regions, only the North American Basin shows a significant diversity-depth trend (r = 0.847, N = 9, $p < 0.01$) reflecting the known tendency of gastropod diversity to increase with depth across the bathyal zone before declining at abyssal depths below 4000 m in the western North Atlantic (Rex 1976). When we performed a partial correlation between diversity and latitude with the effect of depth removed, the relationship remained highly significant (r = −0.835, $p < 0.001$), indicating clearly that the latitudinal diversity gradient shown in figure 6.2 is not a spurious artifact of depth.

As shown for the nonequatorial stations in figure 6.4, local diversity is strongly and positively correlated with regional diversity (r = 0.750, $p < 0.001$, and r = 0.696, $p < 0.001$ with and without the equatorial stations respectively). It is important to note that variation in sampling intensity within basins does not bias the relationship; the sample number per basin is uncorrelated with regional diversity (Spearman's rank order correlation r_s = 0.410, n.s.). When regional and local diversities were estimated by using independent sets of randomly selected samples, the relationship remained significant for all comparisons made (r values ranged from 0.514 to 0.784, $p < 0.03$ to $p < 0.001$, N = 17).

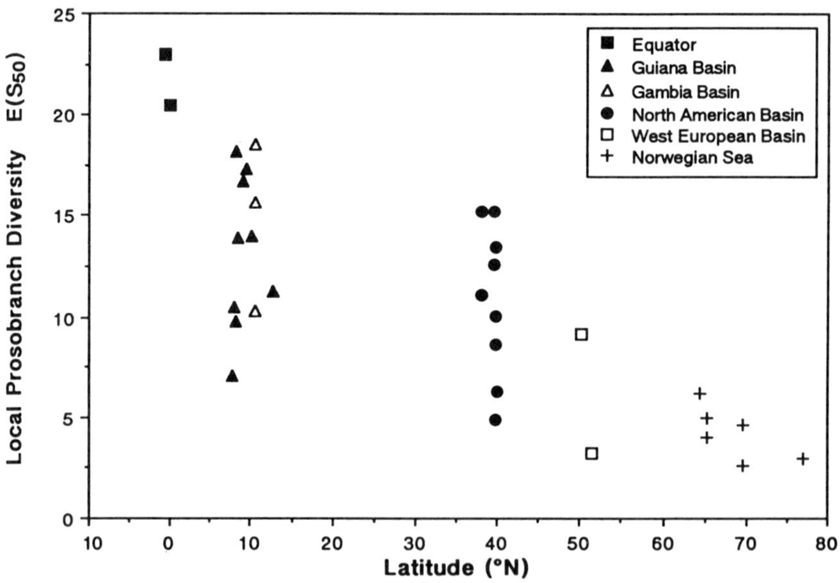

FIGURE 6.2. The relationship between deep-sea prosobranch diversity and latitude for six regions in the Norwegian Sea and North Atlantic. See figure 6.1 and table 6.1 for station localities. Diversity is estimated using Hurlbert's (1971) expected number of species with all samples normalized to 50 individuals.

Regression equation: Y = 16.601 - 0.176X, F = 41.828, r^2 = 0.590, $p < 0.001$, N = 31.

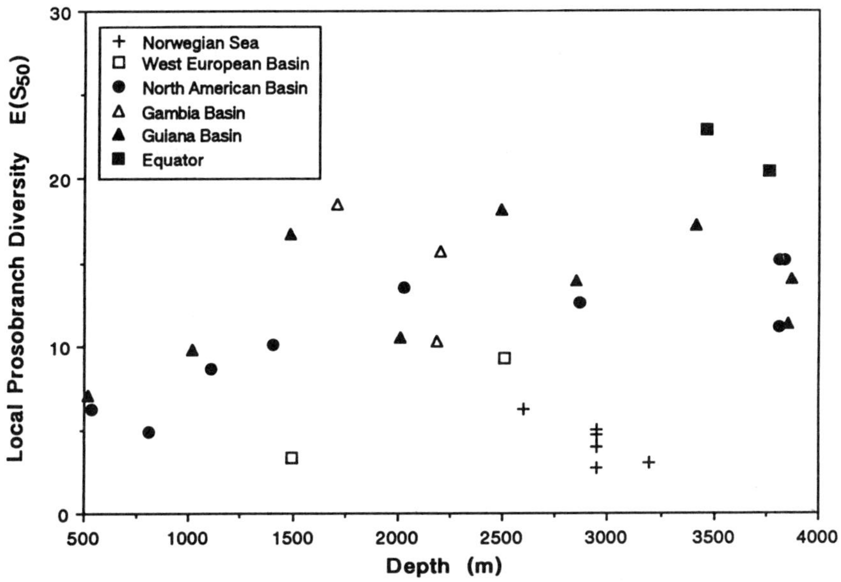

FIGURE 6.3. The relationship between prosobranch species diversity (as in figure 6.1) and depth for all samples used in this study (see figure 6.1, table 6.1). The relationship is not significant ($r^2 = 0.100$, N = 31, n.s.).

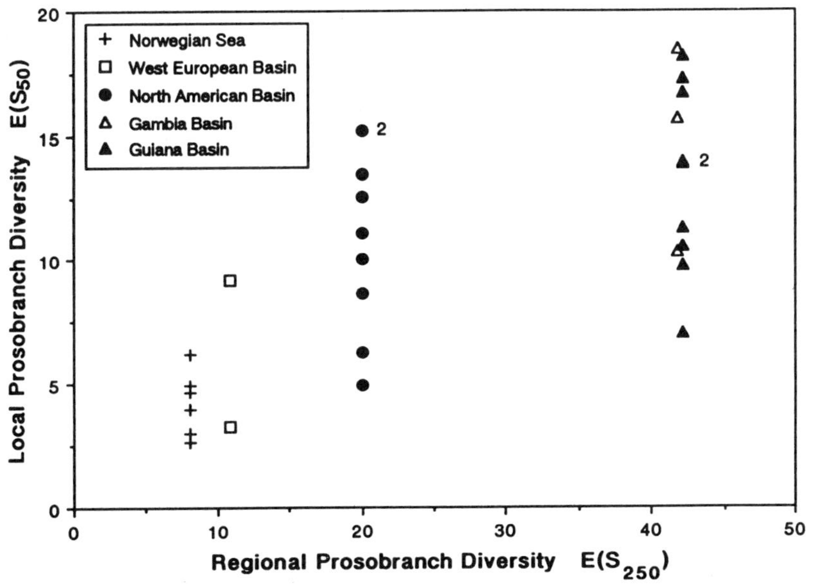

FIGURE 6.4. The relationship between local and regional diversity of deep-sea prosobranchs from five regions in the North Atlantic and Norwegian Sea. Local and regional diversities are normalized to 50 and 250 individuals respectively using Hurlbert's (1971) expected number of species.

Regression equation: $Y = 4.117 + 0.237X$, $F = 25.332$, $r^2 = 0.484$, $p < 0.001$, N = 29.

The relationship between local diversity and the regional incidence of planktotrophic development is shown in figure 6.5. The relationship does seem to be positive over most of the basins (in fact it is weakly significant, r = 0.400, N = 31, $p < 0.05$). However, the western North Atlantic, with the highest percentage of prosobranchs with planktotrophic development, has only intermediate levels of local diversity which makes the overall relationship look somewhat parabolic.

Results of the stepwise multiple regression analysis for predicting local prosobranch diversity are presented in table 6.2. When the equatorial stations are excluded from the analysis, regional diversity enters the equation first, explaining 48 percent of the variance in local diversity. Depth and the percentage of species with planktotrophic development both enter the equation as significant subordinate variables that account for comparable portions of the remaining variance in local diversity. Depth enters second, explaining an additional 9 percent of the residual variance even though depth and local diversity by themselves are uncorrelated (r = 0.208, n.s.). However, an analysis of residuals shows that this significant relationship with depth is primarily due to just three outlying points (one upper slope station from the Guiana Basin [297] with the most negative residual, and two rise stations from the North American Basin [126, 85] with the most positive residuals). When these residuals are removed, the relationship between depth and the residual variation becomes

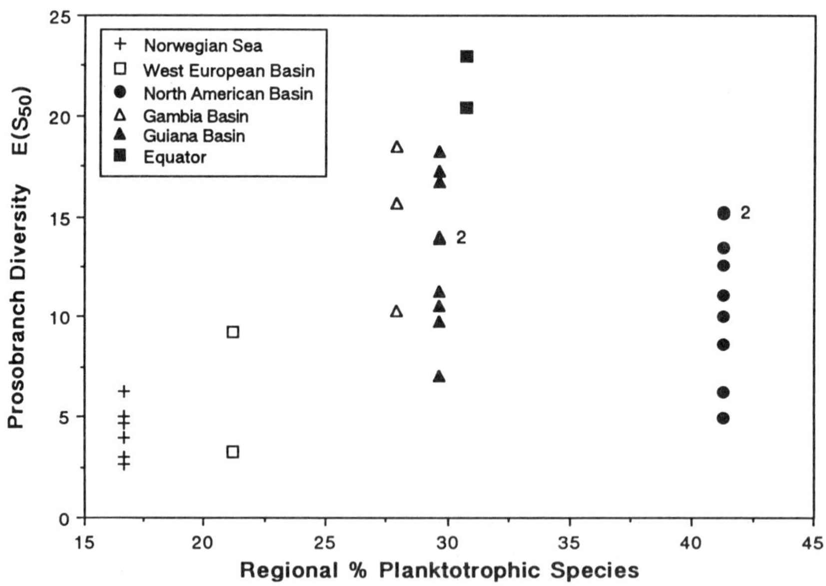

FIGURE 6.5. The relationship between prosobranch diversity and the regional percentage of prosobranch species with planktotrophic larval development in six regions of the Norwegian Sea and North Atlantic. Diversity is estimated using Hurlbert's (1971) expected number of species with all samples normalized to 50 individuals.

Regression equation: Y = 3.574 + 0.250X, F = 5.53, $r^2 = 0.160$, $p < 0.05$, N = 31.

TABLE 6.2 *Results of the Multiple Regression Analysis for Predicting Local Diversity of the Nonequatorial Samples with Three Independent Variables*

Independent Variables	Contribution to R^2	t
X_1 Regional Diversity $E(S_{250})$	0.484	5.629***
X_2 Regional % Planktotrophic Species	0.111	2.990**
X_3 Depth	0.092	3.142**

NOTE: See text for results when equatorial samples are included.
Regression Equation:
$Y = -5.144 + 0.223X_1 + 0.187X_2 + 0.002X_3$; $N = 29$, $F = 18.332$, $R^2 = 0.687$, $p < 0.001$
***$p < 0.001$
**$p < 0.01$

insignificant ($F = 2.479$, $N = 26$, $p = 0.129$). The percentage of species with planktotrophic development enters third, contributing another 11 percent to the variance explained. Here the analysis of residuals (figure 6.6), with the effects of regional diversity and depth removed, reveals a predominance of positive residuals (8 out of 9) for the North American Basin, which is associated with the highest incidence of planktotrophy, and the predominance of negative residuals (5 out of 6) for the Norwegian Sea, which is associated with the lowest incidence of planktotrophy.

Regions with a more even distribution of positive and negative residuals have faunas with an intermediate frequency of planktotrophic development. In other words, regional diversity overpredicts local diversity in the Norwegian Sea, which has a low incidence of planktotrophy and underpredicts diversity in the North American Basin, which has a high incidence of planktotrophy. The residuals are positively and significantly correlated with the frequency of planktotrophic development (figure 6.6), showing clearly why this independent variable contributes significantly to the multiple regression equation (table 6.2).

When the analysis is repeated using the two equatorial stations, results remain virtually identical except that the overall predictive power of the entire equation is improved somewhat ($F = 26.213$, d.f. $= 3,27$, $p < 0.001$, $R^2 = 0.744$). Again, regional diversity is overwhelmingly the most important independent variable ($R^2 = 0.56$), and both depth and incidence of planktotrophy are subordinate variables of very similar explanatory power. Depth enters the equation second accounting for 10 percent of the residual variance. This relationship is attributable to the same three outlying points as before, and equatorial station 156. When all four are removed, the relationship is insignificant ($F = 2.772$, $N = 27$, $p = 0.108$). The incidence of planktotrophy enters last, explaining an additional 8 percent of the variance. An analysis of residuals (with the effect of regional diversity and depth removed) again shows a high proportion of positive (7 out of 9) and negative (5 out of 6) residuals for the

North American Basin and Norwegian Sea respectively. In addition, the two equatorial stations, which together have the second highest regional percentage of species with planktotrophic development (figures 6.5 and 6.7), both have positive residuals. So, again an underprediction of local diversity by regional diversity was associated with a high incidence of planktotrophy and vice versa.

In general, the analysis supports the theory of regional enrichment, including the essential feature that local and regional diversities are linked by dispersal. Normalized estimates of regional and local diversity are highly correlated, and both decrease monotonically with increasing latitude. The proportion of species with planktotrophic development enters into the analysis in a way that is consistent with theory, but its effects are more regionally specific. It accounts for the unusually high local diversities in the North American Basin and equatorial region, and the unusually low local diversities in the Norwegian Sea. The contribution of this variable suggests, in a relative way, that local diversities are augmented in the North American Basin and equatorial region, and limited in the Norwegian Sea by the dispersal potential of their respective regional faunas.

This study provides the first quantitative evidence that species diversity in a deep-sea taxon is related to developmental pattern. As with any comparative approach, and especially one involving this remote environment where so little has been confirmed by direct observation and experimentation, we must exercise caution in evaluating both the data and their interpretation. We have

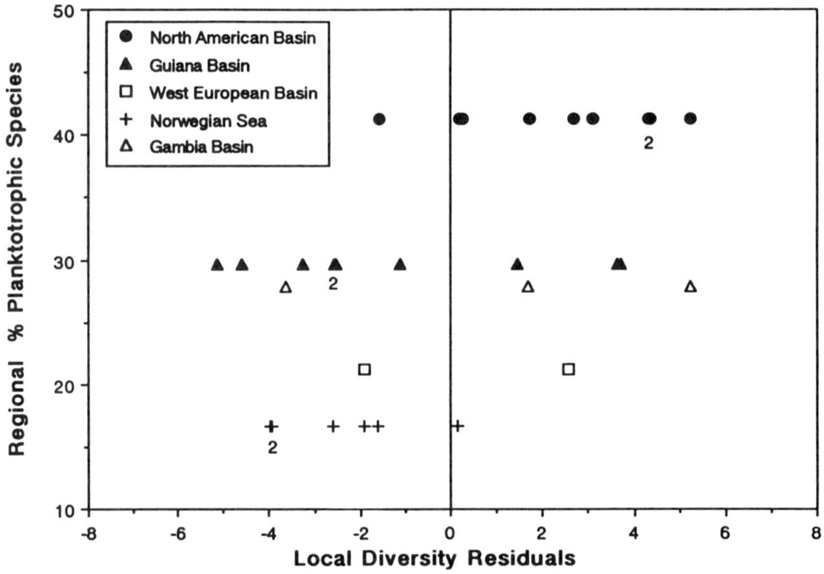

FIGURE 6.6. Analysis of the residuals of local diversity (with the effects of regional diversity and depth removed) against the regional percentage of prosobranchs with planktotrophic development. Equatorial stations are not included. The relationship between the percentage of species with planktotrophic development and the residuals of local diversity is positive and significant ($r = 0.490$, $p < 0.01$, $N = 29$).

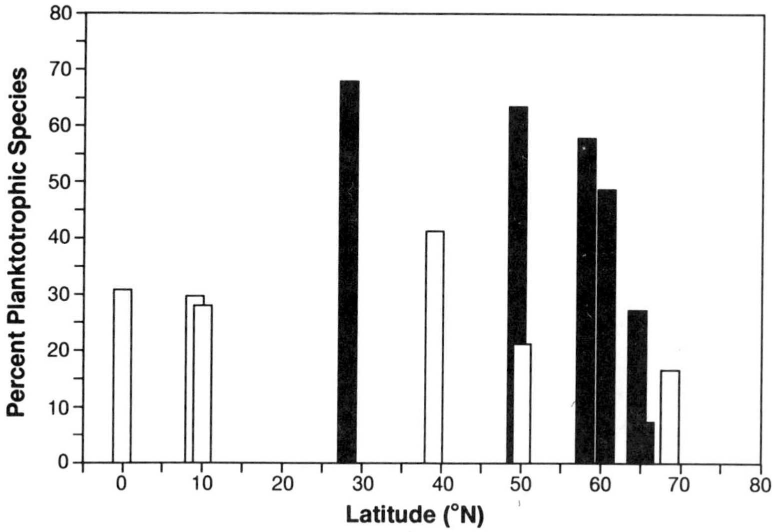

FIGURE 6.7. The relationship between latitude and percentage of prosobranch species with planktotrophic development in deep-sea and coastal environments of the North Atlantic. Solid bars represent coastal prosobranchs (Thorson 1950), and open bars represent deep-sea prosobranchs (table 6.1, figure 6.1). Thorson (1950) reported a frequency of zero for East Greenland at 68°N.

assumed that the proportion of species with planktotrophic development reflects the dispersal potential of the regional species pool. Most deep-sea prosobranchs belong to families and genera in which modes of development appear to be quite dichotomous, that is either planktotrophic with pelagic dispersal or nonplanktotrophic with little or no dispersal (Rex and Warén 1982). However, there are exceptions to the dispersal aspect of this dichotomy in the archaeogastropods. Archaeogastropod larvae do not seem to be truly planktotrophic in the sense that they feed while in the plankton (Strathmann 1978). However, Hadfield and Strathmann (1990) have shown that some archaeogastropod species do have a pelagic nonfeeding phase that is not apparent from protoconch morphology—species with and without a dispersal phase have very similar protoconchs. Thus, it is quite possible that some deep-sea archaeogastropods have a pelagic dispersal phase, but it is not possible to establish the proportion of such species from protoconch morphology. Since pelagic larvae of archaeogastropods do not rely on phytoplankton for food, it is probable that they do not migrate vertically, but disperse demersally in bottom currents.

Many deep-sea archaeogastropods have small protoconchs (ca. 250 μm) and presumably small eggs as well, suggesting that food reserves for development and dispersal are relatively limited. However, recent measurements of metabolic requirements and energy reserves in eggs of an antarctic starfish with small eggs (ca. 200 μm) indicate that larvae could survive about six months without feeding and considerably longer if dissolved organic nutrients are present at even very low concentrations (Hoegh-Guldberg et al. 1992; Shilling and Manahan 1992; Welborn and Manahan 1992). If deep-sea larvae have a similar metabolism, then pelagic larvae of archaeogastropods might undergo large-scale dispersal in cold bottom currents. Some deep-sea archaeogastropods do have Pan-Atlantic distributions (Rex pers. obs.) suggesting long-range dispersal ability. The incidence of planktotrophic development, as determined by protoconch morphology, may underestimate the actual dispersal potential within an assemblage of deep-sea snails (Rex and Warén 1982).

Also, we know nothing specific about duration of planktonic life, dispersal distances and dispersal routes of planktotrophic larvae. The proportion of species with planktotrophic development, while having the predicted relationship to local diversity in the multiple regression, is a subordinate variable that explains only 8–11 percent of the variance and participates in a geographically restricted way. Such relationships can easily be spurious or artifactual. It will be interesting to observe whether the statistical relationships found here are preserved when data on development and diversity become available from additional deep-sea regions and for other taxa.

We also emphasize that the analysis implicates dispersal only on large interregional geographic scales. Only three regions have a sufficient number of samples ($N \geq 6$, table 6.1) to test for a relationship between local diversity and the local proportion of species with planktotrophic development within regions. Of these, only the North American Basin shows a significant, but weak, correlation ($r = 0.713$, $p < 0.05$). Etter and Rex (1990) showed that another aspect of community structure, the distribution of species' bathymetric ranges along a depth gradient, was not associated with developmental mode in deep-sea prosobranchs from the western North Atlantic. The extent to which dispersal activity affects local diversity independent of the potential afforded by regional species pool size may be relatively modest and only detectable at very large interregional scales.

Comparison of Deep-Sea and Coastal Patterns

In figure 6.7, the regional incidence of planktotrophic development is plotted against latitude along with Thorson's (1950) data for coastal prosobranchs. Before comparing the two data bases, several important differences should be pointed out. Thorson's data represent regional collections with a large and variable horizontal component, but little bathymetric coverage (intertidal and subtidal shelf faunas). The deep-sea samples also have large and variable horizontal coverage, but include a broader bathymetric range as well (figure 6.1, table 6.1). The frequency of planktotrophic development can change considerably with depth across the bathyal zone (Rex and Warén 1982); though, as already mentioned, the only region in this study that shows a significant change in developmental type with depth is the North American Basin (see also Rex and Warén 1982). The deep-sea fauna has been assessed for mode of development over a much broader latitudinal range than the coastal fauna, but with fewer regional estimates. Also, it should be remembered that the deep-sea fauna, in general, has been less well sampled than coastal faunas (Grassle and Maciolek 1992).

Though it is not possible to make statistical comparisons in a geographically controlled way, two very clear differences between the faunas are apparent. The most obvious is that the coastal assemblage has a much higher percentage of species with planktotrophic development than has the bathyal deep-sea assemblage, except at high latitudes above 65°N in the Norwegian Sea where the deep sea supports a higher incidence. Thorson (1965) presented additional values for the percentage of prosobranchs with pelagic development in coastal waters from 41° to 71°N, which show a pattern very similar to that illustrated in figure 6.7, but with somewhat higher values overall, possibly because of more intensive collecting. Thorson's (1950) original prediction that planktotrophic development would be suppressed completely in the deep-sea benthos has been refuted in prosobranchs (Bouchet 1976b; Rex and Warén 1982), and in other taxa (see Pearse, this volume). Still, the proportion of species with planktotrophic development (figure 6.7) does remain generally lower than in coastal communities. It has proven very difficult to explain geographic patterns of planktotrophic development as an adaptive response to natural selection in deep-sea species, particularly when it appears to involve ontogenetic vertical migration (Rex and Warén 1982; Potter and Rex 1992). This led Bouchet (1988) to suggest that mode of development simply represents a phylogenetic constraint that requires no selective interpretation. Regardless of the ultimate causes for the existence of planktotrophic development in deep-sea prosobranchs, we have shown here that it does seem to play a role in the assembly of contemporary deep-sea communities (table 6.2), presumably because of its associated dispersal potential (see also Etter and Caswell, this volume).

The other major difference shown in figure 6.7 is that coastal prosobranchs experience a rapid and continuous rank-order decline in the frequency of planktotrophy from about 28° to 68°N. Spight (1977, 1981) pointed out that this decline parallels a latitudinal diversity gradient (see Thorson 1957, 1965). The latitudinal gradient in development type is largely due to a decrease in the diversity of planktotrophic species toward higher latitudes; the diversity of nonplanktotrophic species remains roughly constant throughout this latitudinal span. The deep-sea fauna also shows a decline in the incidence of planktotrophy across the general latitudinal range represented by the coastal data, but the rate of decrease is considerably less. Over the entire range sampled in the deep sea, from the equator to 77°N, there is not a simple continuous drop in the frequency of planktotrophy; again the North American Basin stands out as an anomaly with the highest percentage of species with planktotrophic development at mid latitudes. Apart from this, there is a rank-order decline with latitude, but it is much less steep than for the coastal fauna (figure 6.7) and is not associated in any apparent way with either a shift in the diversity of one developmental mode as in the coastal fauna, or a gradual replacement of one mode by the other. The role of larval development in community structure is regionally specific (figure 6.6), but does not conform to a simple latitudinal gradient. Ecological and environmental factors that shape latitudinal gradients in prosobranch diversity and regional representation of developmental modes may be different and, in any case, appear to be geographically uncoupled.

Acknowledgments

We thank Craig Edwards, Ron Etter, and Andrea Rex for reading versions of the manuscript, and Howard Sanders for providing prosobranch material for the North Atlantic. New unpublished research on the long-range dispersal potential of small larvae at cold temperatures was brought to our attention by Donal Manahan. Warren Dunkle, of the Data Library and Archives at the Woods Hole Oceanographic Institution, provided the map used in figure 6.1. This research is part of a doctoral dissertation presented by Carol Stuart, University of Massachusetts, 1991.

References

Auerbach, M. and A. Shmida. 1987. Spatial scale and the determinants of plant species richness. *Trends in Ecology and Evolution* 2:238–242.

Bouchet, P. 1976a. Mise en évidence d'une migration de larves véligères entre l'étage abyssal et la surface. *Comptes Rendus des Séances de l'Académie des Sciences, Série D* 283:821–824.

Bouchet, P. 1976b. Mise en évidence de stades larvaires planctoniques chez des Gastéropodes Prosobranches des étages bathyal et abyssal. *Bulletin du Museum national d'Histoire naturelle* 277:947–972.

Bouchet, P. 1988. Adaptive and non-adaptive strategies in larval biology of deep-sea snails. *Abstracts of the Fifth Deep-Sea Biology Symposium, Centre de Brest, France, IFREMER*, p. 46.

Bouchet, P. and J.-C. Fontes. 1981. Migrations verticales des larves de Gastéropodes abyssaux: arguments nouveaux dûs à l'analyse isotopique de la coquille larvaire et postlarvaire. *Comptes Rendus des Séances de l'Académie des Sciences, Série 3* 292:1005–1008.

Bouchet, P. and A. Warén. 1979a. Planktotrophic larval development in deep-water gastropods. *Sarsia* 64:37–40.

Bouchet, P. and A. Warén. 1979b. The abyssal molluscan fauna of the Norwegian Sea and its relation to other faunas. *Sarsia* 64:211–243.

Bouchet, P. and A. Warén. 1986. Revision of the northeast Atlantic bathyal and abyssal Aclididae, Eulimidae, and Epitoniidae (*Mollusca, Gastropoda*). *Bollettino Malacologico, Supplemento* 2:300–576.

Chesson, P. L. and R. R. Warner. 1981. Environmental variability promotes coexistence in lottery competitive systems. *American Naturalist* 117:923–943.

Colman, J. G., P. A. Tyler, and J. D. Gage. 1986. Larval development of deep-sea gastropods

(Prosobranchia: Neogastropoda) from the Rockall Trough. *Journal of the Marine Biological Association of the United Kingdom* 66:951–965.

Cornell, H. V. 1985a. Species assemblages of cynipid gall wasps are not saturated. *American Naturalist* 126:565–569.

Cornell, H. V. 1985b. Local and regional richness of cynipine gall wasps on California Oaks. *Ecology* 66:1247–1260.

Cox, G. W. and R. E. Ricklefs. 1977. Species diversity, ecological release, and community structuring in Caribbean land bird faunas. *Oikos* 29:60–66.

Dayton, P. K. and R. R. Hessler. 1972. Role of biological disturbance in maintaining diversity in the deep sea. *Deep-Sea Research* 19:199–208.

Etter, R. J. and M. A. Rex. 1990. Population differentiation decreases with depth in deep-sea gastropods. *Deep-Sea Research* 37:1251–1261.

Grassle, J. F. 1989. Species diversity in deep-sea communities. *Trends in Ecology and Evolution* 4:12–15.

Grassle, J. F. and N. J. Maciolek. 1992. Deep-sea species richness: Regional and local diversity estimates from quantitative bottom samples. *American Naturalist.* 139:313–341.

Grassle, J. F. and L. Morse-Porteous. 1987. Macrofaunal colonization of disturbed deep-sea environments and the structure of deep-sea benthic communities. *Deep-Sea Research* 34:1911–1950.

Gustafson, R. G., D. T. J. Littlewood, and R. A. Lutz. 1991. Gastropod egg capsules and their contents from deep-sea hydrothermal vent environments. *Biological Bulletin* 180:34- 55.

Hadfield, M. D. and M. F. Strathmann. 1990. Heterostrophic shells and pelagic development in trochoideans for classification, phylogeny, and paleoecology. *Journal of Molluscan Studies* 56:239–256.

Hessler, R. R. and H. L. Sanders. 1967. Faunal diversity in the deep-sea. *Deep-Sea Research* 14:65–78.

Hoegh-Guldberg, O., J. R. Welborn, and D. T. Manahan. 1992. Metabolic requirements of antarctic and temperate asteroid larvae. *Antarctic Journal of the United States.* Submitted.

Hurlbert, S. H. 1971. The nonconcept of species diversity: A critique and alternative parameters. *Ecology* 52:577–586.

Jablonski, D. and R. A. Lutz. 1983. Larval ecology of marine benthic invertebrates: Paleobiological implications. *Biological Reviews* 58:21–89.

Josefson, A. B. 1985. Distribution of diversity and functional groups of marine benthic infauna in the Skagerrak (eastern North Sea)—Can larval availability affect diversity? *Sarsia* 70:229–249.

Killingley, J. S. and M. A. Rex. 1985. Mode of larval development in some deep-sea gastropods indicated by oxygen-18 values of their carbonate shells. *Deep-Sea Research* 32:809–818.

Lemche, H. 1948. Northern and arctic tectibranch gastropods. I. The larval shells. II. A revision of the cephalaspid species. *Det Konelige Danske Videnskabernes Selskab, Biologiske Skrifter* 5:1–136.

Menge, B. A. and J. P. Sutherland. 1987. Community regulation: Variation in disturbance, competition, and predation in relation to environmental stress and recruitment. *American Naturalist* 130:730–757.

Potter, E. and M. A. Rex. 1992. Parallel development-depth trends in deep-sea turrid snails from the eastern and western North Atlantic. *Nautilus* 106:72–75.

Rex, M. A. 1976. Biological accommodation in the deep-sea benthos: Comparative evidence on the importance of predation and productivity. *Deep-Sea Research* 23:975–987.

Rex, M. A. 1981. Community structure in the deep-sea benthos. *Annual Review of Ecology and Systematics* 12:331–353.

Rex, M. A. 1983. Geographic patterns of species diversity in the deep-sea benthos. In G. T. Rowe, ed., *Deep-Sea Biology: The Sea,* pp. 453–472. New York: Wiley.

Rex, M. A. and A. Warén. 1982. Planktotrophic development in deep-sea prosobranch snails from the western North Atlantic. *Deep-Sea Research* 29:171–184.

Rex, M. A., M. C. Watts, R. J. Etter, and S. O'Neill. 1988. Character variation in a complex of rissoid gastropods from the upper continental slope of the western North Atlantic. *Malacologia* 29:325–339.

Ricklefs, R. E. 1987. Community diversity: relative roles of local and regional processes. *Science* 235:167–171.

Ricklefs, R. E. 1989. Speciation and diversity: The integration of local and regional processes. In D. Otte and J. A. Endler, eds., *Speciation and Its Consequences,* pp. 599–622. Sunderland, Mass.: Sinaur.

Roughgarden, J., S. Gaines, and H. Possingham. 1988. Recruitment dynamics in complex life cycles. *Science* 241:1460–1466.

Sanders, H. L. 1968. Marine benthic diversity: A comparative study. *American Naturalist* 102:243–282.

Sanders, H. L. 1977. Evolutionary ecology and the deep-sea benthos. In C. E. Goulden, ed., *The Changing Scenes in Natural Sciences 1776–1976,* pp. 223–243. Philadelphia: Academy of Natural Sciences Special Publication.

Shilling, F. M. and D. T. Manahan. 1992. Nutrient transport capacities and metabolic rates scale differently between larvae of an antarctic and a temperate echinoderm. *Antarctic Journal of the United States.* Submitted.

Smith, C. R. 1986. Nekton falls, low-intensity disturbance, and community structure of infaunal benthos in the deep sea. *Journal of Marine Research* 44:567–600.

Smith, C. R. and R. R. Hessler. 1987. Colonization and succession in deep-sea ecosystems. *Trends in Ecology and Evolution* 2:359–363.

Smith, C. R., P. A. Jumars, and D. J. DeMaster. 1986. *In situ* studies of megafaunal mounds indicate rapid sediment turnover and community response at the deep-sea floor. *Nature* 323:251–253.

Spight, T. M. 1977. Latitude, habitat, and hatching type for muricacean gastropods. *Nautilus* 91:67–71.

Spight, T. M. 1981. Latitude and prosobranch larvae: Whose veligers are found in tropical waters? *Ecosynthesis* 1:29–52.

Strathmann, R. R. 1978. The evolution and loss of feeding larval stages of marine invertebrates. *Evolution* 32:894–906.

Terborgh, J. W. and J. Faaborg. 1980. Saturation of bird communities in the West Indies. *American Naturalist* 116:178–195.

Thorson, G. 1950. Reproduction and larval ecology of marine bottom invertebrates. *Biological Reviews* 25:1–45.

Thorson, G. 1957. Bottom communities (sublittoral or shallow shelf). *Geological Society of America Memoir* 67:461–534.

Thorson, G. 1965. The distribution of benthic marine Mollusca along the N.E. Atlantic shelf from Gibraltar to Murmansk. *Proceedings of the First European Malacological Congress 1962,* 5–23.

Underwood, A. J. and P. G. Fairweather. 1989. Supply-side ecology and benthic marine assemblages. *Trends in Ecology and Evolution* 4:16–20.

Vinogradova, N. G. 1962. Vertical zonation in the distribution of the deep-sea benthic fauna in the ocean. *Deep-Sea Research* 8:245–250.

Welborn, J. R. and D. T. Manahan. 1992. Seasonal changes in concentrations of amino acids and sugars in seawaters by McMurdo Sound, Antarctica: Uptake of amino acids by asteroid larvae. *Antarctic Journal of the United States.* Submitted.

CHAPTER SEVEN

Ultrastructural Features of Gonads and Gametes in Deep-Sea Invertebrates

Kevin J. Eckelbarger

ABSTRACT

Ultrastructural studies of the gonads and gametes of deep-sea invertebrates are relatively rare and have been confined to a few species of anthozoans, molluscs, echinoderms, and vestimintiferans. Echinoderms have received the greatest attention and recent studies of 50 bathyal and abyssal species have revealed a variety of unusual ultrastructural features, primarily involving the male gamete. These include highly aberrant sperm morphology in an abyssal concentricycloid, the presence of elongated heads, modified nuclei and acrosomes, and lipid droplets in the middlepieces of bathyal echinoids, dimorphic sperm in an abyssal echinoid, and pigmented sperm in a bathyal asteroid. The ovarian follicle cells of some bathyal and abyssal holothuroids are unusual in that they resemble podocytes, cells generally associated with excretory organs and involved in ultrafiltration. There is currently no evidence that the ultrastructural modifications observed in bathyal and abyssal species is directly correlated with deep-sea habitats.

Biologists have only recently shifted their attention to investigations of reproduction and development in deep-sea invertebrates whereas extensive research has been conducted on shallow-water species. Studies dealing with the reproductive biology of deep-sea organisms have emphasized the frequency of breeding, larval dispersal potential (as inferred from egg size), and species fecundity (reviewed in Gage and Tyler 1991).

Histological studies of gonads and gametes of deep-sea animals are required to determine the gametogenic strategy employed by any given organism. However, published studies of the histology of deep-sea invertebrate gonads have been restricted to a few species of cnidarians, polychaetes, crustaceans, molluscs, brachipods, vestimentiferans, and echinoderms. Nearly all of these studies were conducted at the light microscopic level using material collected by dredge or trawl (see Gage and Tyler 1991 for review). These studies have provided us with useful information on the organization of the gonad and some indication of reproductive periodicity.

The ultrastructural features of invertebrate gonads and gametes have been thoroughly documented for many common, shallow-water species (see Adiyodi and Adiyodi 1983a, b) while studies dealing with deep-sea species have been rare and are limited to only a few species of anthozoans, molluscs, echinoderms, and vestimentiferans. This is unfortunate because ultrastructural studies can provide useful reproductive and systematic information that cannot be obtained through light microscopic studies alone. For example, egg and sperm morphology is species-specific and can have substantial systematic value (Franzen 1977; Eckelbarger and Çrassle 1983; Wirth 1984; Bacetti 1985, 1986) but subtle differences between germ cells cannot be ascertained using light microscopy alone. Ultrastructural studies also provide information on the kinds of yolk materials stored in the egg, the probable role of accessory cells in sperm or egg growth and nutrition, and the possible mechanisms of yolk synthesis (vitellogenesis). Knowledge of vitellogenic mechanisms sometimes provides an indication of the patterns of egg production in a given species (e.g., semelparity vs. iteroparity) without having to conduct laborious and expensive seasonal sampling of populations.

Despite the value of ultrastructural studies, there are several reasons why they have not been extensively employed with deep sea invertebrates: (1) obtaining living, undamaged specimens often requires expensive ships and manned submersibles; (2) it is difficult to obtain seasonal ship time in order to study gametogenesis in reproductively synchronized species; (3) there is no guarantee that reproductive specimens of both sexes will be obtained simultaneously; (4) morphologists have not participated to a great extent in deep-sea research; (5) experience has shown that standard ultrastructural techniques for tissue preservation are often inadequate for deep-sea organisms and new techniques must be developed (unpubl. obs.).

Review of Gonad and Gamete Morphology

Echinoderms

Echinoderms are frequently the most abundant and diverse megafaunal organisms encountered in the abyssal and bathyal zones of the ocean (Billet and Hansen 1982; Pawson 1982). The reproductive biology of many deep-sea echinoderms have been investigated including the light microscopic features of their gonads (Tyler 1988; Gage and Tyler 1991). Over the last five years, the author has conducted an ultrastructural survey of echinoderm gonad and gamete morphology in 44 bathyal species from the Bahamas Islands, one abyssal species from Hawaii, and 4 abyssal species from the Rockall Trough in the North Atlantic (see table 7.1). In the majority of species examined, gonad and gamete morphology conforms to that reported from shallow-water species but there are a number of notable exceptions. These studies serve to demonstrate the variability observed among related taxa and suggest possible reproductive adaptations in response to unique selective pressures in the deep sea.

Males Until recently, ultrastructural studies of echinoderm gametes have been biased toward the male gamete, perhaps because sperm are readily obtained and are relatively easy to preserve and describe quantitatively. Sperm morphology has been described for more than 100 species of echinoderms of which about 25 percent were described at the ultrastructural level. Approximately 80 percent of all sperm described are from shallow water species.

Metazoan sperm are morphologically categorized as being "primitive" or "modified" and their morphology is influenced by the environment into which they are released prior to fertilization (Franzen 1956, 1970; Afzelius 1979). Primitive sperm which characterize the majority of echinoderm species, are usually encountered in species having external fertilization in sea water. The primitive spermatozoon is a small cell with a round or conical head (nucleus), a small terminal acrosome, a middlepiece containing a few mitochondria, and a single posterior flagellum of the 9 + 2 type. Modified sperm show a wide variety of morphological alterations and are generally observed in species having a modified mode of fertilization. Outside of the Insecta, echinoderm sperm are probably better documented than for any other invertebrate group. Until recently, echinoderm sperm were believed to be highly conservative in morphology and, with rare exception, followed the primitive model (Field 1895; Tyler and Tyler 1966; Chia et al. 1975). Exceptions to this trend were observed in the tabloid-headed sperm of the crinoid *Isometra vivipara* (Holland 1976a), the conical-headed sperm of the crinoid *Antedon bifida* (Holland 1976b), the cylindrical-headed sperm of the sea cucumber *Cucumaria lubrica* (Atwood and Chia 1974), and the tabloid-headed sperm of the sea cucumber *Cucumaria pseudocurata* (Atwood 1975).

Studies of the sperm of abyssal echinoderms have been limited to two

TABLE 7.1 *Summary of Bathyal and Abyssal Echinoderms Examined by Electron Microscopy*

Taxa	Sex	Collection Depth (m)	Unusual Features
Class Concentricycloidea			
F. Xyloplacidae			
Xyloplax turnerae	M	2060	modified sperm
Class Echinoidea			
F. Cidaridae			
Calocidaris micans	M, F	507	
Cidaris blakei	M, F	730	
Cidaris rugosa	M, F	803	
Stylocidaris lineata	M, F	600	
F. Echinothuridae			
Araeosoma belli	M, F	460	modified sperm
Araeosoma fenestratum	M, F	800	modified sperm
Hygrosoma sp.	M	600	modified sperm
Phormosoma placenta	M, F	710	modified sperm
Sperosoma antillense	M, F	720	modified sperm
F. Aspidodiadematidae			
Aspidodiadema jacobyi	M, F	689	modified sperm
F. Saleniidae			
Salenia goesiana	M, F	630	
F. Arbaciidae			
Coelopleurus floridanus	M, F	425	
F. Echinolampadidae			
Conolampus sigsbei	M, F	475	
F. Brissidae			
Brissopsis atlantica	M, F	560	
F. Palaeopuesidae			
Linopneustes longispinus	M, F	657	
Palaeopneustes cristata	M, F	554	
Palaeopneustes tholoformis	M, F	554	
Archaeopneustes hystrix	M	513	
Palaeobrissus hilgardi	M, F	1025	
F. Asterostomatidae			
Phrissocystis multispina	M, F	1800	modified sperm
Class Ophiuroidea			
F. Ophiacanthidae			
Ophiomitra valida	M, F	588	
F. Ophiuridae			
Ophiomusium sp.	M	834	
F. Ophiomyxidae			
Ophiomyxa tumida	M, F	770	
F. Gorgonocephalidae			
Gorgonocephalus sp.	F	?	
Class Holothuroidea			
F. Holothuriidae			
Holothuria occidentalis	M, F	420	modified ovary
F. Stichopodidae			
Eostichopus regalis	F	490	
F. Synallactidae			
Amphigymnas bahamensis	F	680	
Bathyplotes natans	M, F	900	modified ovary
Mesothuria sp.	M, F	915	modified ovary
Hansenothuria benti	M, F	888	modified ovary
Pseudostichopus sp.	M	4300	
Benthothuria funebris	M, F	4300	

Taxa	Sex	Collection Depth (m)	Unusual Features
F. Pelagothuriidae			
Enypniastes eximia	M, F	900	
F. Deimatidae			
Deima sp.	M, F	5050	
F. Molpadiidae			
Molpadia sp.	M	5050	
Class Asteroidea			
F. Luidiidae			
Luidia sp.	M	?	
F. Caymanostellidae			
Caymanostella sp.[a]	M	1000	
F. Goniasteridae			
Peltaster placenta	M, F	690	
Peltaster nidarosiensis	F	740	
Anthenoides piercei	M, F	?	
Nymphaster sp.	M, F	540	
Pseudarchaster sp.	F	640	
F. Ophidiasteridae			
Ophidiaster alexandri	M, F	2717	
Linckia nodosa	M, F	621	
F. Asterinidae			
Asterinopsis pilosa	M, F	?	
F. Brisingidae			
Novodinia antillensis	M, F	817	modified sperm

[a]Healy et al. 1988.

species. The sperm of *Xyloplax turnerae,* a deep-sea representative of the newly erected class of echinoderms, the Concentricycloidea (Rowe et al. 1988), are highly aberrant relative to other echinoderms and most metazoans. The sperm consist of an elongate, internally segmented acrosome, a long, tapering nucleus, a single flagellum attached by a centriolar rootlet to the anterior portion of the nucleus, and a single elongate mitochondrion located posterior to the nucleus (Healy et al. 1988). In the abyssal echinoid *Phrissocystis multispina,* two types of spermatozoa are produced including a euspermatozoon typical of echinoids (figures 7.11, 7.12), and a paraspermatozoon (figures 7.9, 7.10), which is bipolar-tailed (Eckelbarger et al. 1989b). The mature paraspermatozoon bears an apical acrosome morphologically similar to the euspermatozoon (figures 7.1 and 7.2). A single primary flagellum lies parallel to the long axis of the nucleus and extends equidistant both anteriorly and posteriorly to form a bipolar-tailed cell (figure 7.4). Longitudinal sections of the distal region of the flagellum show that it is composed of disorganized microtubular doublets that taper near the terminal region (figure 7.3). The primary flagellum appears to originate from the distal centriole but is joined just posterior to the acrosome by a secondary axoneme originating from the proximal centriole at the base of the nucleus (figure 7.3). As a result, two 9 + 2 axonemes originating from separate centrioles lie on opposite sides of the nucleus (figure 7.8). Microtubules associated with the distal centriole fuse with the primary flagellum soon after emerging from the posterior region of the cells (figures 7.4–7.6). This is the first recorded occurrence of dimorphic sperm in the Echinodermata and one of the few reports of bipolar-tailed sperm in the Metazoa.

The sperm of shallow-water asteroids, holothuroids, and ophiuroids are similar in having a spherical acrosomal vesicle and subacrosomal material both accommodated in a deep anterior invagination of the spherical nucleus, centrioles usually oriented at an angle to each other, a ring-shaped mitochondrion, and a flagellum sometimes offset at an angle relative to the long axis of the cell (Chia and Bickell 1983). Recently examined sperm of bathyal asteroids, ophiuroids, and holothuroids are similar to those reported from shallow-water species, as exemplified by *Bathyplotes natans* (figure 7.14). However, there are two exceptions. The head of the sperm of the swimming holothuroid *Enypniastes exima* is about twice the size of other reported holothuroid sperm and fertilizes a large, yolky egg (Eckelbarger et al. 1989c). Although the sperm of the brissingid asteroid *Novodinia antillensis* is superfically similar to other asteroid sperm (figure 7.13), to our knowledge, it is unique among the Metazoa in possessing a small, eccentric pigment spot in the middlepiece (figure 7.15). Due to the presence of this pigment, a concentrated solution of spawned sper-

---

FIGURES 7.1–10. Paraspermatozoon of *Phrissocystis multispina*.
From Eckelbarger et al. 1989a.

FIGURE 7.1. Apical region of nucleus (N) showing acrosome (*) and anteriorly directed axoneme (A).

FIGURE 7.2. Apical region of nucleus (N) showing acrosome (*) and lateral components of two lateral axonemes which fuse into a single axoneme (A) just anterior to the acrosome. Insert: cross section through two 9 + 2 axonemes just anterior to acrosome.

FIGURE 7.3. Longitudinal section through terminal region of anteriorly directed flagellum showing tapering tip.

FIGURE 7.4. Longitudinal section through showing anteriorly and posteriorly directed flagellum (F) and two adjacent centrioles (C). Arrows indicate microtubular components extending from distal centriole which join the posteriorly directed portion of flagellum. M, mitochondrion. Insert: cross section through terminal region of anteriorly directed axoneme shown in figure 7.3. Note disorganized arrangement of microtubules.

FIGURE 7.5. Spermatid with superfluous nuclear chromatin inclusion (*). Arrow indicates branching of posterior flagellum (F) as it extends anteriorly alongside the nucleus (N). M = mitochondrion.

FIGURE 7.6. Middlepiece region showing the association of the microtubular elements of the primary flagellum (arrow heads) with those emanating from the distal centriole (DC). PC = proximal centriole; M = mitochondrion; N = nucleus.

FIGURE 7.7. Cross section through the middlepiece showing the circular mitochondrion (M), the parallel centrioles (C), and the oblique section through the laterally positioned axoneme (arrow).

FIGURE 7.8. Cross section through the central nuclear region (N) showing axonemes on both sides of the nucleus (arrow heads).

FIGURE 7.9. SEM of mature paraspermatozoon showing posteriorly (PF) and anteriorly directed (AF) components of the flagellum.

FIGURE 7.10. SEM of nuclear region (N) showing anteriorly (AF) and posteriorly directed (PF) components of flagellum. Arrow heads indicate primary axoneme lateral to nucleus.

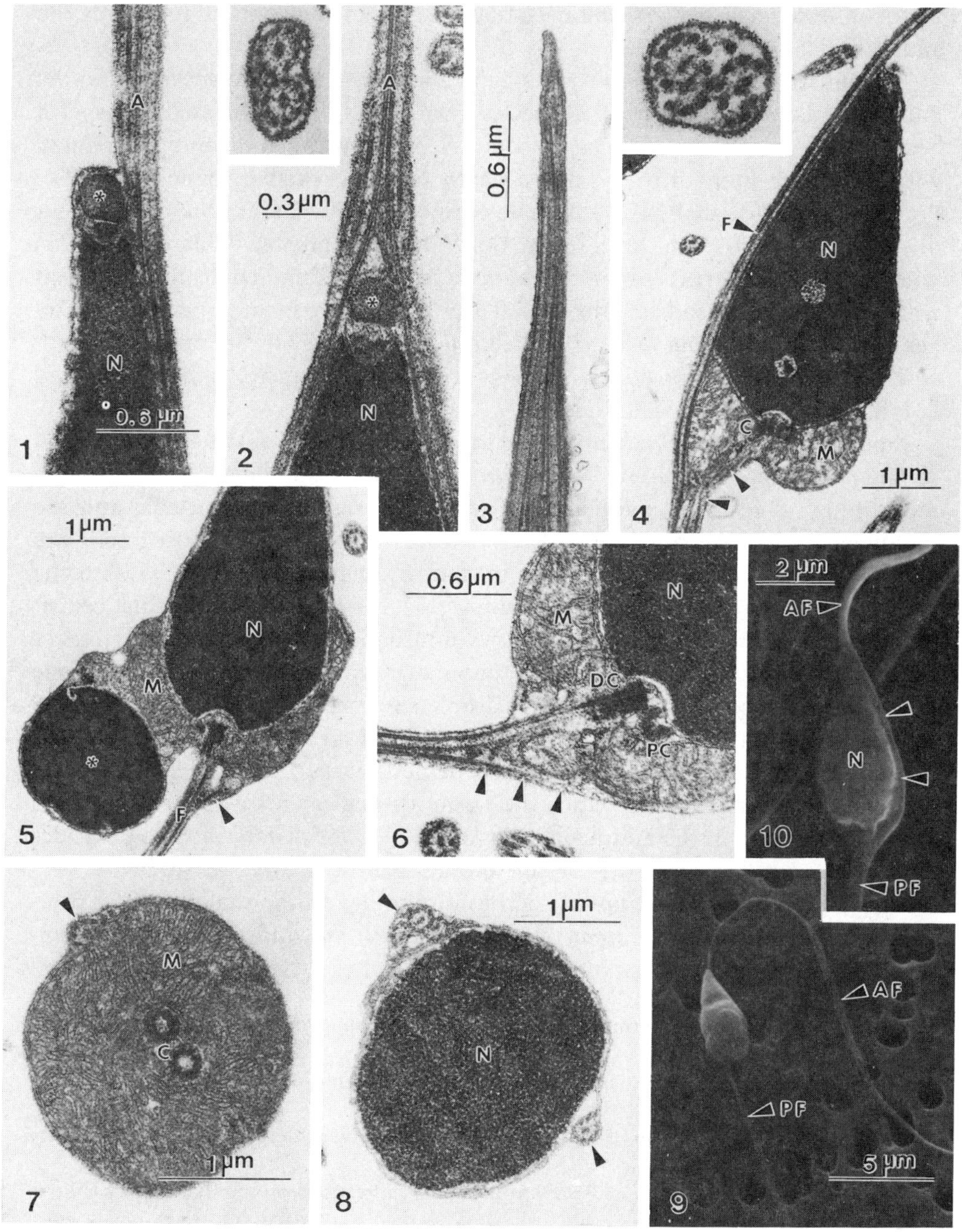

1
A
N
0.6 µm
2
A
N
0.3 µm
3
0.6 µm
4
F
N
C
M
1 µm
5
1 µm
N
M
F
6
0.6 µm
M
N
DC
PC
7
M
C
1 µm
8
1 µm
N
9
AF
PF
5 µm
10
2 µm
AF
N
PF

matozoa is reddish-pink in color, similar to that of the somatic tissues of the adult animal.

Recent studies of sperm morphology in 23 bathyal echinoid species from the Bahamas Islands and one abyssal species from Hawaii demonstrated that sperm head length (acrosome, nucleus, and middlepiece) was significantly correlated with the upper limit of their known depth ranges (Eckelbarger et al. 1989a). Previous studies of shallow-water male echinoids showed that about 19 percent of the species had sperm head lengths of 9.0 μm or greater (Chia et al. 1975) whereas about 44 percent of the bathyal species examined from Bahamian waters had sperm head lengths of 9.0 μm or greater, including the 26.0 μm head of *Aspidodiadema jacobyi* (Eckelbarger et al. 1989a). Sperm head length was significantly and positively correlated with egg diameter in echinoids and holothuroids but not in asteroids.

Sperm of the bathyal echinothuriid echinoids *Phormosoma placenta, Sperosoma antillense, Araeosoma fenestratum,* and *A. belli,* have several unique morphological features involving the acrosomal vesicle, the nucleus, and the middlepiece that are not typical for shallow water species. Sperm nuclei are highly elongated and abruptly taper posteriorly (figures 7.16–7.17). With the exception of *Sperosoma antillense* (figure 7.17), the nuclei lack a distinct centriolar fossa. The acrosomal vesicle shows unusual regional staining differences including a densely staining central region surrounded by an electron-opaque component (figure 7.18). Intracellular droplets resembling lipid extend from the extreme posterior region of the middlepiece of all four species and form a collar around the proximal portion of the axoneme (figures 7.19–22, 26). A single circular mitochondrion surrounds the posterior nuclear fossa (figure 7.23) and the two centrioles in the middlepiece (figure 7.24). The distal centriole possesses a centriolar satellite consisting of nine radiating arms (figure 7.25).

In an ultrastructural study of sperm morphology in the shallow water echinothuriids *Asthenosoma ijimai, Araeosoma owstoni,* and *Hapalosoma gem-*

FIGURE 7.11. Mature euspermatozoon of *Phrissocystis multispina.* A = acrosome; N = nucleus; M = mitochondrion.

FIGURE 7.12. Acrosome (A), subacrosomal fossa (*), and nucleus (N) of *P. multispina* euspermatozoon.

FIGURE 7.13. Mature sperm of *Novodinia antillensis* showing acrosome (A), nucleus (N), and mitochondrion (M).

FIGURE 7.14. Mature sperm of *Bathyplotes natans* showing acrosome (A), nucleus (N), and mitochondrion (M).

FIGURE 7.15. Middlepiece region of mature sperm of *N. antellensis* showing pigment granule (*) adjacent to mitochondrion (M). N = nucleus.

FIGURE 7.16. SEM of mature sperm of *Sperosoma antillense.*

FIGURE 7.17. Longitudinal section through mature sperm of *S. antellense.* N = nucleus; M = mitochondrion.

FIGURE 7.18. Anterior region of mature sperm of *S. antellense* showing acrosomal vesicle with two regions of differing electron densities and subacrosomal fossa (*).

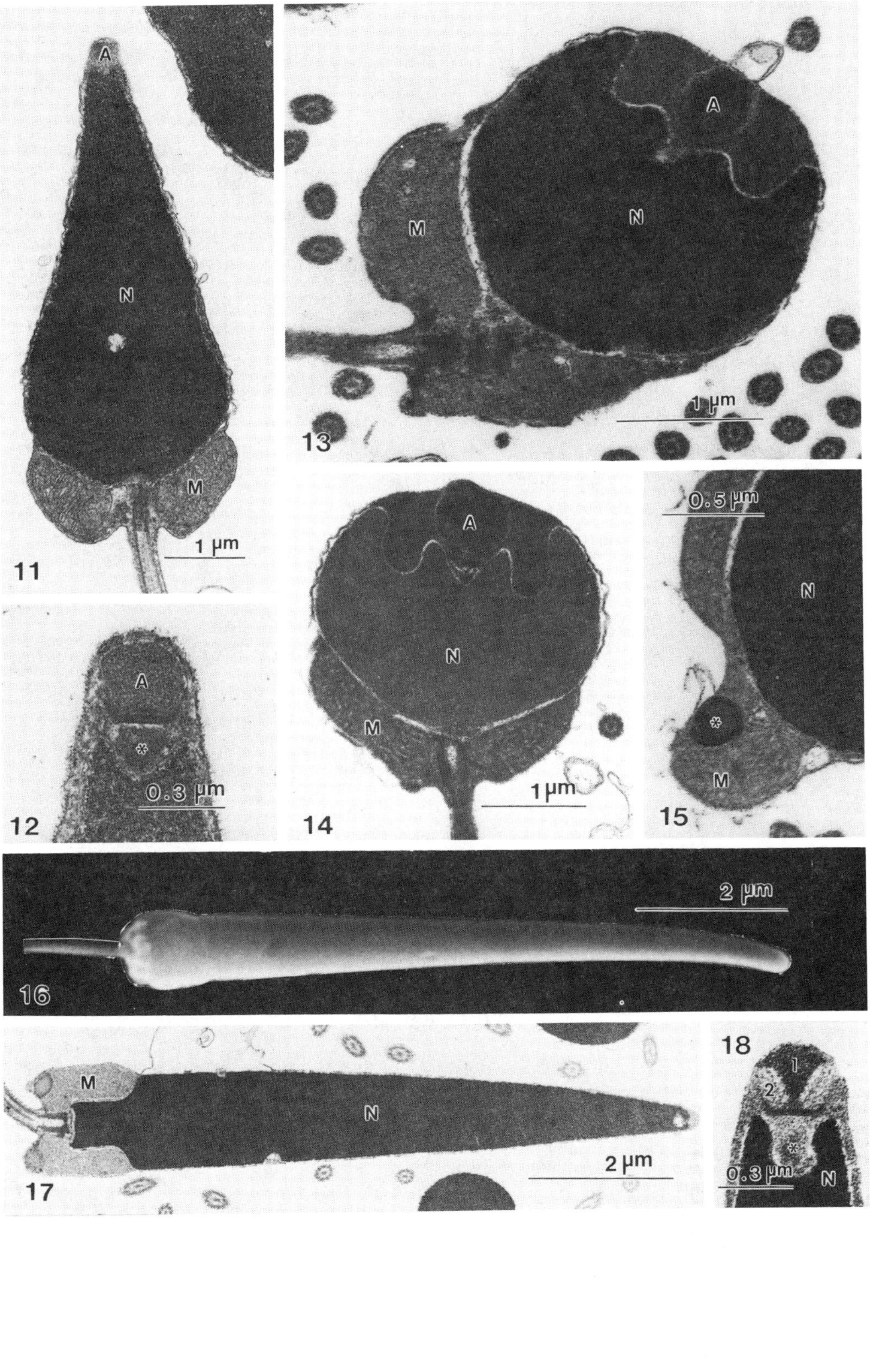
A
N
M
1 μm
11
A
0.3 μm
12
A
M
N
1 μm
13
A
N
M
1 μm
14
0.5 μm
N
M
15
2 μm
16
M
N
2 μm
17
18
1
2
0.3 μm
N

miferum from Japanese waters, very similar features were found to those described above for the bathyal species (Amemiya et al. 1980). However, the sperm from shallow water species had shorter head lengths. Sperm head length was estimated to be 7 μm for *A. ijimai,* 6 μm for *A. owstoni,* and 4 μm for *H. gemmiferum.* In contrast, head lengths of 12 μm for *Phormosoma placenta,* 8.5 μm for *A. fenestratum,* and 9 μm for *A. belli* and *Sperosoma antillense* were noted from these bathyal species (Eckelbarger et al. 1989c).

Females Ultrastructural studies of the ovary and oogenesis in shallow water holothurians are relatively abundant (see Giese et al. 1991) but only a single study has been published on any bathyal species (Eckelbarger and Young 1992). This recent investigation dealt with comparative holothurian ovarian morphology and oogenesis in six shallow water species from the northeast Pacific and four bathyal species from the Bahamas. Ovarian morphology was similar in all ten species except for selected features of the peritoneal cells of the outer layer of the ovary and the follicular inner epithelial cells surrounding the developing oocytes in some species. While the follicular inner epithelial cells of the abyssal species *Enypniastes eximia* and *Benthothuria* sp. are squamous as in shallow water holothuroids (figure 7.27), those of *Mesothuria* sp. (figure 7.28), *Hansenothuria benti* (figure 7.29), *Holothuria occidentalis* (figure 7.30) and *Bathyplotes natans* (Tyler and Eckelbarger in press) resemble podocytes, fenestrated epithelial cells usually associated with excretory organs.

The process of vitellogenesis has been intensively studied in shallow water echinoderms (see Giese et al. 1991 for review) but has only recently been examined in any bathyal or deep sea species. A comparative study of bathyal holothuroids (Eckelbarger and Young 1992) shows that vitellogenesis appears to be very similar in both shallow water and deep-sea echinoderms. The Golgi complex and rough endoplasmic reticulum appear to play a role in the

---

FIGURES 7.19–26. Sperm of echinothuriid echinoids (from Eckelbarger et al. 1989b).
FIGURES 7.19–22. Longitudinal sections through the posterior region of the mature sperm of *Phormosoma placenta* (figure 7.19), *Sperosoma antillense* (figure 7.20), *Araeosoma fenestratum* (figure 7.21), and *A. belli* (figure 7.22). Horizontal lines in figure 7.19 indicate level of cross sections through sperm of *P. placenta* shown in figures 7.23–7.26. N = nucleus; M = mitochondrion; CF = centriolar fossa; DC = distal centriole; PC = proximal centriole; L = lipid-like deposit.
FIGURE 7.23. Cross section through posterior nucleus (N) and surrounding mitochodrion (M) of sperm of *P. placenta.*
FIGURE 7.24. Cross section through centrioles (C) and surrounding mitochondrion (M) in middlepiece of *P. placenta* sperm.
FIGURE 7.25. Cross section through centriolar satellite showing nine radiating arms (arrow heads) in middlepiece of sperm of *P. placenta.*
FIGURE 7.26. Cross section through middlepiece of sperm of *P. placenta* showing ring of lipid-like bodies (L) surrounding proximal portion of axoneme (A).

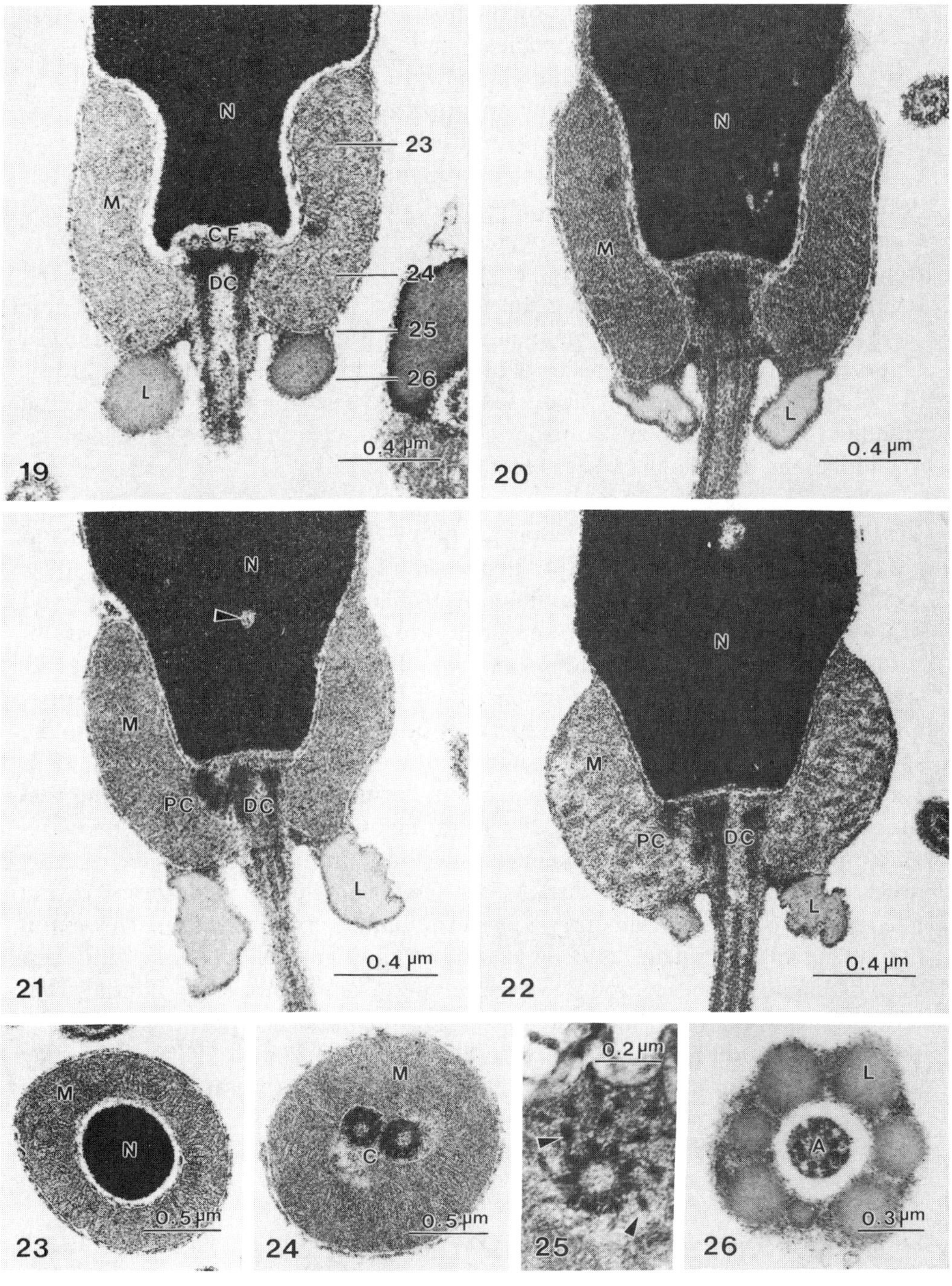
19
N
M
CF
DC
L
23
24
25
26
0.4 μm
20
N
M
L
0.4 μm
21
N
M
PC
DC
L
0.4 μm
22
N
M
PC
DC
L
0.4 μm
23
M
N
0.5 μm
24
M
C
0.5 μm
25
0.2 μm
26
L
A
0.3 μm

autosynthetic production of yolk by the oocytes themselves. However, seven of the ten species studied showed significant endocytotic activity during vitellogenesis suggesting at least the possibility of an extraovarian, heterosynthetic contribution to yolk synthesis. Our investigation of ovarian structure in bathyal asteroids and echinoids showed no unusual features although we were able to gather additional ultrastructural information on the relationship between developing oocytes and nutritive phagocytes in the bathyal echinoid *Brissopsis atlantica*. In early vitellogenic oocytes, thin, highly branching cell processes from the surrounding phagocytes penetrate into the cortical ooplasm of the cell (figures 7.31–7.32) and maintain this intimacy throughout vitellogenesis. Some of the nutritive phagocytes are squamous in form and resemble follicle cells observed in many other invertebrate ovaries (figure 7.33). Although most nutritive phagocytes contain few organelles, some possess abundant electron dense granules (figures 7.31–7.32), a prominent nucleus, cisternae of rough endoplasmic reticulum, and Golgi complexes (figure 7.33).

Vestimentiferans Vestimentiferans are deep-sea worms that lack a mouth and gut and are currently placed within their own phylum, the Vestimentifera (Jones 1988). Unlike many deep-sea animals, vestimentiferans have received a good deal of attention due to their often great size and prominence as members of many hydrothermal vent communities. Ultrastructural features of spermiogenesis in *Riftia pachyptila* have been described (Gardiner and Jones 1985; Jones and Gardiner 1985) and show the sperm to be at least superficially similar to the sperm of perviate pogonophores such as *Siboglinum ekmani* (Franzen 1973). Spermatids are associated with cytophores during differentiation and the mature spermatozoon is comprised of a head region with a helical acrosome and an elongated helical nucleus integrated with one mitochondrial helix, a short neck region, and a tail. The nucleus consists of an elongated posterior region filled with condensed chromatin and a shorter anterior region lacking chromatin but having microfilaments associated with the nuclear envelope. Following their detachment from the cytophore, mature sperm are arranged in bundles that lack distinct walls (Gardiner and Jones 1985). Ultrastructural observations of these sperm bundles have been recently described (Cary et al. 1989). This study demonstrated that each sperm bundle is composed of several hundred sperm with acrosomes that are completely detached from the rest of the sperm cell. Following spawning, coordinated beating of sperm flagellae propel the sperm

FIGURE 7.27. Ovary of *Enypniastes eximia* showing early vitellogenic oocyte (OC) and overlying follicle cell (FC). FB = fibroblast cell.
FIGURE 7.28. ovary of *Mesothuria* sp. showing vitellogenic oocyte (OC) and overlying podocyte-like follicle cell (arrow heads). BL = basal lamina; FB = fibroblast cell.
FIGURE 7.29. podocyte-like follicle cell (arrow heads) covering the surface of an early vitellogenic oocyte (oc) of *Bathyplotes natans*. A = amoebocyte.
FIGURE 7.30. Podocyte-like follicle cell from ovary of *Holothuria occidentalis*.

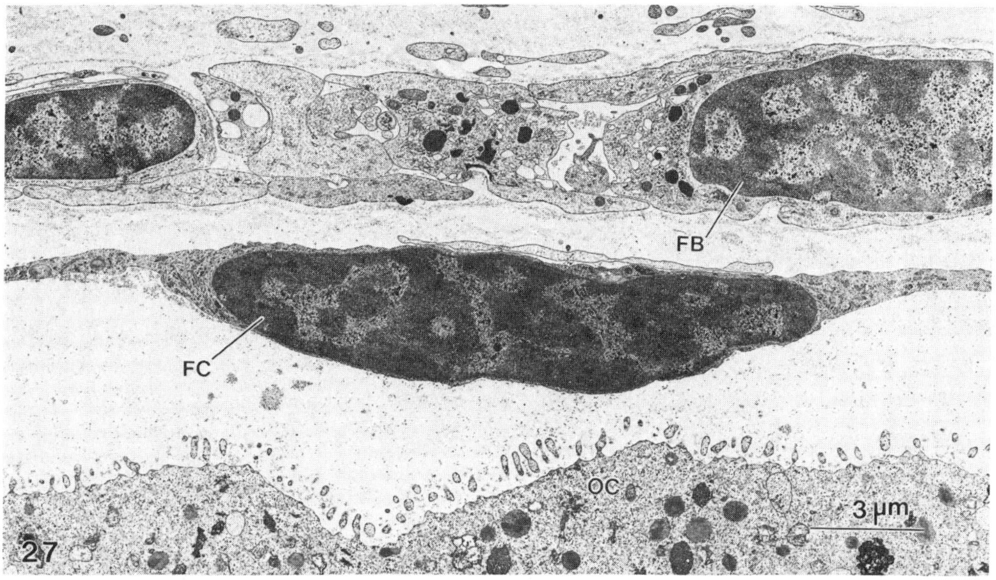
FB
FC
OC
3 µm
27

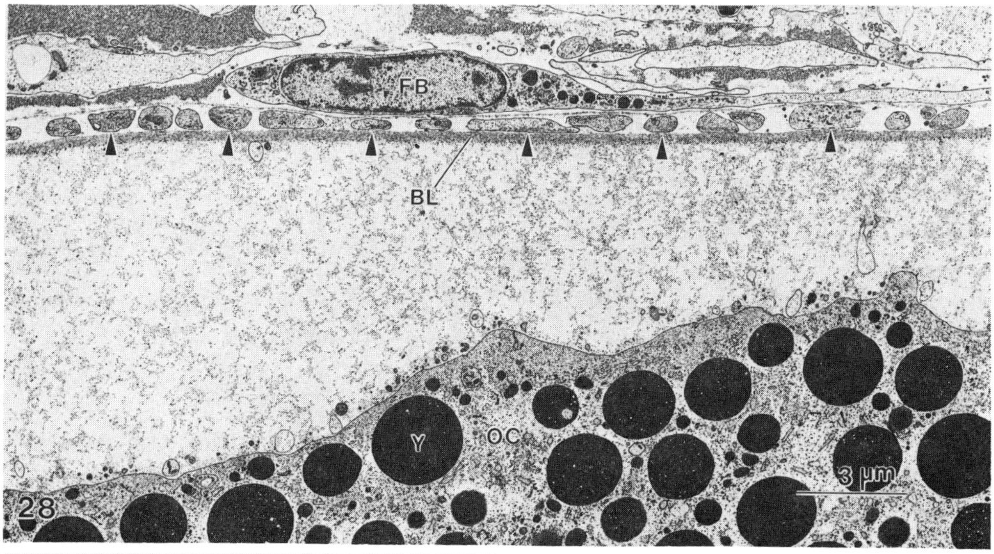
FB
BL
Y
OC
3 µm
28

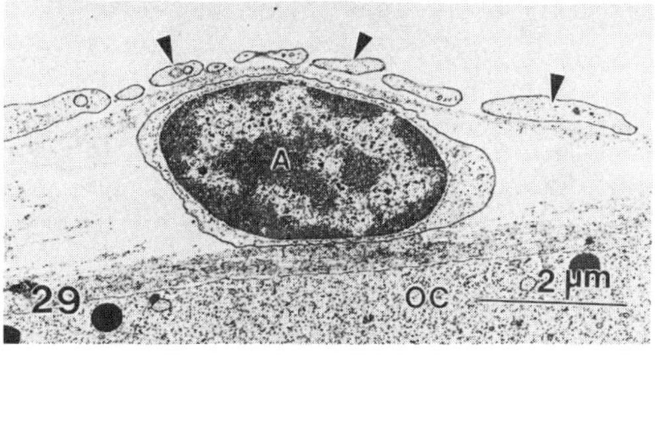
A
OC
2 µm
29

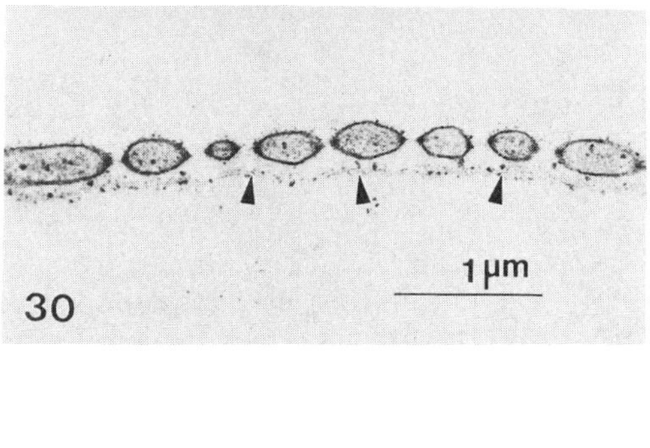
1 µm
30

bundles through sea water for 15 minutes before releasing individual sperm that become relatively immobile.

Preliminary observations of oogenesis in vestimintiferans have only recently been conducted (Gardiner et al. 1992). Oogenesis in *Riftia pachyptila* is intra-ovarian in that oocytes remain in the ovary for most of vitellogenesis. A network of blood vessels is present in the ovary and oocytes form portions of the blood vessel walls in a manner remarkably similar to that described in some polychaete species (see Eckelbarger 1988 for review). Many features of oocyte differentiation have been described although the mechanisms of yolk synthesis are still unclear. The presence of all stages of oocyte development within the ovary suggests that gametogenesis occurs continuously in mature individuals. Limited ultrastructural observations of the mature, spawned egg of *R. pachyptila* have been made using both scanning and transmission electron microscopy (Cary et al. 1989).

Anthozoans Information on the ultrastructure of anthozoan gonads and gametes in shallow water species is rather limited (reviewed by Fautin and Mariscal 1991) so it is difficult to make meaningful comparisons to the equally limited studies of deep-sea species. Some ultrastructural observations of the gonad and selected stages of sperm and egg differentiation have been made in the anemone *Paracalliactis stephensoni* trawled from 2000 meters in the Bay of Biscay (Van-Praet 1990a), and *Phelliactis hertwigi* and *P. robusta* trawled at depths of 719–1448 meters and 1600–2173 meters, respectively, from the Porcupine Seabight in the northeast Atlantic Ocean (Van-Praet 1990b). As in other sea anemones, spermatogenesis takes place within intramesoglean cysts. No significant morphological deviation from most shallow-water species was reported although the middlepiece mitochondria do not appear to fuse during differentiation and the middlepiece mitochondria and lipid droplets are symmetrical rather than asymmetrical as in many shallow-water species. Mature sperm were structurally similar to those of shallow-water species and eggs developed while in close association with trophonema cells as in most other anthozoans. Oogenesis was similar to that of other anthozoans although the mature eggs lack the large microvilli (cytospines) that usually characterize sea anemones.

FIGURES 7.31–33. Ovary of *Brissopsis atlantica*.

FIGURE 7.31. Cytoplasmic processes (double arrow heads) of highly convoluted nutritive phagocytes (NP) penetrating into early vitellogenic oocyte (OC). Single arrow heads indicate invaginations of oocyte surface. Note large electron dense granules with nutritive phagocyte cytoplasm. N = nucleus; Nu = nucleolus.

FIGURE 7.32. Higher magnification of figure 7.31 showing invaginations (arrow heads) of oocyte (OC) oolemma, adjacent intercellular space (*), and electron dense granules (G) within cytoplasm of nutritive phagocyte (NP). N = nucleus.

FIGURE 7.33. Nucleus (N) of nutritive phagocyte adjacent to developing oocyte (OC). ER = endoplasmic retirculum; G = Golgi complex.

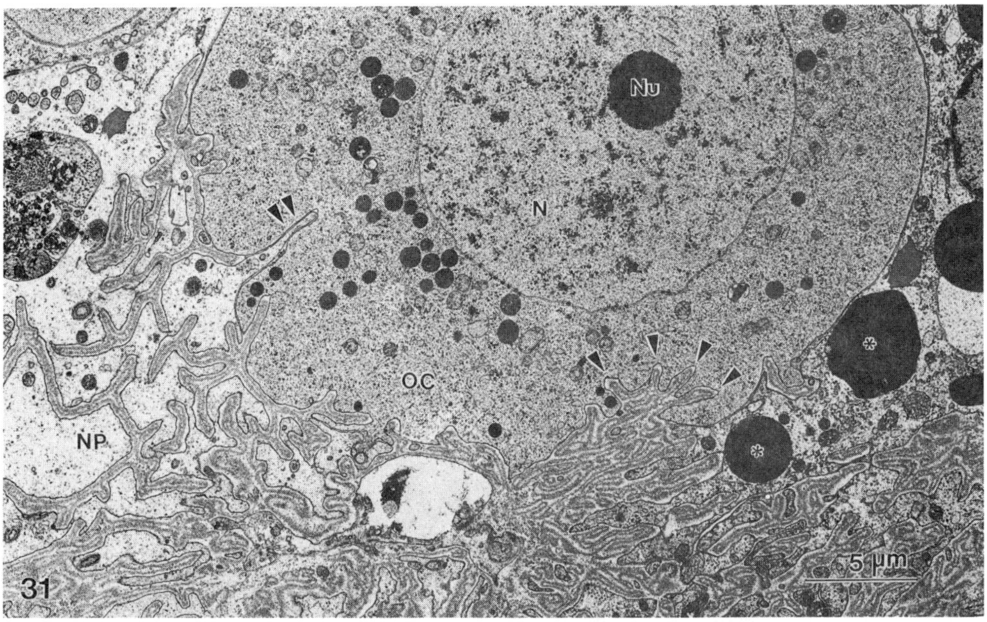
Nu
N
OC
NP
*
5 µm
31

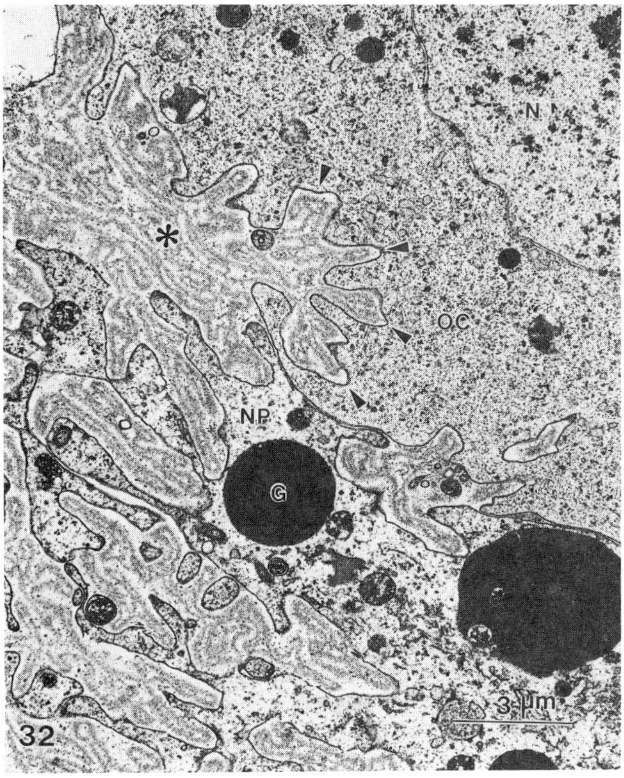
N
*
OC
NP
G
3 µm
32

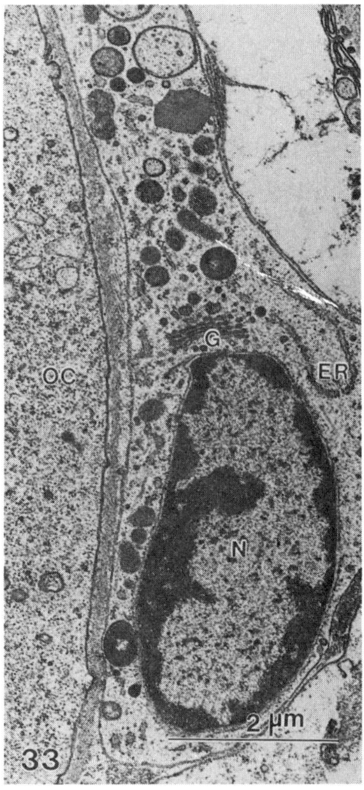
OC
G
ER
N
2 µm
33

Molluscs Ultrastructural studies dealing with gonad and gamete morphology in shallow-water molluscs are extensive (Maxwell 1983). Similar studies on deep-sea species are few in number and include sperm development in the cephalopod *Vampyroteuthis infernalis* (Healy 1989a) collected from 640 meters off New South Wales, Australia, the spermatozeugmata of the gastropod *Abyssochrysos* cf. *eburneum* (Healy 1989b) collected from 960 meters off Brazil, the sperm of *Pleurotomaria africana* trawled from 450–500 meters off Western Australia (Healy 1988), and the euspermatozoon of the hydrothermal vent caenogastropod *Provanna* n. sp. (Healy 1990) recovered from 2615 meters on the East Pacific Rise. The ultrastructural features described in these sperm were considered characteristic of the group and were not viewed as adaptations attributable to specialized deep sea habitats.

Discussion

The deep-sea environment appears, by shallow-water standards, to present some very unusual selective pressures to resident organisms including high pressure, low temperature, complete darkness, relatively low food availability, and often low population densities. However, there is currently no evidence that any of the unusual ultrastructural features observed in the gonads or gametes of deep-sea species (in comparison to shallow-water species) is directly correlated with this unique habitat or that they offer any special selective advantage. This points out the general enigma that reproductive biologists have faced since they began documenting gamete diversity. While animal gametes show enormous morphological and physiological variation, there is usually no obvious biological advantage or apparent adaptive significance to any given modification. Do some morphological features of gametes offer a species a selective advantage in a specific environment or are gamete variations neutral or without adaptive significance?

Variability in the male gamete has been far better recorded than in the female. There is extensive evidence that sperm morphology is highly correlated with the environment into which they are released at spawning and the particular mechanism of fertilization (Franzen 1956, 1977). Sperm morphology is known to be influenced by many factors including: (1) the presence or absence of external or internal fertilization; (2) whether or not eggs and sperm are released in close proximity to each other; (3) whether or not hypodermic impregnation occurs; (4) the presence or absence of a jelly mass around the spawned eggs; (5) the presence or absence of spermatophores or sperm storage prior to fertilization; (6) the presence or absence of unusually large eggs or thick egg envelopes; or (7) whether or not the sperm encounter high current velocities on their way to the eggs.

Perhaps the most important factor to recognize in gamete evolution is ancestry. Like whole organisms, gametes reflect their evolutionary history. Therefore, similarities in gamete morphology between related species is not surprising. When attempting to assess the selective pressures responsible for a particular gamete modification, one must first take into account the features of sperm morphology that are shared by closely related species. That is, is the feature common within the taxon or is it unique? It is especially useful to compare gamete morphology in closely related species from significantly different habitats. Unfortunately, many deep-sea species have no close relatives in shallow water habitats so it is often impossible to make meaningful comparisons.

The significance of elongated nuclei (sperm heads) in many Bahamian echinoids (Eckelbarger et al. 1989c) is unclear but the study demonstrated a significant and positive correlation between egg size and sperm head length in both echinoids and holothuroids. Franzen (1983) reported a strong correlation between the evolution of elongated sperm morphology and large, yolk-rich eggs in bivalve molluscs although no correlations were found in comparative studies of capitellid polychaetes (Eckelbarger and Grassle 1987) and gastropod molluscs (Hodgson and Bernard 1988). Although the Bahamian study demonstrated that larger eggs were often fertilized by larger sperm, this does not imply that large eggs predominate in the bathyal zone. Many bathyal species have small eggs that develop into planktotrophic larvae (Young and Cameron 1989; Young et al. 1989).

The presence of lipid droplets in the sperm of four species of Bahamian echinothuriid echinoids is an unusual feature observed in only a few other echinoid species (Eckelbarger et al. 1989b). While preliminary studies have shown that the sperm of these echinothuriids swim far longer after spawning than the sperm of shallow-water species (see Young, this volume), it is unclear if the lipid stores serve as an extra energy reserve. Bathyal echinothuriid population densities are often very low and it is tempting to suggest that the presence of additional energy reserves in the sperm reflects a need for greater swimming ability due the great spatial separation between males and females. On the other hand, the sperm of echinothuriids living in shallow Japanese waters have similar lipid deposits despite the fact that the population densities are presumably higher (Amemiya et al. 1980). This suggests that this unusual sperm modification is a taxonomic feature resulting from a common ancestry rather than a modification in fertilization biology.

The sperm head of the Bahamian bathyal echinoid *Aspidodiadema jacobyi* is the longest recorded from the Echinodermata (Eckelbarger et al. 1989c). The nucleus, at 27 μm in length, is sickle-shaped and about 25 percent the diameter of the mature egg. Recent studies of *A. arcitum* from abyssal waters near Hawaii show that the sperm are similar in morphology (Young unpubl. obs.) suggesting that this unusual shape is characteristic of the genus. On the other hand, the eggs of *A. jacobyi* are released within a unique, sticky mucus mass through which the sperm probably must swim to reach the eggs. The egg masses appear to become entangled in the fine barbs that adorn the spines of the adult female so it is possible that fertilization occurs here (Young, this volume). Sperm of the holothuroids *Cucumaria pseudocurata* (Atwood 1975) and *C. lubrica* (Atwood and Chia 1974) both have elongated heads, a morphological feature correlated with the viscous, mucus-covered egg masses.

Based on limited observations of the unique dimorphic spermatozoa of the abyssal echinoid *Phrissocystis multispina* it was concluded that the euspermatozoa may become entangled in the bipolar-tailed paraspermatozoa, causing clumping of the sperm and thus reducing diffusion during a spawning event (Eckelbarger et al. 1989a). Similar conclusions were made in a series of experiments with the dimorphic sperm of the fresh water prosobranch mollusc *Goniobasis* (Woodard 1940). However, recent observations of sperm behavior in *P. multispina* demonstrated that both sperm types swim normally in a forward direction and do not appear to entangle themselves (Young unpubl. obs.). Additional observations and fertilization experiments will be required to elucidate sperm function in this unusual animal.

The spermatozoon of the abyssal concentricycloid *Xyloplax turnerae* is undoubtedly the most highly modified echinoderm sperm yet reported (Healy et al. 1988). This species is unique in that it is sexually dimorphic and copulation very likely occurs between males and females (Rowe et al. 1988). Although it was acknowledged that sperm structure can undergo extensive modification during evolution when the biology of fertilization is highly modified, the unusual morphology of *X. turnerae* sperm was largely attributed to the genetic distance believed to exist between the Concentricylcoidea and other echinoderm classes (Healy et al. 1988). The fact that the sperm of the abyssal asteroid *Cayamostella* sp. were unmodified and yet this species shares the same habitat as *X. turnerae,* is consistent with the idea that sperm modifications are not an obligate correlate of abyssal environments.

The deep sea provides marine scientists with an exciting laboratory for studying reproductive evolution. Recent work has demonstrated that meaningful experimental studies can be conducted at great depths to examine fertilization ecology, larval biology, and links between nutrition and gametogenesis in many deep-sea species. With greater access to this environment through the use of submersibles, investigators can begin to better understand the life histories of deep-sea invertebrates and how species have adapted to extreme environments. It is clear from more than a century of study on shallow-water species that gametes carry a rich evolutionary history and one that is strongly influenced by environmental conditions. Continued examination of gonad and gamete morphology in deep-sea organisms should prove useful in elucidating the factors that shape gamete morphology and physiology and how this relates to species fitness.

Acknowledgments

The author wishes to thank the ship and submersible crews of the R/V *Seward Johnson* and R/V *Edwin Link* for their assistance at sea. Thanks are also extended to Pamela Blades-Eckelbarger, who provided assistance with tissue preparation, ultramicrotomy, and darkroom work, and to S. Tyler and K. Edwards for use of the Electron Microscopy Center in the Dept. of Zoology, University of Maine, Orono. Paul Tyler provided abyssal echinoderm specimens from the Rockall Trough. This work was supported by NSF grant OCE-8916264 to Young and Eckelbarger. This represents contribution number 270 of the Darling Marine Center, University of Maine.

References

Adiyodi, K. G. and R. G. Adiyodi. 1983. *Reproductive Biology of Invertebrates*. Vol. 1: *Oogenesis, Oviposition, and Oosorption*; vol. 2: *Spermatogenesis and Sperm Function*. New York: Wiley.

Afzelius, B. A. 1979. Sperm structure in relation to phylogeny in lower metazoa. In D. W. Fawcett and J. M. Bedford, eds., *The Spermatozoon: Maturation, Motility, Surface Properties, and Comparative Aspects*. Baltimore: Urban and Schwarzenberg.

Amemiya, S., T. Suyemitsu, and I. Uemura. 1980. Morphological observations on the spermatozoa of echinothurid sea urchins. *Development Growth and Differentiation* 22:327–335.

Atwood, D. G. 1975. Fine structure of an elongated dorso-ventrally compressed echinoderm (Holothuroidea) spermatozoon. *Journal of Morphology* 145:189–207.

Atwood, D. G. and F.-S. Chia. 1974. Fine structure of an unusual spermatozoon of a brooding sea cucumber *Cucumaria lubrica*. *Canadian Journal of Zoology* 52:519–523.

Baccetti, B. 1985. Evolution of the sperm cell. In C. B. Metz and A. Monroy, eds., *Biology of Fertilization*, vol. 2: *Biology of the Sperm*, pp. 3–58. New York: Academic Press.

Baccetti, B. 1986. Evolutionary trends in sperm structure. *Comparative Biochemistry and Physiology* 85A:29–36.

Billett, D. M. S. and B. Hansen. 1982. Abyssal aggregations of *Kolga hyalina* Danielssen and Koren (Echinodermata: Holothuroidea) in the northeast Atlantic Ocean: a preliminary report. *Deep-Sea Research* 29:799–818.

Cary, S. C., H. Felbeck, and N. D. Holland. 1989. Observations on the reproductive biology of the hydrothermal vent tube worm *Riftia pachyptila*. *Marine Ecology Progress Series* 52:89–94.

Chia, F.-S. and L. R. Bickell. 1983. Echinodermata. In K. G. and R. G. Adiyodi, eds., *Reproductive Biology of Invertebrates*, vol. 2: *Spermatogenesis and Sperm Function*, pp. 545–620. New York: Wiley.

Chia, F.-S., D. G. Atwood, and B. Crawford. 1975. Comparative morphology of echinoderm sperm and possible phylogenetic implications. *American Zoologist* 15:533–565.

Eckelbarger, K. J. 1988. Oogenesis and female gametes. In W. Westheide and C. O. Hermans, eds., *The Ultrastructure of Polychaeta*, pp. 281–307. New York: Gustav Fischer Verlag. [P. Ax, ed., *Microfauna Marina*, vol. 4].

Eckelbarger, K. J. and J. P. Grassle. 1983. Ultrastructural differences in the eggs and ovarian follicle cells of *Capitella* (Polychaeta) sibling species. *Biological Bulletin* 165:379–393.

Eckelbarger, K. J. and J. P. Grassle. 1987. Spermatogenesis, sperm storage, and comparative sperm morphology in nine species of *Capitella, Capitomastus,* and *Capitellides* (Polychaeta: Capitellidae). *Marine Biology* 95:415–429.

Eckelbarger, K. J. and C. M. Young. 1992. Ovarian ultrastructure and vitellogenesis in ten species of shallow-water and bathyal sea cucumbers (Echiinodermata: Holothuroidea). *Journal of the Marine Biological Association of the United Kingdom.* 72:759–781.

Eckelbarger, K. J., C. M. Young, and J. L. Cameron. 1989a. Ultrastructure and development of dimorphic sperm in the abyssal echinoid *Phrissocystis multispina* (Echinodermata: Echinoidea): implications for deep-sea reproductive biology. *Biological Bulletin* 176:257–271.

Eckelbarger, K. J., C. M. Young, and J. L. Cameron. 1989b. Modified sperm ultrastructure in four species of soft-bodied echinoids (Echinodermata: Echinothuriidae) from the bathyal zone of the deep sea. *Biological Bulletin* 177:230–236.

Eckelbarger, K. J., C. M. Young, and J. L. Cameron. 1989c. Modified sperm in echinoderms from the bathyal and abyssal zones of the deep sea. In J. Ryland and P. A. Tyler, eds., *23d European Marine Biology Symposium: Reproduction, Genetics, and Distributions of Marine Organisms,* pp. 67–74. Fredensborg, Denmark: Olsen and Olsen.

Fautin, D. G. and R. N. Mariscal. 1991. Cnidaria: Anthozoa. In F. W. Harrison and J. A. Westfall, eds., *Microscopic Anatomy of Invertebrates,* vol. 2: *Placozoa, Porifera, Cnidaria, and Ctenophora,* pp. 267–358. New York: Wiley-Liss.

Field, G. W. 1895. On the morphology and physiology of the echinoderm spermatozoan. *Journal of Morphology* 11:235–270.

Franzen, A. 1956. On spermiogenesis, morphology of the spermatozoan and biology of fertilization among invertebrates. *Zoolologiska Bidrag Fran Uppsala* 31:355–482.

Franzen, A. 1970. Phylogenetic aspects of the morphology of spermatozoa and spermiogenesis. In B. Baccetti, ed., *Comparative Spermatology.* New York: Academic Press.

Franzen, A. 1973. The spermatozoon of *Siboglinum* (Pogonophora). *Acta Zoologica* 54:179–192.

Franzen, A. 1977. Sperm structure with regard to fertilization biology and phylogenetics. *Verh. Dtsch. Zool. Ges.* 1977:123–138.

Franzen, A. 1983. Ultrastructural studies of spermatozoa in three bivalve species with notes on evolution of elongated sperm nucleus in primitive spermatozoa. *Gamete Research* 7:199–214.

Gage, J. D. and P. A. Tyler. 1991. *Deep-Sea Biology: A Natural History of Organisms at the Deep-Sea Floor.* New York: Cambridge University Press.

Gardiner, S. L. and M. L. Jones. 1985. Ultrastructure of spermiogenesis in the vestimentiferan tube worm *Riftia pachyptila* (Pogonophora: Obturata). *Transactions of the American Microscopical Society* 104:19–44.

Gardiner, S. J., E. Shrader, and M. L. Jones. 1992. Preliminary observations on oogenesis in the tube worm *Riftia pachyptila* Jones (Vestimentifera). *American Zoologist* 32:124A.

Giese, A. C., J. S. Pearse, and V. B. Pearse. 1991. *Reproduction of Marine Invertebrates,* vol. 6: *Echinoderms and Lophophorates.* Pacific Grove, Calif: Boxwood Press.

Healy, J. M. 1988. Ultrastructural observations on the spermatozoa of *Pleurotomaria africana* Tomlin (Gastropoda). *Journal of Molluscan Studies* 54:309–316.

Healy, J. M. 1989a. Spermatozoa of the deep-sea cephalopod *Vampyroteuthis infernalis* Chun: Ultrastructure and possible phylogenetic significance. *Philosophical Transactions of the Royal Society of London* 323:589–608.

Healy, J. M. 1989b. Spermatozeugmata of *Abyssochrysos:* Ultrastructure, development and relevance to the systematic position of the Abyssochrysidae (Prosobranchia, Caenogastropoda). *Bulletin Museum Natural History Paris* 11:509–533.

Healy, J. M. 1990. Taxonomic affinities of the deep-sea genus *Provanna* (Caenogastropoda): New evidence from sperm ultrastructure. *Journal Molluscan Studies* 56:119–122.

Healy, J. M., F. W. E Rowe, and D. T. Anderson. 1988. Spermatozoa and spermiogenesis in *Xyloplax* (Class Concentricycloidea): A new type of spermatozoon in the Echinodermata. *Zoologica Scripta* 17:297–310.

Hodgson, A. N. and R. T. F. Bernard. 1988. A comparison of the structure of the spermatozoa and spermatogenesis of 16 species of patellid limpet (Mollusca: Gastropoda). *Journal of Morphology* 195:205–223.

Holland, N. D. 1976a. Morphologically specialized sperm from the ovary of *Isometra vivipara* (Echinodermata: Crinoidea). *Acta Zoologica* 57:147–152.

Holland, N. D. 1976b. The fine structure of the conical-headed sperm of the crinoid *Antedon bifida*. *Experientia* 32:101–102.

Jones, M. L. 1988. The Vestimentifera, their biology, systematic, and evolutionary patterns. *Oceanologica Acta* 4:69–82.

Jones, M. L. and S. L. Gardiner. 1985. Light and scanning electron microscopic studies of spermatogenesis in the vestimentiferan tube worm *Riftia pachyptila* (Pogonophora: Obturata). *Transactions of the American Microscopical Society* 104:1–18.

Maxwell, W. L. 1983. Mollusca. In K. G. and R. G. Adiyodi, eds., *Reproductive Biology of Invertebrates*, pp. 275–342. New York: Wiley.

Pawson, D. L. 1966. Ecology of holothurians. In R. A. Boolootian, ed., *Physiology of Echinodermata*, pp. 63–71. New York: Interscience-Wiley.

Rowe, F. W. E., A. N. Baker, and H. E. S. Clark. 1988. The morphology, development, and taxonomic status of *Xyloplax* Baker, Rowe, and Clark (1986) (Echinodermata: Concentricycloidea), with the description of a new species. *Proceedings of the Royal Society of London* B 233:431–459.

Tyler, P. A. and K. J. Eckelbarger. In press. Reproduction in the holothurian *Bathyplodes natans* (Holothuroidea: Synallactidae) from bathyal depths in the northeast and western Atlantic. *Journal of the Marine Biological Association of the United Kingdom.*

Tyler, A. and B. S. Tyler. 1966. The gametes: Some procedures and properties. In R. A. Boolootian, ed., *Physiology of Echinoderms*, pp. 639–682. New York: Interscience.

Van-Praet, M. 1990a. Gametogenesis and reproductive cycle in the deep-sea anemone *Paracalliactis stephensoni* (Cnidaria: Actiniaria). *Journal of the Marine Biological Association of the United Kingdom* 70:163–172.

Van-Praet, M. 1990b. Reproduction in two deep-sea anemones (Actiniaria); *Phelliactis hertwigi* and *P. robusta*. *Progress in Oceanography* 24:207–222.

Wirth, U. 1984. Die Struktur der Metazoen-Spermien und ihre Bedeutung fur die Phylogenetik. *Verh. Natur. Ver. Hamb.* 27:295–362.

Woodard, T. M. 1940. The function of the apyrene spermatozoa of Goniobasis laqueata Say. Journal of Experimental Zoology 85:103–123.

Young, C. M. and J. L. Cameron. 1989. Developmental rate as a function of depth in the bathyal echinoid *Linopneustes longispinus*. In J. S. Ryland and P. A. Tyler, eds., *Reproduction, Genetics, and Distributions of Marine Organisms*, pp. 225–234. Fredensborg, Denmark: Olsen and Olsen.

Young, C. M., J. L. Cameron, and K. J. Eckelbarger. 1989. Extended pre-feeding period in a planktotrophic echinoid larva from the bathyal zone of the deep sea. *Journal of the Marine Biological Association of the United Kingdom*. 69:695–702.

CHAPTER EIGHT

Environmental Control of Quasi-Continuous and Seasonal Reproduction in Deep-Sea Benthic Invertebrates

Paul A. Tyler, Lucia S. Campos-Creasey, and Lucy A. Giles

Abstract

In this paper we examine the reproduction and diet of three of the dominant taxa in the deep-sea—protobranch bivalves, echinoids, and asteroids. For each, previously published data are reviewed to provide a conceptual framework in which to describe the diet and its possible control on reproduction. We have examined species pairs from each of these taxa, one of the pair being a quasi-continuous breeder and the other a seasonal breeder. The diet of these species is examined in light of the supply of surface derived organic matter that arrives at the deep-sea bed as a marked pulse in late spring and early summer. All the quasi-continuous species examined conformed to the pattern predicted by showing no seasonal variation in their diet. In the seasonal breeders the pattern of reproduction did not vary with the supply of phytodetritus from the surface. In the protobranch there appeared to be a cessation of feeding coincident with the arrival of phytodetritus whereas in the echinoid there was an increase in feeding activity but *not* a concomitant increase in percent organic matter in the stomach. In the seasonally breeding asteroid there was variation in the sedimentary organic matter in the stomach with season but there was also an additional seasonal signal in the flux of fish carcasses from the mesopelagic zone. These data support the concept that seasonal reproduction occurs in the deep sea but that its control is not necessarily related to surface derived flux. Thus, the concept of "typical" patterns in the deep sea is further eroded.

During the last ten years the traditional view of the deep sea being a tranquil, conservative environment (Menzies 1965) has been challenged. This challenge has resulted in a reevaluation of the physical and chemical parameters that control the ecology of benthic organisms within this environment. Two broad groups of ecological controls can be recognized. The first consists of large-scale aperiodic, often dramatic, events such as benthic storms that erode the sediment surface under certain conditions and redeposit this sediment in thick layers elsewhere (Hollister and McCave 1984; Gross et al. 1988), and turbidity currents and debris flows that can smother benthic invertebrates (Leeder 1985).

The second group of ecological controls consists of predictable and unpredictable input of organic material to the deep-sea which, although nowhere near as dramatic as those events of benthic storms and turbidity currents, may have a profound effect on the ecology of organisms at the deep-sea floor. Rowe and Staresinic (1979) conceptualized the pathways, sources, and sinks of food into the deep sea. Large food falls may be in the form of animal carcasses (Bruun 1957) and although these falls do not appear to have a seasonal component, they do provide up to 10 percent of the energy requirement of the deep-sea benthos (Smith 1985). Large food falls of animal origin are quickly consumed by scavenging species at the deep-sea bed. It is likely that this food source enhances reproductive activity but evidence is not yet available. A second form of large food fall are macroalgae and seagrasses. These have been observed regularly in deep-sea photographs and appear to attract specific benthic species (Young et al. 1993; J. Pearse pers. comm.). This form of large food fall is known to be ingested by a variety of deep-sea benthos (Mortensen 1935; Wolff 1976, 1979; Suchanek et al. 1986) but, as with food falls of animal origin, there is no evidence that this food enhances reproductive activity. Large unpredictable falls of wood are colonized by xylophagous molluscs which grow rapidly then direct considerable energy toward reproductive activity (Turner 1973).

A more predictable supply of food to the deep-sea bed is in the form of phytodetritus and other particulate organic material from surface waters. The first indication that such input was not a slow, constant "rain of particles" came from sediment trap studies in the Sargasso Sea (reviewed by Deuser 1986). These studies were later complemented by photographic observations of surface derived phytodetrital material arriving at the seabed (Lampitt 1985; Rice et al. 1986, 1991). During descent, and particularly after arrival at the deep-sea bed, this phytodetrital material is colonized by bacteria and various protists (Lochte and Turley 1988; Gooday 1988; Gooday and Turley 1990; Thiel et al. 1990).

Evidence of a seasonal reproductive cycles of deep-sea invertebrates was established at the same time as predictable seasonal environmental events were being identified. Early evidence of reproductive seasonality provided by George and Menzies (1967, 1968) was supplemented by data on seasonal recruitment (Schoener 1968). The first sampling program to test for seasonal cycles of reproduction revealed a seasonal cycle in 2 of 11 species examined; the remaining species had quasi-continuous (nonseasonal) gametogenic pattern (Rokop 1974, 1977). Rokop's approach demonstrated the benefit of long times series in establishing patterns of life history in the deep-sea. In the 1970s the Scottish Marine Biological Association established two stations ("Permanent" and "M") at 2900 m and 2200 m respectively in the NE Atlantic (Gage and Tyler 1991). These stations have been sampled as regularly as ship scheduling would allow, and have demonstrated seasonal life history processes in a limited number of deep sea invertebrate species. The data on seasonal reproduction were reviewed by Tyler (1988) and the data set as a whole by Gage and Tyler (1991) and Tyler et al. (1992).

As a result of these observations, there arose a suspicion that the downward vertical flux of phytodetritus was a significant controlling factor in both the gametogenic cycle and recruitment of seasonally breeding species (Tyler and Gage 1984a; Thiel et al. 1990). However, it remains an enigma how two, or more, sympatric species, under the same environmental conditions, can have very different reproductive patterns.

The purpose of this paper is to review what is known of the diet of three common deep-sea taxa of different feeding types and to compare these data to the diet of species from a time series of samples in the NE Atlantic to see if, and how, surface derived organic matter might control reproductive patterns in deep-sea invertebrates.

Materials and Methods

We examined species pairs (each including a quasi-continuous and a seasonal breeder) that were (1) as closely related taxonomically as possible, and (2) from samples collected as close as possible in both space and time. To this end we selected three suitable species pairs represented in our archived material: the protobranch bivalves *Malletia cuneata* (a continuous breeder) and *Ledella pustulosa* (a seasonal breeder); the echinoids *Phormosoma placenta* (continuous) and *Echinus affinis* (seasonal); and the asteroids *Bathybiaster vexillifer* (continuous) and *Plutonaster bifrons* (seasonal).*

Samples of all species examined were taken by either a 3-m-wide Agassiz trawl, a Woods Hole pattern or IOS pattern epibenthic sledge, or an USNEL box-corer. Samples came either from the Scottish Marine Biological Association

*Throughout this paper we shall use the convention of referring to the quasi-continuous species before the seasonal species in text, tables, and figures.

time series or from the IOSDL Deep Sea Benthic Sampling Programme. Samples were fixed in 10 percent buffered seawater formalin and transferred to 70 percent alcohol, after one week, for long-term storage.

Malletia Cuneata/Ledella Pustulosa

To document the reproductive cycles of these two species, at least ten specimens from each sample were examined. Where this sample size proved inadequate an additional ten specimens were examined. Specimens were decalcified in RDC decalcifying solution, dehydrated through graded alcohols, cleared in histoclear, and embedded in paraffin wax. All ten specimens were embedded in a single block with the same orientation. The blocks were serial-sectioned and every other slide was stained with haematoxylin and eosin. The stained slides were examined and the size of the largest oocyte was noted. Individual fecundity was too low for sensible oocyte-size frequency counts. For each sample, the mean maximum oocyte diameter was plotted against time.

For the determination of diet, the shell and digestive gland were removed from other fixed samples to reveal the stomach. This was opened and the stomach contents were removed with a fine pasteur pipette. For grain size analysis, the stomach contents $< 63\ \mu m$ were counted on a coulter counter. Because each stomach contained a limited quantity of material, the stomach contents of 10 individuals from each sample were combined. The stomach contents were examined under a light microscope and a scanning electron microscope (SEM) to determine the main food items present.

Phormosoma Placenta/Echinus Affinis

After measuring the test diameter of each echinoid the test was opened to reveal the digestive system. The gut was divided into oesophagus, stomach and intestine. From each of these portions the contents were removed and their dry weight, ash-free dry weight (AFDW) and percent total organic matter determined by drying at 70°C, reweighing and burning in a muffle furnace at 550°C. For the determination of organ indices the gut wall fractions were dried at 70°C and weighed. Organ indices were determined according to:

$$\text{organ index} = \text{organ AFDW/test diameter or arm radius.}$$

Subsamples of the gut contents were examined by both light microscopy and SEM.

Bathybiaster Vexillifer/Plutonaster Bifrons

The maximum arm radius (R) and interambulacral radius (r) (sensu Mortensen 1927) were measured for all seastar individuals. The dorsal surface was removed by incising between the dorsal marginal plates and the paxillar area to expose the digestive system whilst leaving the gonads attached to the underside of the dorsal surface.

The gut contents were removed from the stomach and stored in 70 percent alcohol. After staining overnight with 0.5 percent rose bengal in 70 percent alcohol the gut contents were examined under a dissecting microscope. All the recognizable, as well as stained unrecognizable, organic matter was removed and stored in 70 percent alcohol. The volume of the remaining sediment was noted. A portion of this sediment was dried at 70°C overnight and ashed at 550°C for 6 h to determine the percent of organic matter content. Dietary species diversity was calculated by the Shannon-Weiner Index taking the unrecognizable organic matter as a single category.

Review and Results

Mollusca: Bivalvia: Protobranchia

Many studies of protobranch bivalves have emphasized taxonomy and shell morphology (Verrill and Bush 1898; Knudsen 1970). More recently Howard Sanders and John Allen have provided details of soft part anatomy, diet, and reproduction (particularly egg number and size), and prodissoconch length for several families (Allen and Sanders 1973, 1982; Sanders and Allen 1973, 1977, 1985; Allen and Hannah 1989). Scheltema (1972) was the first to look at reproduction in deep-sea protobranchs. He detailed the egg size and suggested quasi-continuous reproduction for five species of *Nucula* from slope regions off the NE United States. Reproduction formed the basis of a one-year, five-sample study in the San Diego Trough, California (Rokop 1977). The three protobranch species Rokop examined demonstrated quasi-continuous breeding thus supporting the previous concept of continuous breeding in this deep-sea group. This conclusion was challenged by Lightfoot et al. (1979) who found seasonal breeding in *Ledella pustulosa* and *Yoldiella jeffreysi* in a short time series of samples from the NE Atlantic.

Diet in deep-sea protobranchs is even more poorly known than reproduction. Allen (1979) suggested the digestive physiology of deep-sea protobranchs allows them to utilize refractory organic matter. This led to an increase in the length of the intestine although Yonge (1939) did not believe the intestine played a part in the digestive processes of shallow-water protobranchs. More recently Allen and his co-workers have noted the presence of diatom, silicoflagellate, and foraminifera fragments in the stomachs of the protobranchs they have examined (Allen and Sanders 1973; Sanders and Allen 1977; Allen 1978). The long, often coiled, hind gut of many deep-sea protobranch species has been considered an adaptative response to the refractory nature of food in the deep-sea (Allen and Hannah 1989).

With this background we examined the diet and its relation to reproduction in:

MALLETIA CUNEATA/LEDELLA PUSTULOSA

Reproduction The gonads in both species of protobranch are situated adjacent to the digestive glands. The oocytes develop from oogonia located close to the intestine wall. The oocytes develop in a similar way with the well-developed oocytes having an eosinophilic central cytoplasm with a paler, fine granular, peripheral, cortical granule layer some 5 to 15 μm thick. In *M. cuneata* the maximum fecundity is 30 oocytes per individual and the maximum oocyte size observed was 240 μm diameter, whereas in *L. pustulosa* the fecundity was a maximum of 240 and the maximum oocyte size was 120 μm diameter.

Comparison of the mean maximum oocyte size for each sample (figure 8.1a, b) suggests there is no seasonal cycle of oocyte development in *M. cuneata* although there is considerable variation between different samples. Conversely *L. pustulosa* shows a seasonal variation confirming the preliminary observations of Lightfoot et al. (1979) that this species was a seasonal breeder. Gametogenesis in *L. pustulosa* starts in April/May of each year and there is a release of gametes in the spring of the following year.

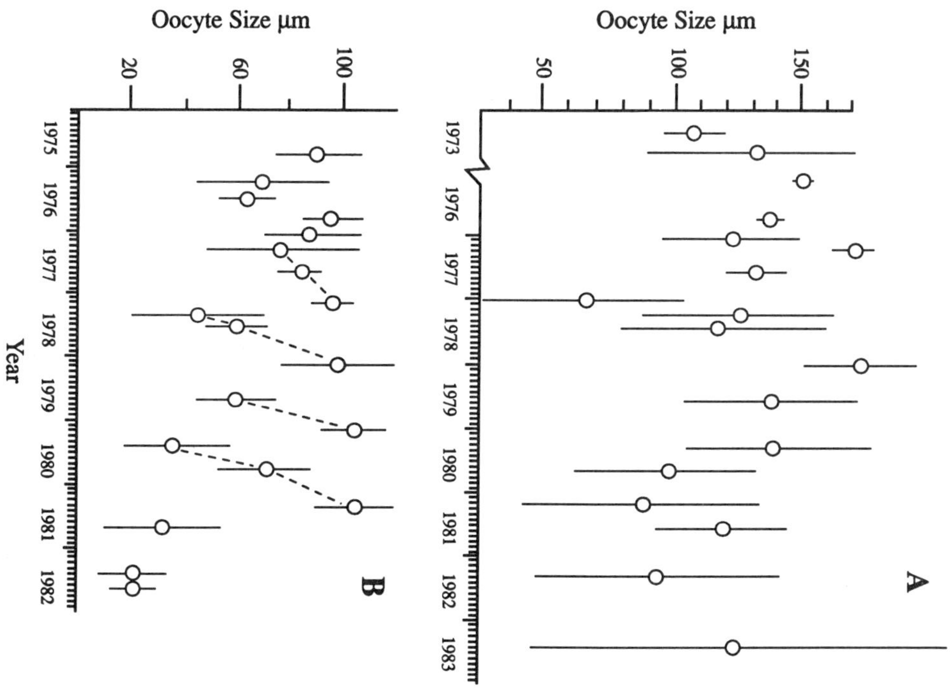

FIGURE 8.1. Mean maximum oocyte size (+/- 1 sd) for: A. *Malletia cuneata*; B. *Ledella pustulosa*, from samples taken at 2900 m in the Rockall Trough, NE Atlantic

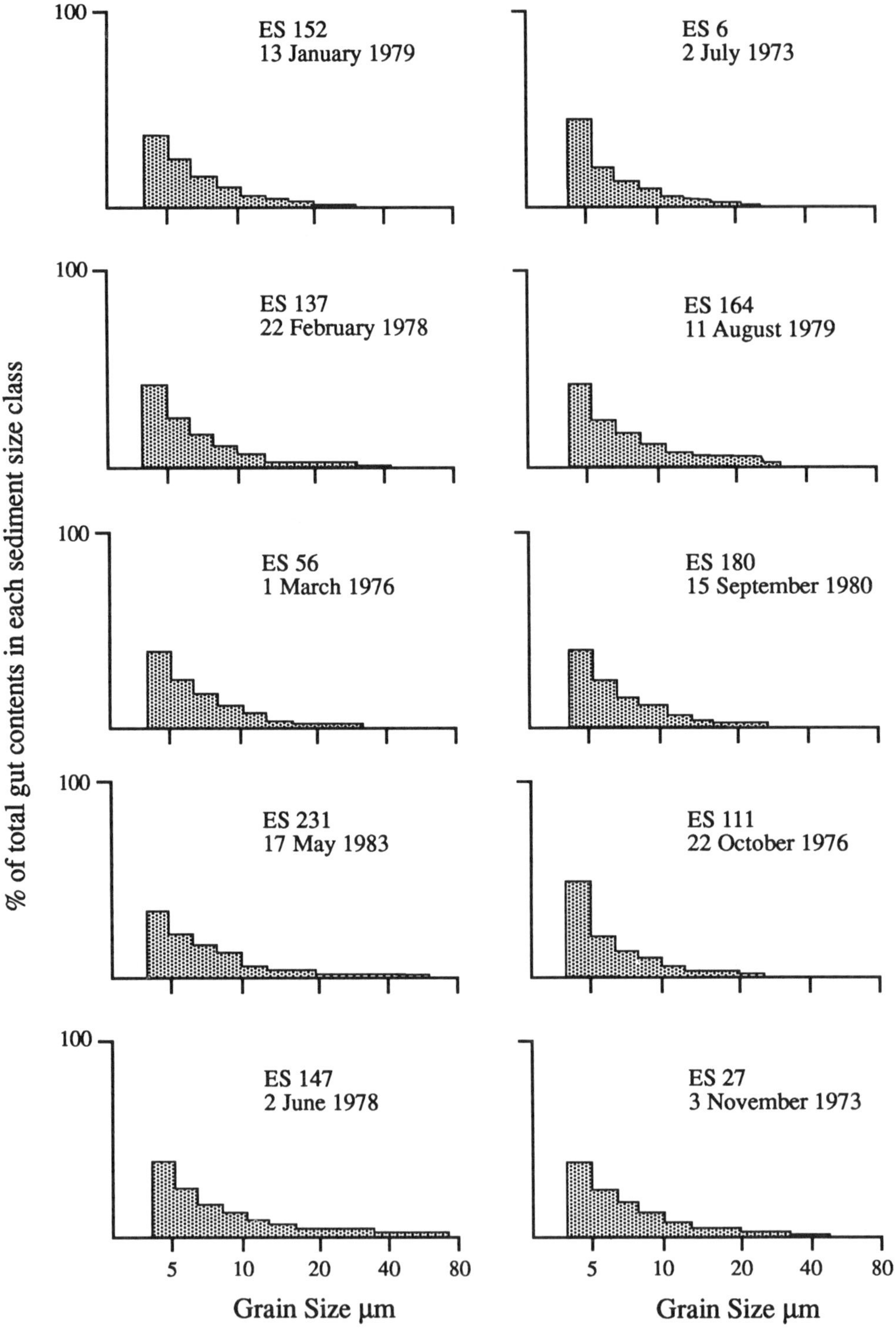

FIGURE 8.2. Particle size distribution of the stomach contents of *Malletia cuneata*.

Nutritional Biology *Malletia cuneata* has a single loop digestive tract (Sanders and Allen 1985) and *Ledella pustulosa* has a double loop gut (Allen and Hannah 1989). There was no significant variation in the grain size distribution in the stomach of *M. cuneata* throughout the year (figure 8.2), whereas in *L. pustulosa* (figure 8.3) there was an increase in the mean grain size in the February samples and a marked peak in the fraction < 5 μm in the June samples. Samples taken in the other months conform to the grain size distribution found in *M. cuneata.*

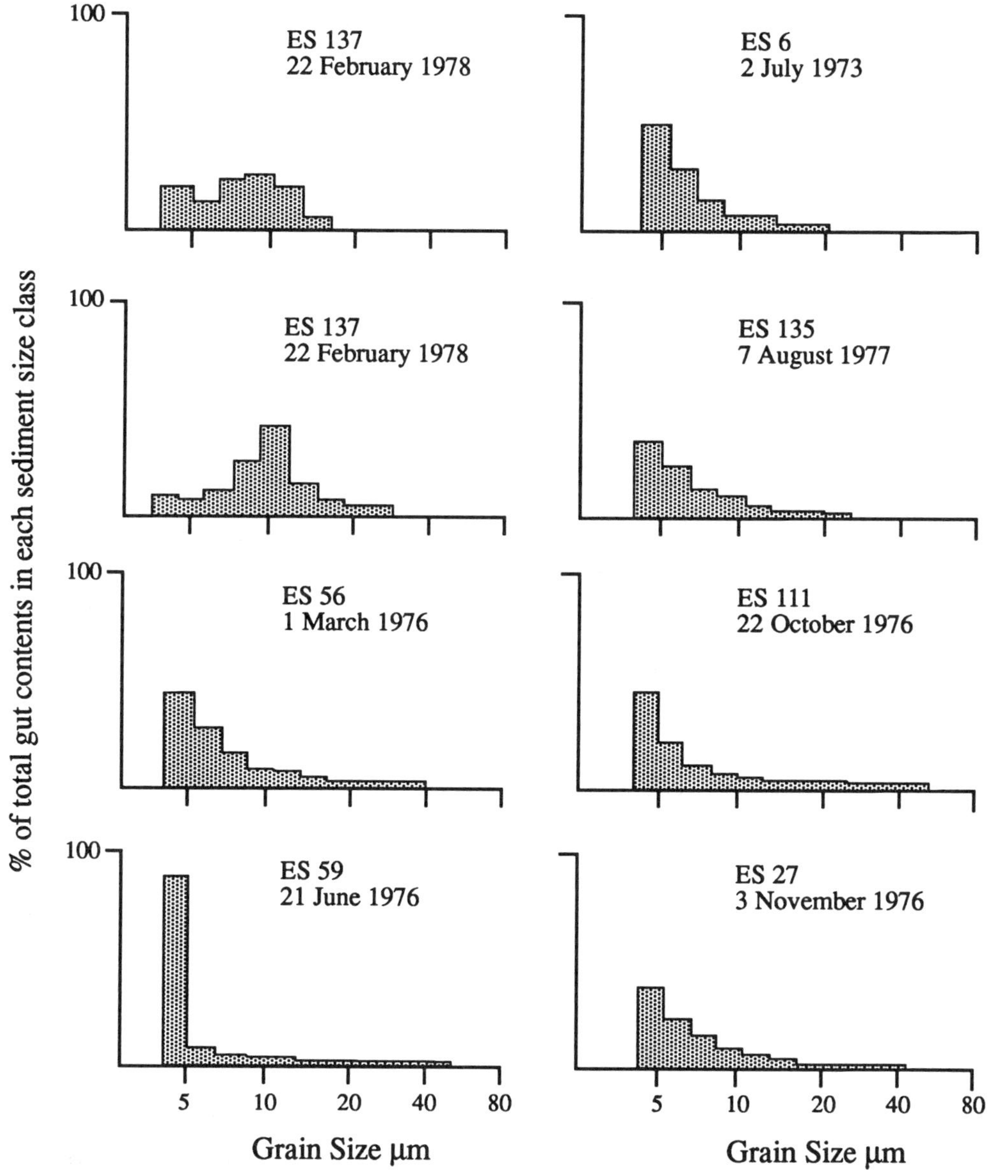

FIGURE 8.3. Particle size distribution of the stomach contents of *Ledella pustulosa.*

Light microscope examination of the stomach contents of *M. cuneata* (Tyler et al. 1993a) shows this material to be dominated by diatoms (both intact and fragments), coccoliths and foraminiferans. There appeared to be no seasonal variation in the occurrence of this material. Scanning electron microscope observations of the gut contents shows this material to be dominated by either the entire coccosphere or coccoliths of the species *Cyclococcolithus leptoporus, Coccolithus pelagicus* and *Emiliana huxleyi*. Because of the action of the style in the stomach of protobranchs, much of the diatom material appears as fragments. Whole specimens of species of the genera *Thalassiosira, Coscinodiscus, Skeletonema* and *Thalassionema* can be recognized as well as spores of *Chaetoceros*. In addition there are a few pennate diatoms, possibly *Rhizosolenia*, and both benthic and pelagic foraminifera.

Light microscope observations of the stomach contents of *L. pustulosa* suggest a similar pattern to that found in *M. cuneata* with the diet being composed of whole and fragmented diatoms including *Chaetoceros* spores, and coccoliths (Tyler et al. in 1993a). The variety of the diet is less than that observed in *M. cuneata* and there is evidence that there is a reduction in intake in *L. pustulosa* during the months of April, May, and June. These data are equivocal when taken in conjunction with the grain size and SEM observations. The grain size data suggest that in February and June we would expect to find a different population of material to reflect the grain size distribution. The only material we find in the February samples that may represent this increase in mean particle size are clumps of diatom spores. The other material present appears to be similar to that found in the stomachs of *M. cuneata*. It is possible that the peak in the grain size frequency in the June sample of *L. pustulosa* represents the disaggregated coccoliths from the surface bloom or fine inorganic material from surficial sediments. The SEM observations show species similar to those found in the stomach of *M. cuneata* but a greater number of *Chaetoceros* spores were found in the August sample of *L. pustulosa*.

Echinodermata: Echinoidea

Deep-sea echinoids show the two dominant patterns of reproduction described for deep-sea species. Species that breed on a quasi-continual basis include the deep-water echinothuriids and *Poriocidaris purpurata* (Tyler and Gage 1984b) whilst seasonal breeders include *Echinus affinis, E. alexandri* and *E. acutus* from the NE Atlantic (Tyler and Gage 1984a; Gage et al. 1986) and *Linopneustes longispinus* and *Stylocidaris lineata* from the Bahamas (Young and Cameron 1989; Young et al. 1992). This last species shows a seasonal pairing behavior, coincident with the spawning season which, presumably, enhances fertilization success.

The diet of deep-sea echinoids has been relatively well studied compared to other deep-sea taxa. The great Danish echinoid taxonomist Theodor Mortensen kept a careful record of dietary material he found in the guts of echinoids he examined, and as long ago as 1935 he showed that deep-water echinothurids take in macroalgal remains (Mortensen 1935, 1938). More recent observations (Lewis 1963) have shown that bathyal echinoids from the Caribbean feed on seagrasses and *Sargassum*.

Phormosoma Placenta/Echinus Affinis

Reproduction The gametogenic biology and reproductive periodicity of these two species was described by Tyler and Gage (1984a, b). *Phormosoma placenta* breeds quasi-continually and *Echinus affinis* spawns annually in the late winter.

Nutritional Biology Although the data show a considerable degree of scatter the ash-free dry weight of the stomach contents shows no seasonal variation in *Phormosoma placenta* but there was an increase in the stomach contents AFDW of *Echinus affinis* in the early June samples (figure 8.4a, b). This pattern is reflected in the total stomach content dry weight of both species. If, however,

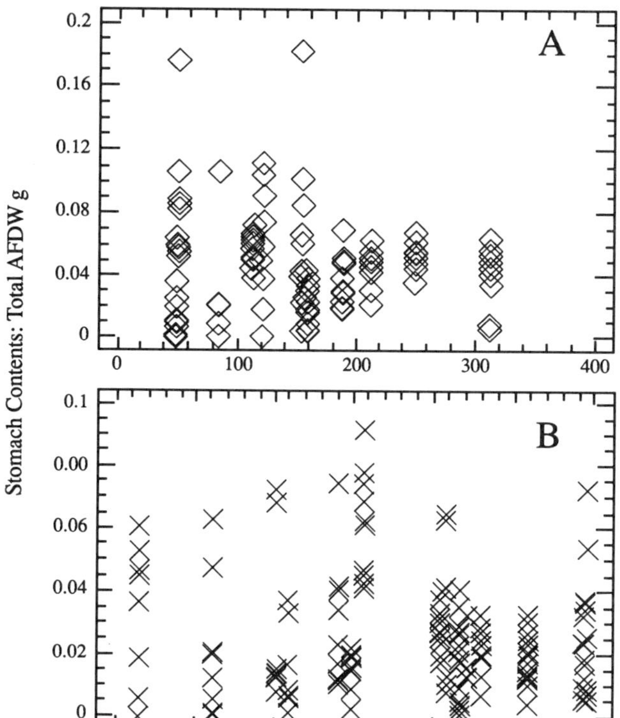

FIGURE 8.4. Scatterplot of the ash-free dry weight of the stomach contents of: A. *Phormosoma placenta*; B. *Echinus affinis* against Julian day.

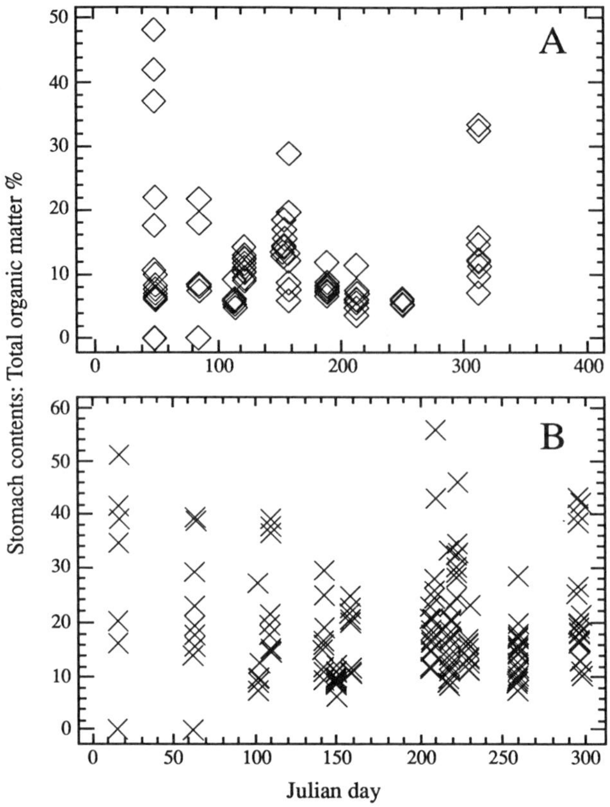

FIGURE 8.5. Scatterplot of the % total organic matter of the stomach contents of: A. *Phormosoma placenta*; B. *Echinus affinis* against Julian day.

the organic matter (as a percent of the total stomach contents) is plotted against Julian day there is no obvious seasonal variation in either species (figure 8.5a, b) although the intra-sample variation is high in most samples. The percent organic matter content of *E. affinis* is greater than that of *P. placenta*.

These data suggest that although there is no variation in the percent organic matter in the stomach of *Echinus affinis* between samples taken throughout the year, there is an increase in the actual amount of both total gut contents and AFDW in June. We propose that *E. affinis* feeds more actively when the phytodetritus arrives at the deep-sea floor.

Light microscopic and SEM observations of the stomach contents of both species suggest this material is dominated by diatom species, especially spores of *Chaetoceros*. Other common diatom genera include *Thalassionema, Nitzschia* and *Rhizosolenia*. Benthic and pelagic foraminifera are also found.

Observations of the behavior of *Echinus affinis* during the residence of phytodetritus at the deep-sea bed suggests they forage in the hollows where this material tends to accumulate (Campos et al. in submission). No such foraging behavior has been observed in *Phormosoma placenta*.

Echinodermata: Asteroidea

The asteroids have provided considerable data on both reproductive and dietary patterns in the deep sea. The majority of species examined in the NE Atlantic Ocean are quasi-continuous breeders (Tyler 1986) although two species, *Plutonaster bifrons* and *Dytaster grandis,* show a distinct seasonal cycle with spawn-out in the early months of the year (Tyler and Pain 1982; Tyler et al. 1990).

The diet of asteroids from sublittoral to abyssal depths was examined by Carey (1972) who showed a progression from a dominance of carnivory in more shallow waters to omnivory in the deep-sea species. In the NE Atlantic Khripounoff (1979) elucidated, in detail, the diet of all the asteroid species collected by the French BIOGAS program (Laubier and Monniot 1985). This latter study showed a diet varying both between and within species from carnivory to almost purely sediment feeding. Our own initial observations in the NE Atlantic suggested that most asteroid species took a variety of prey; the most notable exception being *Porcellanaster ceruleus* which appeared to be exclusively a mud-swallower.

Bathybiaster Vexillifer/Plutonaster Bifrons

Reproduction The histology of gamete development and reproductive periodicity of these two species was described by Tyler et al. (1982) and Tyler and Pain (1982). To compare the reproductive with nutritional biology we present here data for the weights of gonads and pyloric caecae.

The relationship of gonad ash-free dry weight with time confirms that *Bathybiaster vexillifer* is a continuous breeder and *Plutonaster bifrons* is a seasonal breeder (Tyler et al. 1982; Tyler and Pain 1982). To determine if gonad size differences are related to season and or age we have plotted the gonad ash-free dry weight against Julian day and maximum radius (R) (figure 8.6a, b). In *Bathybiaster* the gonad AFDW shows a significant power regression (ANOVA F = 168.04; n = 131) with size (R) but no variation with season, suggesting reproductive effort increases with age. In *Plutonaster* the gonad weight increases with R but is not significant (ANOVA F = 14.57; n = 155) whereas there is a marked variation in the gonad AFDW with season (Tyler et al. in press b).

Nutritional Biology The pyloric caecum AFDW shows a similar pattern to that of the gonad AFDW in both species. *Bathybiaster* shows a strong variation with size, whereas the greatest variation in *Plutonaster* is with season (figure 8.7a, b). If we compare the gonad index and the pyloric caecum indices of both species, as representative of reproductive effort and nutritive storage respectively, on an annual basis *Bathybiaster* puts double the effort into reproduction as *Plutonaster* (t = 11.86; $p < 0.01$) whereas *Plutonaster* puts signifi-

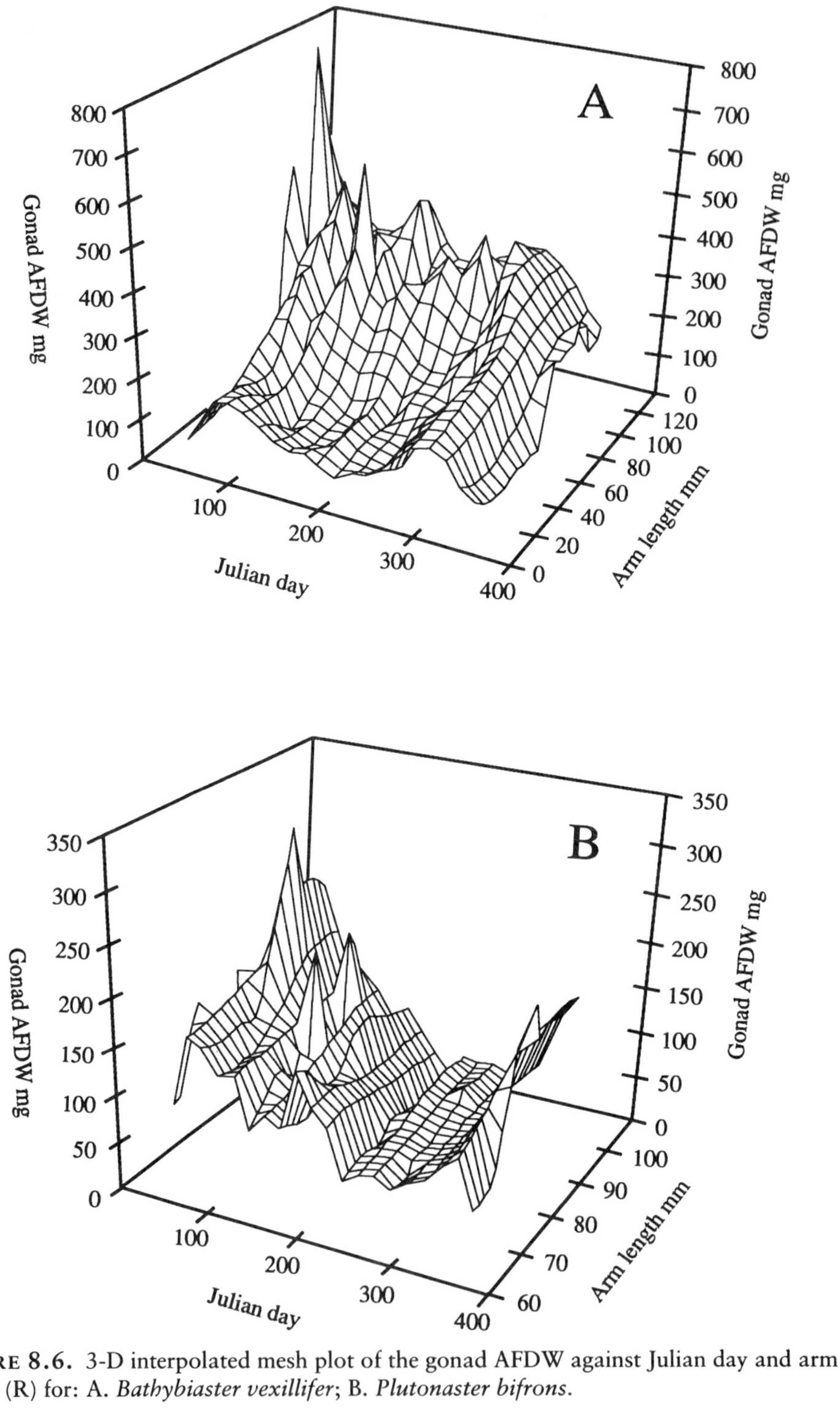

FIGURE 8.6. 3-D interpolated mesh plot of the gonad AFDW against Julian day and arm radius (R) for: A. *Bathybiaster vexillifer*; B. *Plutonaster bifrons*.

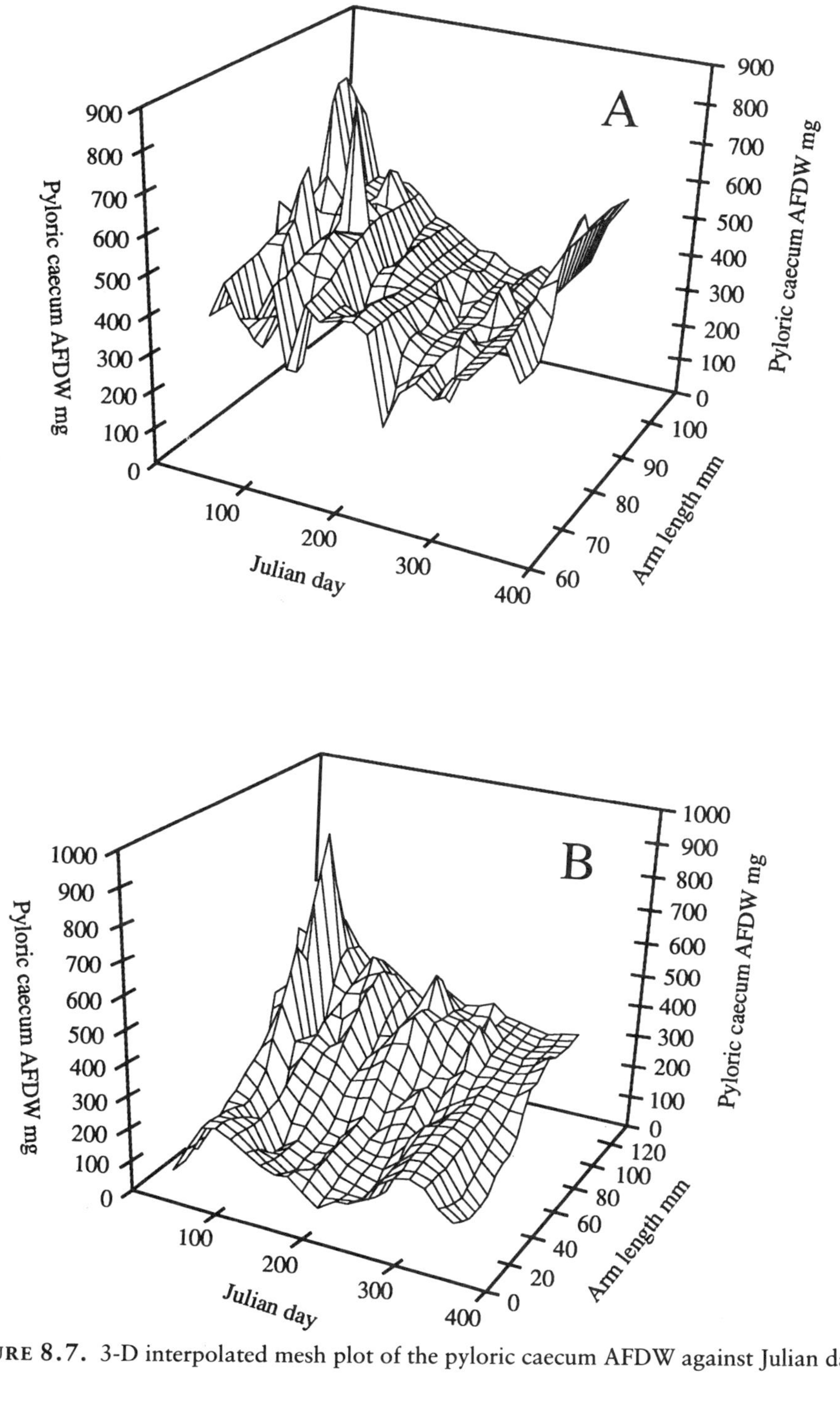

FIGURE 8.7. 3-D interpolated mesh plot of the pyloric caecum AFDW against Julian day

cantly more effort into nutrient storage than does *Bathybiaster* (t = 7.42; $p < 0.01$).

These data confirm that there is a seasonal signal in both the gonad and the pyloric caecum of *Plutonaster bifrons* but not in *Bathybiaster vexillifer.*

Although stomach eversion is usually prevented by the gastric ligaments, a number of the specimens had empty stomachs. We do not know if this is the natural condition or whether the stomach contents were voided during capture.

The dominant food item in the stomach of *Bathybiaster vexillifer* was the irregular echinoid *Hemiaster expurgitus* (Tyler et al. 1993b). Individuals of this echinoid found in the gut were up to 30 mm test length although the dominant size range was < 20 mm. Gastropods found in the stomach were dominated by *Natica* sp. and a variety of protobranch bivalves including *Yoldiella jeffreysei, Ledella pustulosa,* and *Malletia johnstoni.* In addition to these common species, other noteworthy prey were the polychaete worms *Poecilochaetus* cf. *fulgens,*

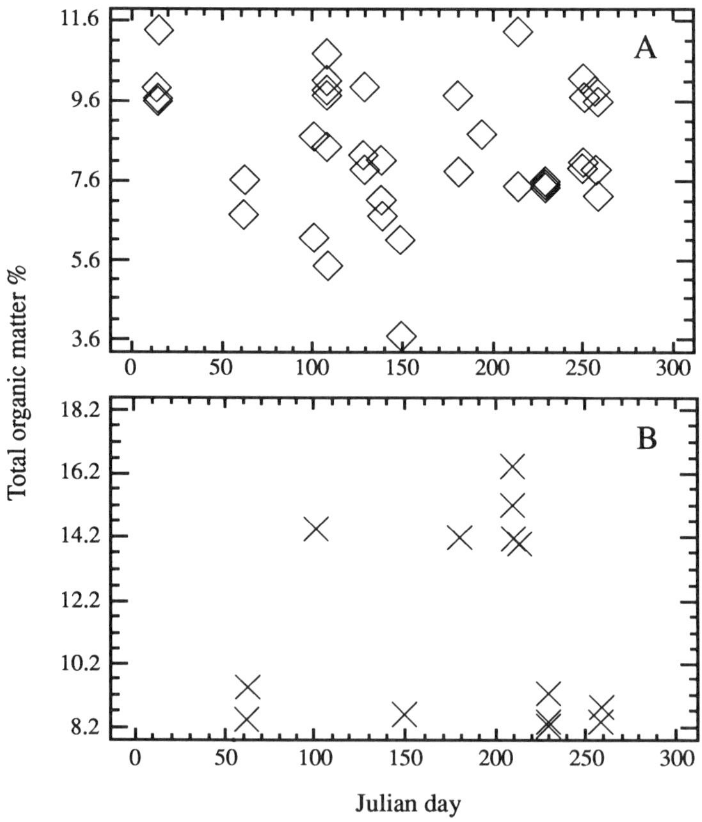

FIGURE 8.8. Scatterplot of % organic matter against Julian day for: A. *Bathybiaster vexillifer*; B. *Plutonaster bifrons.*

Terebellides sp. A, *Diplocirrus* sp. and at least three species of maldanids. Also found was the holothurian *Ypsilothuria bitentaculata* and, although peracarid fragments (particularly amphipods and isopods) were found in a number of samples, these were generally rare. Fragments of fish vertebrae and scales as well as unrecognizable organic matter were found in a number of specimens. The prey species diversity for each individual sample ranged from 1.65 to 3.63 (mean = 2.55; SD = 0.61) (Tyler et al. 1993b).

As with *Bathybiaster vexillifer*, the diet of *Plutonaster bifrons* was dominated by *Hemiaster expergitus* (particularly fragments) and unrecognizable organic matter (Tyler et al. 1993b). The size range of *H. expergitus* differed significantly from those in the stomach of *B. vexillifer* ($\chi = 31.97$; $p < 0.01$) (Tyler et al. 1993b). Protobranch bivalves (an undescribed *Nuculoma* sp.), gastropods and polychaetes formed a very minor element in the diet of *P. bifrons*. The most notable feature in the stomach contents was the occurrence of a partially ingested and badly decomposed specimen of blue whiting (*Micromestistius poutassou*) embedded in the stomach of two specimens of *P. bifrons*. In addition, fish scales and vertebrae were found in other specimens of *P. bifrons*. The dietary species diversity of *P. bifrons* varied from 0.89 to 2.16 (mean = 1.49; SD = 0.36).

The gut species diversity was significantly different between the two seastar species ($t = 4.82$; $p < 0.001$).

When all the recognizable food and stained organic matter had been removed from the stomachs of both species, only a limited number of individuals contained residual sediment. Analysis of the organic carbon content of this material from *Bathybiaster* revealed values of 3.6 to 11.3 percent but with no apparent seasonal variation (figure 8.8a). Conversely, in *Plutonaster*, the values ranged from 8.2 to 16.2 percent and showed a possible seasonal cycle (figure 8.8b) (Tyler et al. 1993b).

Examination of seabed photographs taken with the IOS epibenthic sledge suggests that *Bathybiaster vexillifer* burrows "several centimeters" into the sediment (Rice et al. 1991) whereas *Plutonaster bifrons* feeds very close to the sediment surface and rarely is the dorsal surface covered by sediment (Tyler et al. 1993b).

Conclusions

During the last ten years some of the basic tenets of deep-sea biology have been questioned. The discovery of hydrothermal vents has shown that certain areas of the deep-sea contain small patches of high biological production. On a less dramatic, but no less important, scale was the discovery that the apparently unchanging environment of the deep-sea could have imposed on it a seasonal signal by the rapid downward flux of surface-derived organic material (Billett et al. 1983; Deuser 1986; Rice et al. 1986) and that a small proportion of deep-sea invertebrate species were breeding on a predictable seasonal basis and showing seasonal growth patterns (Gage and Tyler 1991). Those species that reproduce on a seasonal basis produce a small egg indicative of planktotrophic development. The period of larval development is linked to the downward flux of particulate material in late spring. By contrast, those species that reproduce quasi-continually produce a large egg indicative of lecithotrophy and are independent of the downward flux.

As a result of the discovery of seasonal processes in the deep-sea there was an intuitive feeling that the seasonally varying gametogenic processes observed were linked, through diet, to the vertical flux of phytodetritus (Tyler and Gage 1984a; Thiel et al. 1990). An appropriate way to test if this flux had an influence on the reproductive processes in benthic invertebrates at the deep-sea bed was to examine pairs of closely related species, both taxonomically and spatially, each pair consisting of a quasi-continuous breeder and a seasonal breeder.

The results of the data from all the quasi-continuous breeders were that the seasonal pulse of phytodetritus to the deep-sea floor was "damped out" in the diet of these species. In *Malletia cuneata* there appeared to be no seasonal variation in the number of particles of phytodetrital material in the stomach whilst in *Phormosoma placenta* the organic matter content of the stomach showed no seasonal variation. The organ indices of *P. placenta* also showed no seasonal variation. The diet of *Bathybiaster vexillifer* suggested this species was a carnivore and these data together with photographic observations suggest this species feeds "several centimeters" down in the sediment (Rice et al. 1991) where any sediment ingested would no longer bear a seasonal signal. This lack of seasonal variation in the diet was reflected in the lack of variation in the pyloric caecum and gonad AFDW.

The data on diet for the seasonal breeders remain equivocal. In *Ledella pustulosa* the diet is similar to that of *Malletia cuneata,* except there appears to be a reduction in feeding coincident with the arrival of phytodetritus at the deep-sea bed. In *Echinus affinis* there is an increase in the total stomach content and total organic matter (as AFDW) coincident with the phytodetritus deposition which, together with the photographic observations of feeding, suggests a seasonal benefit to *E. affinis* metabolism. This seasonal signal appears to be transmitted directly to the gonad as the stomach wall shows an increase associated with the deposition of phytodetritus (Campos-Creasey et al. in submission). The principal dietary component of *Bathybiaster vexillifer,* namely *Hemiaster expergitus,* was also the main element in the diet of *Plutonaster bifrons.* However, other than this echinoid, the diet of *P. bifrons* was composed mainly of unrecognizable organic matter including fish remains. This suggests that *P. bifrons* is a scavenger at the sediment water interface and thus any sediment it takes in with its food will possess a seasonal signal. In addition, the occurrence of two specimens scavenging a blue whiting suggests the possibility of an additional seasonal signal. The blue whiting spawns in the top 1000 m over the Rockall Trough in March/April of each year before migrating north (Bailey 1982). Mauchline and Gordon (1984) have shown that the blue whiting forms an important element of the diet of deep water benthopelagic fish during these months. It is possible that *P. bifrons* may also benefit from the sinking of dead blue whiting that provides a food focus and rich source of energy prior to the downward flux of phytodetritus.

Acknowledgments

This study was completed during the tenure of NERC Grant GR3/7275 to P.A.T and a Brazilian Government Scholarship to L.S.C. Both are gratefully acknowledged. Dr. J. D. Gage (SMBA) is thanked for his valuable comments during the preparation of this paper. Dr. M. Rex (Universiity of Massachusetts), Mr. G. J. L. Paterson and Mr. N. Merrett (Natural History Museum, London) are thanked for identifications of gastropods, polychaetes, and fish, respectively.

References

Allen, J. A. 1978. Evolution of deep-sea protobranch bivalves. *Proceedings of the Royal Society of London,* Series B 284:387–401.

Allen, J. A. 1979. Adaptations and radiation of deep-sea bivalves. *Sarsia* 64:19–27.

Allen, J. A. and F. J. Hannah. 1989. Studies on the deep-sea Protobranchia: The subfamily Ledellinae (Nuculanidae). *Bulletin of the British Museum Natural History* (Zoology) 55:123–171.

Allen, J. A. and H. L. Sanders. 1973. Studies on the deep-sea Protobranchia (Bivalvia): The families Silicilidae and Lametilidae. *Bulletin of the Museum of Comparative Zoology of Harvard College* 145:263–310.

Allen, J. A. and H. L. Sanders. 1982. Studies on the deep-sea Protobranchia (Bivalvia): The subfamily Spinulinae (family Nuculanidae). *Bulletin of the Museum of Comparative Zoology of Harvard College* 150:1–30.

Bailey, R. S. 1982. The population of the blue whiting in the North Atlantic. *Advances in Marine Biology* 19:257–355.

Billett, D. S. M., R. S. Lampitt, A. L. Rice, and R. F. C. Mantoura. 1983. Seasonal sedimentation of phytoplankton to the deep-sea benthos. *Nature* 302:520–522.

Bruun, A. F. 1957. Deep-sea and abyssal depths. *Geological Society of America Memoir* 67(1):641–672.

Campos, L. S., P. A. Tyler, J. D. Gage, and D. L. Harbour. In submission. Vertical flux coupled to the diet and seasonal life history in the deep-sea echinoid *Echinus affinis.*

Carey, A. G., Jr. 1972. Food resources for sublittoral, bathyal, and abyssal asteroids in the northeast Pacific. *Ophelia* 10:35–47.

Deuser, W. G. 1986. Seasonal and interannual variations in deep water particle fluxes in the Sargasso Sea and their relation to surface production. *Deep-Sea Research* 33:225–246.

Gage, J. D. and P. A. Tyler. 1991. *Deep-Sea Biology: A Natural History of Organisms at the Deep-Sea Floor.* Cambridge: Cambridge University Press.

Gage, J. D., P. A. Tyler, and D. Nichols. 1986. Reproduction and growth of *Echinus acutus* var. *norvegicus* and *E. elegans* on the continental slope off Scotland. *Journal of Experimental Marine Biology and Ecology* 101:41–53.

George, R. Y. and R. J. Menzies. 1967. Indication of cyclic reproductive activity in abyssal organisms. *Nature* 215:878.

George, R. Y. and R. J. Menzies. 1968. Further evidence for seasonal breeding cycles in the deep-sea. *Nature* 220:80–81.

Gooday, A. J. 1988. A response by benthic foraminifera to the deposition of phytodetritus in the deep sea. *Nature* 332:70–73.

Gooday, A. J. and C. M. Turley. 1990. Responses by benthic organisms to inputs of organic material to the ocean floor: A review. *Proceedings of the Royal Society of London* 331:119–138.

Gross, T. F., A. J. Williams, and A. R. M. Nowell. 1988. A deep-sea sediment transport storm. *Nature* 331:518–521.

Hollister, C. D. and I. N. McCave. 1984. Sedimentation under deep-sea storms. *Nature* 309:220–225.

Khripounoff, A. 1979. Relations trophiques dans l'ecosysteme benthique abyssal Atlantique: Descriptions et bilan energetique. Diplome de Docteur de 3e Cycle L'Universite Pierre et Marie Curie.

Knudsen, J. 1970. The systematics and biology of abyssal and hadal Bivalvia. *Galathea Report* 11:7–241.

Lampitt, R. S. 1985. Evidence for the seasonal deposition of detritus to the deep-sea floor and its subsequent resuspension. *Deep-Sea Research* 32:885–897.

Laubier, L. and C. Monniot. 1985. *Peuplements Profonds du Golfe de Gascogne* Campagnes BIOGAS. IFREMER, Brest, France.

Leeder, M. R. 1985. *Sedimentology: Processes and Product.* London: Allen and Unwin.

Lewis, J. B. 1963. The food of some deep water echinoids from Barbados. *Bulletin of Marine Science* 13:360–363.

Lightfoot, R., P. A. Tyler, and J. D. Gage. 1979. Seasonal reproduction in deep-sea bivalves and brittlestars. *Deep-Sea Research* 26:967–973.

Lochte, K. and C. M. Turley. 1988. Bacteria and cyanobacteria associated with phytodetritus in the deep sea. *Nature* 333:67–69.

Mauchline, J. and J. D. M. Gordon. 1984. Feeding and bathymetric distribution of the gadoid and morid fish of the Rockall Trough. *Journal of the Marine Biological Association of the United Kingdom* 64:657–665.

Menzies, R. J. 1965. Conditions for the existence of life on the abyssal sea floor. *Oceanography and Marine Biology: An Annual Review* 3:195–210.

Mortensen, T. 1927. *Handbook of the Echinoderms of the British Isles.* Oxford: Oxford University Press.

Mortensen, T. 1935. *A Monograph of the Echinoidea:* II. Copenhagen: Reitzel.

Mortensen, T. 1938. On the vegetarian diet of some deep-sea echinoids. *Annotationes zoologicae japonenses* 17:225–228.

Rice, A. L., D. S. M. Billett, J. Fry, A. W. G. John, R. S. Lampitt, R. F. C. Mantoura, and R. J. Morris. 1986. Seasonal deposition of phytodetritus to the deep sea floor. *Proceedings of the Royal Society of Edinburgh* 88B:265–279.

Rice, A. L., D. S. M. Billett, M. H. Thurston, and R. S. Lampitt. 1991. The Institute of Oceanographic Sciences Biology Programme in the Porcupine Seabight: Background and General Introduction. *Journal of the Marine Biological Association of the United Kingdom* 71:281–310.

Rokop, F. J. 1974. Reproductive patterns in deep-sea benthos. *Science* 186:743–745.

Rokop, F. J. 1977. Year-round reproduction in deep sea bivalve molluscs. In S. E. Stancyk, ed., *Reproductive Ecology of Marine Invertebrates,* pp. 189–198. Columbia: University of South Carolina Press.

Rowe, G. T. and N. Staresinic. 1979. Sources of organic matter to the deep-sea benthos. *Ambio Special Report* 6:19–24.

Sanders, H. L. and J. A. Allen. 1973. Studies on the deep-sea Protobranchia (Bivalvia): Prologue and Pristoglomidae. *Bulletin of the Museum of Comparative Zoology* 145:237–262.

Sanders, H. L. and J. A. Allen. 1977. Studies on the deep-sea Protobranchia (Bivalvia): The family Tindariidae and the genus *Pseudotindaria. Bulletin of the Museum of Comparative Zoology* 148:23–59.

Sanders, H. L. and J. A. Allen. 1985. Studies on the deep-sea Protobranchia (Bivalvia): the family Malletiidae. *Bulletin of the British Museum Natural History (Zoology)* 49: 195–238.

Scheltema, R. 1972. Reproduction and dispersal of bottom dwelling marine invertebrates: A speculative summary. In R. W. Bauer, ed., *Barobiology and Experimental Biology of the Deep Sea,* pp. 58–68. University of North Carolina Press.

Schoener, A. 1968. Evidence for a reproductive periodicity in the deep sea. *Ecology* 49:81–87.

Smith, C. R. 1985. Food in the deep-sea: Utilization, dispersal, and flux of nekton falls at the Santa Catalina Basin floor. *Deep-Sea Research* 32:417–442.

Suchanek, T. H., S. L. Williams, J. C. Ogden, D. K. Hubbard, and I. P. Gill. 1986. Utilization of shallow water seagrasses detritus by Caribbean deep-sea macrofauna: ^{13}C evidence. *Deep-Sea Research* 32:201–214.

Thiel, H., O. Pfannkucke, G. Schriever, K. Lochte, A. J. Gooday, C. Hemleben, R. F. G. Mantoura, C. M. Turley, J. W. Patching, and F. Riemann. 1990. Phytodetritus on the deep-sea floor in the central oceanic region of the northeast Atlantic. *Biological Oceanography* 6:203–239.

Turner, R. 1973. Wood-boring bivalves, opportunistic species in the deep sea. *Science* 180:1377–1379.

Tyler, P. A. 1986. Studies of a benthic time series: Reproductive biology of benthic invertebrates in the Rockall Trough. *Proceedings of the Royal Society of Edinburgh* 88B:175–190.

Tyler, P. A. 1988. Seasonality in the deep sea. *Oceanography and Marine Biology: An Annual Review* 26:227–258.

Tyler, P. A. and J. D. Gage. 1984a. Seasonal reproduction of *Echinus affinis* (Echinodermata: Echinoidea) in the Rockall Trough, northeast Atlantic Ocean. *Deep-Sea Research* 31:387–402.

Tyler, P. A. and J. D. Gage. 1984b. The reproductive biology of echinothurid and cidarid sea urchins from the deep sea (Rockall Trough, North East Atlantic Ocean). *Marine Biology* 80:63–74.

Tyler, P. A. and S. L. Pain. 1982. The reproductive biology of *Plutonaster bifrons, Dytaster insignis* and *Psilaster andromeda* (Asteroidea: Astropectinidae) from the Rockall Trough. *Journal of the Marine Biological Association of the United Kingdom* 62:869–887.

Tyler, P. A., S. L. Pain, and J. D. Gage. 1982. The reproductive biology of the deep-sea asteroid *Bathybiaster vexillifer. Journal of the Marine Biological Association of the United Kingdom* 62:57–69.

Tyler, P. A., D. S. M. Billett, and J. D. Gage. 1990. Seasonal reproduction in the seastar *Dytaster grandis* from 4000 m in the north-east Atlantic Ocean. *Journal of the Marine Biological Association of the United Kingdom* 70:173–180.

Tyler, P. A., J. D. Gage, and D. S. M. Billett. 1992. Reproduction and recruitment in deep-sea invertebrate populations in the NE Atlantic Ocean: A review of the options. In G. Colombo, I. Ferrari, V. U. Ceccherelli, and R. Rossi, eds., *Marine Eutrophication and Population Dynamics*, pp. 257–262. Fredensborg, Denmark: Olsen and Olsen.

Tyler, P. A., R. Harvey, L. A. Giles, and J. D. Gage. In 1993a. Reproductive strategies and diet in the deep-sea nuculanid protobranchs (Bivalvia: Nuculoidea) from the Rockall Trough. *Marine Biology* 114:571–580.

Tyler, P. A., J. D. Gage, G. J. L. Paterson, and A. L. Rice. 1993b. Dietary constraints on the reproductive periodicity in two sympatric deep-sea astropectiniid seastars. *Marine Biology* 115:267–277.

Verrill, A. E. and K. J. Bush. 1898. Revision of the deep-water Mollusca of the Atlantic coast of North America, with descriptions of new genera and species. *Proceedings of the United States National Museum* 20:775–901.

Wolff, T. 1976. Utilization of seagrass in the deep-sea. *Aquatic Botany* 2:161–174.

Wolff, T. 1979. Macrofaunal utilization of plant remains in the deep-sea. *Sarsia* 64:117–136.

Yonge, C. M. 1939. The Protobranchiate Molusca: A functional interpretation of their structure and evolution. *Proceedings of the Royal Society of London* Series B 230: 79–147.

Young, C. M. and J. L. Cameron. 1989. Developmental rate as a function of depth in the bathyal echinoid *Linopneustes longispinus*. In J. S. Ryland and P. A. Tyler, eds., *Reproduction, Genetics, and Distributions of Marine Organisms*, pp. 225–231. Fredensborg, Denmark: Olsen and Olsen.

Young, C. M., P. A. Tyler, J. L. Cameron, and S. G. Rumrill. 1992. Seasonal breeding aggregations in low-density populations of the bathyal echinoid *Stylocidaris lineata*. *Marine Biology* 113:603–612.

Young, C. M., P. A. Tyler, R. H. Emson, and J. D. Gage. 1993. Selection and perception of macrophyte detrital falls by the bathyal sea urchin *Stylocidaris lineata*. *Deep-Sea Research* 40:1475–1486.

CHAPTER NINE

The Biology of External Fertilization in Deep-Sea Echinoderms

Craig M. Young

ABSTRACT

External fertilization by free spawning is much more common than internal fertilization among bathyal and abyssal echinoderms. A few species of echinoids may pseudocopulate by exchanging gametes in masses of mucus, which are retained on the adult spines. In deep-sea species of *Aspidodiadema,* elongate sperm nuclei may be an adaptation for swimming through the thick mucus that holds the eggs. Denny's (1988) model of fertilization success predicts that, all else being equal, external fertilization should be a more viable strategy on smooth sedimentary bottoms in the deep sea than in the more turbulent flow conditions found in shallow water. This is contrary to Thorson's Rule, which predicts greater incidence of brooding (and, by inference, internal fertilization) with depth. Reproduction at low population density is sometimes achieved by echinoids, holothuroids, and ophiuroids by aggregation behaviors. Some species aggregate year round and others remain with spawning partners only during the reproductive season. Gametes of echinothuriids contain lipid reserves that apparently allow sperm to remain motile more than three times as long as typical shallow-water echinoid sperm. Functional significance of some gamete modifications, including biflagellate, dimorphic, and pigmented sperm, remains unknown.

Despite early opinions to the contrary (Thorson 1966), many marine organisms waste a substantial portion of their reproductive effort at the time of fertilization (Pennington 1985; Brazeau and Lasker 1990; Levitan et al. 1992). This loss is minimized most effectively by copulation and internal fertilization. External fertilization, in which eggs and sperm are spawned freely into the water, is a gamble of gametes under even the best of conditions. Shallow-water animals that succeed at this lottery often do so by precise synchrony of gametogenesis and spawning (reviewed by Giese and Kanatani 1987), and by producing so many gametes that substantial losses can be tolerated (Thorson 1950). Most of the cues used by shallow-water organisms for synchronizing gametogenesis and spawning are not available to organisms living in the continuously dark habitats of the deep sea (table 9.1). One might expect, therefore, that copulation, internal fertilization, and postcopulatory sperm storage mechanisms as well as brooding should be common in deep-sea animals. However, seasonal reproduction is surprisingly common in the deep sea (reviewed by Tyler 1988) indicating that cues must exist for synchronizing reproductive processes in the absence of light. Moreover, the generalization that deep-sea animals should brood their young (Thompson 1878; Thorson 1950), is rapidly losing acceptance with the discovery that many deep-sea species produce planktotrophic or pelagic lecithotrophic larvae (see chapters by Pearse, Gustafson and Lutz, and Bouchet, this volume). These recent changes in our thinking about reproductive frequency and modes in the deep sea raise the possibility that external fertilization may be much more common than previously thought.

In this review, I present the existing evidence for free spawning in deep-sea echinoderms and discuss some advantages and disadvantages of external fertilization in the deep sea. I also review behavioral, morphological, and physiological mechanisms that may increase the probability of fertilization. Echino-

Table 9.1 *Environmental Cues Used by Shallow Water Animals to Synchronize Reproductive Processes, and Their Availability in the Deep Sea*

Reproductive Process	Cue	Shallow-Water Example	Available in Deep Sea
Gametogenesis	lunar cycles	echinoids	no
	day length	asteroids	no
	temperature	echinoids	maybe
	salinity	mussels	maybe
	food availability	echinoids	yes
Spawning	phytoplankton	mussels	yes (with lag)
	temperature	clams	maybe
	bright sunlight	echinoderms	no
	lunar period	corals	no
	sunrise	ascidians	no
	sunset	cnidarians	no
	rain storms	mussels	no
	wave actions	mussels	no

derms are among the most abundant and diverse megafaunal organisms in the deep sea. Because the reproductive biology and development of shallow-water echinoderms has been studied extensively, the phylum is ideal for broad comparative studies (e.g., Emlet et al. 1987; Eckelbarger et al. 1989a) and for comparisons between shallow-water and deep-sea species.

How Common Is External Fertilization in the Deep Sea?

Very few deep-sea animals have been observed to spawn in situ. A notable example is the observation by Van Dover (in press) of spawning behavior in the vestimentiferan tube worm *Riftia pachyptila*. Some echinoderms release gametes spontaneously after surviving the physiological stress associated with transport to the surface by trawl or submersible. For example, Mortensen (1914) recorded gamete release by dredged individuals of a deep-sea species of *Ophiothrix*. Likewise, apparently healthy individuals of the echinothuriid echinoids *Araeosoma fenestratum* and *Phormosoma placenta* and the goniasterid asteroid *Peltaster placenta* have sometimes released gametes in shipboard cold rooms within a few hours of collection (personal observations).

Spawning is induced by KCl injection in many deep-sea echinoids as easily as in shallow-water forms. For example, Pawson and Miller (1979) induced gamete release in the bathyal arbaciid sea urchin *Coelopleurus floridanus* from the Florida continental slope and described long genital papillae in the males of this species. More recently, we have spawned numerous other bathyal species and a few shallow abyssal species in shipboard laboratories and cold rooms (table 9.2). Potassium chloride injection does not induce spawning in the soft-

TABLE 9.2 *Laboratory Observations of Spawning in Deep-Sea Echinoderms from the Bahamas and the Rockall Trough*

Bahamas (Bathyal)	Rockall Trough (Bathyal)	Rockall Trough (Abyssal)
Echinoidea:		
Stylocidaris lineata	*Cidaris cidaris*	*Echinus affinus*
Cidaris blakei	*Echinus acutus norvegicus*	*Phormosoma placenta*
Coelopleurus floridanus	*Echinus elegans*	
Conolampas sigsbei	*Spatangus raschi*	
Aspidodiadema jacobyi		
Salenia goesiana		
Archaeopneustes histrix		
Palaeobrissus hilgardi		
Brissopsis atlanticus		
Phormosoma placenta		
Araeosoma fenestratum		
Lytechinus euerces		
Linopneustes longispinus		
Asteroidea:		
Peltaster placenta	*Pontaster tenuispinus*	

bodied echinothuriids, but spontaneous spawning has been observed in two species (table 9.2). Although the species coverage is smaller, bathyal and abyssal animals in the northeast Atlantic Ocean show a similar pattern. Seven species have been induced to spawn in the laboratory, and there is no evidence for brooding or internal fertilization in any of the other echinoid or asteroid species in this region. Most of the deep-sea species in the North Atlantic are now known to produce pelagic lecithotrophic larvae (reviewed by Pearse, this volume), so free spawning is likely to be the predominant mechanism of fertilization.

The sperm of some deep-sea echinoderms are spawned in thick mucus which remains attached for a time to the spines or to the aboral surface of the male adult. For example, the tiny bathyal echinoid *Salenia goesiana* releases strands of sperm which quickly become entangled on the spines (figure 9.1A, B). The sperm of this species, which have small, circular heads (Eckelbarger et al. 1989a) not unlike the sperm of some sea cucumbers, are retained in the mucus strands for at least several hours. The sticky sperm strands are not easily removed by pipette. Spines of *S. goesiana* are covered with small spinelets which accumulate debris (figure 9.1C) and to which the sperm masses attach readily.

We have also observed the spawning of mucus-bound gametes in two species of the genus *Aspidodiadema, A. jacobyi* in the Bahamas and *A. arcitum* off Hawaii. Both released their sperm and eggs in a cloud of viscous mucus that was held in a tight mass on the aboral surface of the body between the spines. The surface microstructure of the spines of *Aspidodiadema* spp. is shown in figure 9.1E, F. Spinelets roughen the spine surface and seem to assist in the retention of the viscous gamete masses. Elsewhere, we have described development of *Aspidodiadema jacobyi,* noting that the mouth does not develop in the larva until 21 days after settlement (Young et al. 1989). We assumed at that time that development of these small eggs took place in the plankton and speculated that the extended pre-feeding period could provide nutritional flexibility needed during planktonic development in a food-poor environment. However, our repeated observations of mucus-bound gametes now raise the possibility that eggs are actually brooded on the spines during the early developmental stages. If this is the case, then fertilization may occur by a form of pseudocopulation when the mucus masses of adjacent individuals come into contact.

The few echinoids known to brood their young are mostly from relatively shallow Antarctic waters (Thompson 1878; David and Mooi 1990), and the mode of fertilization is not known for any of these species. It seems likely that males release sperm into the water, and that fertilization of eggs retained on or in the female takes place without any form of copulation. Copulation is suspected, however, in the tiny Concentricycloidea, which inhabit deep-sea wood. These animals are sexually dimorphic; males have pairs of marginal spines thought to function in sperm transfer. Ultrastructural studies of the sperm support the hypothesis that fertilization takes place internally (Healy et al. 1988).

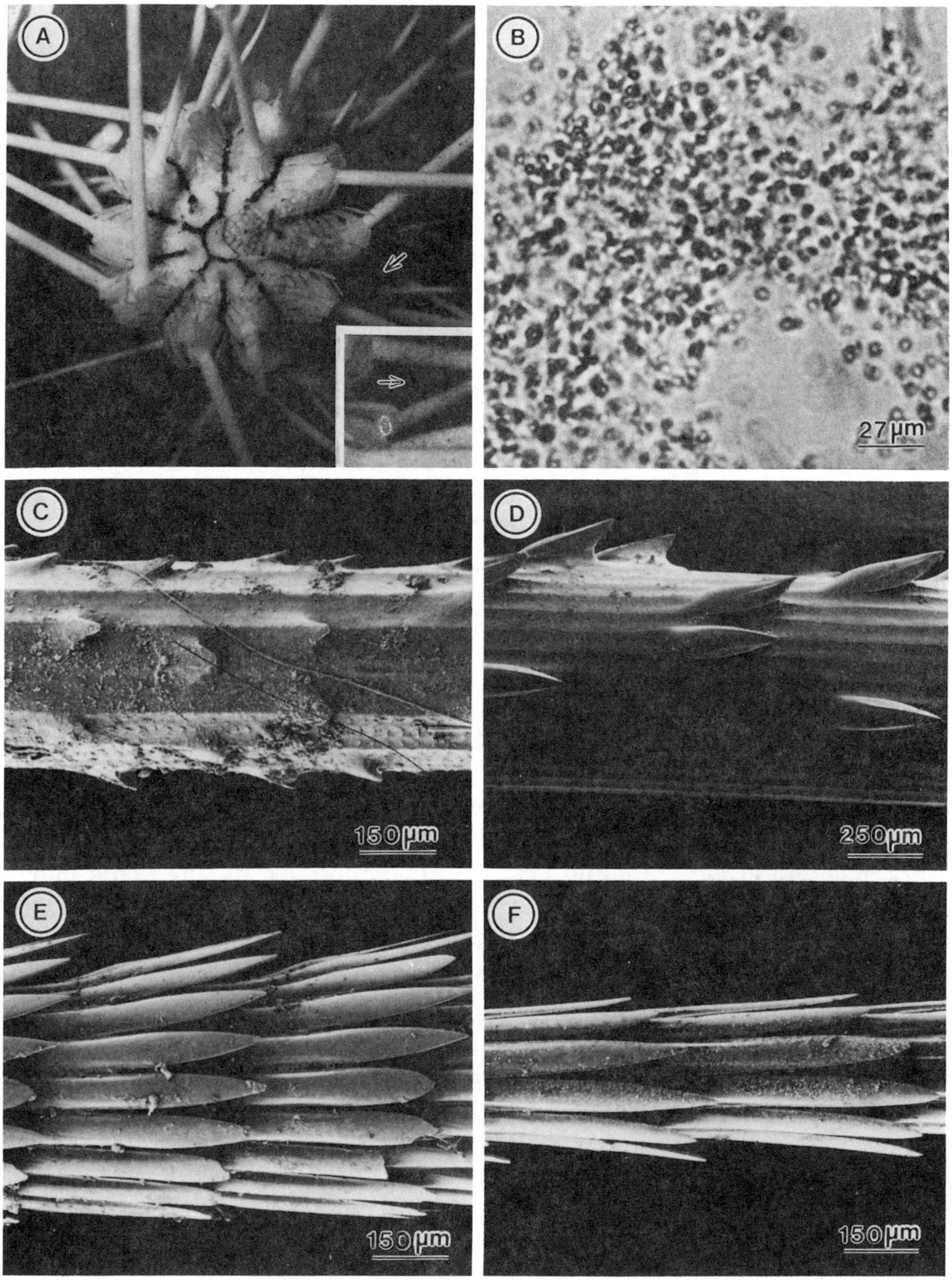

FIGURE 9.1. A: Male *Salenia goesiana* with strands of mucus-bound sperm attached to spines (arrow). Inset: close-up of sperm strands. B: Live spermatozoa of *Salenia goesiana* bound in mucus. C: Scanning electron micrograph (SEM) of a spine of *Salenia goesiana*, showing spinelets. D: SEM showing spinelets on a spine of *Phrissocystis multispina*. E: SEM showing spinelets on a spine of *Aspidodiadema arcitum*. F. SEM showing spinelets on a spine of *Aspidodiadema jacobyi*.

Numerous asteroids are known to brood their young either externally or internally (reviewed by Hyman 1955), though most of these are from shallow temperate and polar waters. Deep-sea members of the family Pterasteridae probably brood their young in an internal nidamental chamber as do some (but not all: cf. Chia 1966; McEdward 1992) shallow-water species (Fisher 1940). Freely spawned sperm are probably passed into the chamber with respiratory currents similar to those described in a nonbrooding species by Nance and Braithwaite (1979). The deep-sea Antarctic brisingid *Odinella nutrix* (Fisher 1940) forms a brood chamber by interlocking spines at the bases of the arms. External fertilization probably occurs even in those asteroid species that retain their embryos. Internal brooding has been documented in at least two deep-sea holothurians, *Oneirophanta mutabilis* (Hansen 1968) and *Ocnus sacculus* (Pawson 1983). Details of fertilization remain completely unknown for all brooding asteroids and holothuroids from the deep sea.

Taking into account all of the species for which reproductive mode has now been described or inferred, it is apparent that external fertilization by free spawning is the rule rather than the exception among deep-sea echinoderms.

Population Densities Required for External Fertilization in the Deep Sea

Nutrients have commonly been assumed to limit population size in the deep sea (reviewed by Rowe 1983; Gage and Tyler 1991). This assumption is supported by data showing an inverse relationship between biomass and depth (Rowe 1971; Lampitt et al. 1986) and between population density and depth (Smith and Hamilton 1983). Many species of macrofauna and meiofauna are represented by single individuals in box core samples, and rarefaction curves often fail to reach asymptotes, indicating that the majority of deep-sea species are rare (Gage and Tyler 1991). On the basis of these observations, one might assume that external fertilization would be problematic for many deep-sea species. However, there have been no attempts to estimate the population size that would be required for successful fertilization in a deep-sea animal. One approach to this problem is to consider natural population densities in light of hydrodynamic models that allow us to predict the likelihood that sperm will encounter eggs under flow regimes characteristic of deep-sea habitats.

Denny (1988) and Denny and Shibata (1989) have advanced a useful time-averaged hydrodynamic model which predicts the downstream distribution and density of gametes on the basis of gamete release rate and various flow parameters. The concentration of sperm at any set of x, y, and z coordinates with respect to a spawning male at coordinates 0, 0, 0 can be predicted with the equation:

$$C_{(x,y,z)} = \frac{Q\bar{u}}{2\pi\alpha_y\alpha_z u_*^2 x^2} - \left[\frac{y^2\bar{u}^2}{2\alpha_y^2 u_*^2 x^2} + \frac{z^2\bar{u}^2}{2\alpha_z^2 u_*^2 x^2}\right]$$

in which Q is the sperm release rate, $\bar{u}$ is the mainstream current velocity, a_x and a_y are directional diffusivity values in the x (downstream) and y (lateral) directions, and u_* is friction velocity.

According to the equation, sperm concentration at a given point should increase with a higher rate of sperm release or with a higher current velocity. Sperm concentration should be reduced by increased bottom roughness which contributes to turbulence (measured as friction velocity, u_*) and by downstream distance. The two squared terms in the denominator have the largest effects in reducing the sperm concentration.

For all echinoid species studied to date, a high level of fertilization is attained only when sperm are above a threshold concentration ranging from 10^2 to 10^5 sperm per milliliter (Vogel et al. 1982). We (Will Jaeckle, Brian Bingham, Paul Tyler, and I) have compared the fertilization thresholds of several bathyal echinoderms from subtropical and temperate seas with those of shallow-water species from the same regions. Some representative examples chosen from larger data sets are shown in figure 9.2. The position of the threshold along the *x* axis varies from species to species, but there is no evidence that deep-water species are capable of successful fertilization at significantly lower sperm concentrations than their shallow-water counterparts. Knowing the approximate range of concentrations required for fertilization, we can use the Denny equation for estimating the size and shape of the region downstream from a spawning male where sperm concentration is adequate.

Several different physical processes drive water currents and determine current velocities in the deep sea. Continental slopes, rises, and nearby abyssal regions are often influenced by boundary currents (e.g., the Gulf Stream and Kuroshio Current), and by the filament currents, meanders and eddies associated with these currents. Coastal regions may also experience currents driven by tides and by density effects associated with upwelling or downwelling. Flows in such regions may be very high. For example, at the HEBBLE site on the Nova Scotia Rise, a maximum current velocity of 73 cm/sec was recorded during a benthic storm, and sustained speeds in excess of 35 cm/sec were recorded for several days (Richardson et al. 1981). Flow velocities near 40 cm/sec have been reported at bathyal depths on the Blake Plateau (Pratt 1963) and underlying the Florida Current (Weatherly 1972). The currents in abyssal troughs and basins (Elliott and Thorpe 1983; Saunders 1988; Thistle and Eckman 1990) and in the centers of major oceanic gyres (Hayes 1980; Warren and Owens 1988; Klein and Mittelstaedt 1992) tend to be much weaker (table 9.3). Flows are known to accelerate over underwater pinnacles, seamounts and ridges (Genin et al. 1986; Mullineaux and Butman 1990). Buoyant plumes emanating from hydrothermal vents entrain surrounding water and cause local currents to flow toward vents (Mullineaux, this volume).

Wimbush (1976) and Wimbush and Munk (1970) have suggested as a generalization that values of both current velocity and friction velocity are about one order of magnitude lower on abyssal plains ($\bar{u}$ = 3 cm/sec, u_* = 0.1 cm/sec) than on continental shelves ($\bar{u}$ = 30 cm/sec; u_* = 1.0 cm/sec). The same flow parameters are about 1 order of magnitude higher on some wave-swept rocky shores and shallow subtidal reefs (e.g., $\bar{u}$ = 110 cm/sec and $\bar{u}$ = 10 cm/sec; Denny 1988) than at greater depths on continental shelves. For the purpose of illustration, I will use these generalized values in Denny's equation. However, it should be kept in mind that current velocities vary both spatially and temporally, so the generalizations will have many exceptions. For the other flow parameters in the equation (e.g., directional diffusivity values, sperm release rates), I use values suggested by Denny (1988). One potential source of

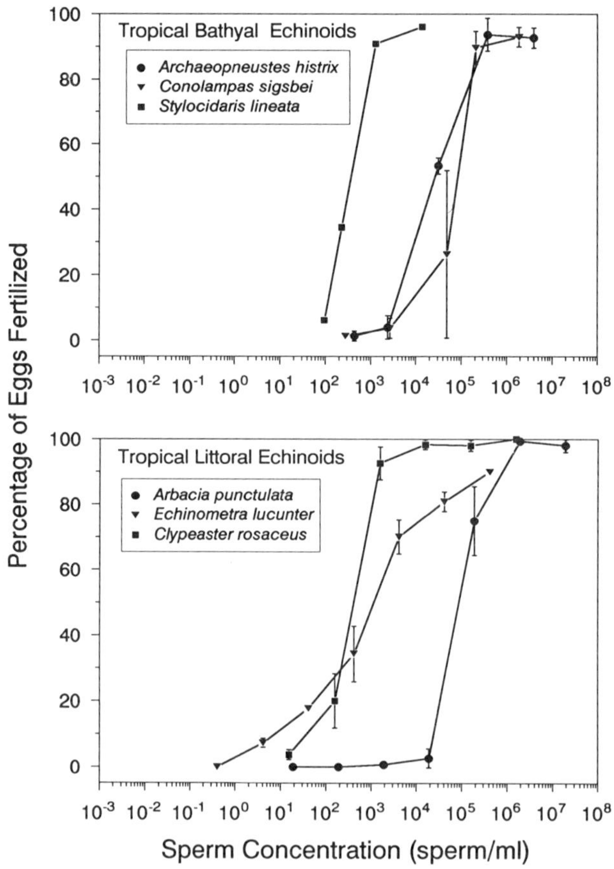

FIGURE 9.2. Percentage of eggs fertilized as a function of sperm concentration for representative tropical bathyal echinoids (upper panel) and tropical echinoids from shallow water (lower panel). With the exception of unreplicated data on *Stylocidaris lineata* (after Young et al. 1992), each point represents mean of 3 samples with standard deviation.

TABLE 9.3 *Examples of Current Speeds in the Deep-Sea Benthic Boundary Layer*

Location	Depth (m)	Current Speeds (cm/s)	Reference
San Diego Trough	1050	0–10	Thistle and Eckman 1990
Gay Head-Bermuda Transect	1800	5–25	Grassle et al. 1975
Florida Current	780	0–38	Weatherly 1971
Eastern Tropical Pacific	5200	0–15	Hayes 1980
Blake Plateau	870	0–41	Pratt 1963
Madeira Abyssal Plain	5300	0–7.7	Elliott and Thorpe 1983
Cross Seamount	410	9–25	Mullineaux and Butman 1990
North Atlantic	3584	7.5–14	Knauss 1965
North Atlantic	5337	0–2	Knauss 1965
Northeast Atlantic	4500	3–5	Klein and Mittelstaedt 1992
Subarctic Pacific	4000	1.5	Warren and Owens 1988
Madeira Abyssal Plain	5000	1–2	Saunders 1988
Central North Pacific	4000	4.7–7.1	Schmitz 1987
Tongue of the Ocean, Bahamas	490	0–18	Young et al. 1993

error is in the estimation of sperm release rates. In the model, it is assumed that sperm are released at some constant rate into the water column, where they disperse as though they were water molecules. However, the sperm of some deep-water echinoids pool on the aboral surface of the spawning male (figure 9.3A). The rate at which sperm are removed from the adult is probably a function of bottom shear, which will vary with current speed.

Sperm concentrations directly downstream from a spawning male sea urchin in "typical" shelf, deep-sea, and intertidal habitats are shown in figure 9.4. The model predicts that sperm may remain at a high enough concentration to fertilize eggs more than 7 meters away from a spawning male in the deep sea, but only for about 2.5 meters on the shelf and less than 1 meter in a shallow intertidal habitat. The dramatic difference is largely a consequence of differing bottom roughness. On a smooth deep-sea bottom, low turbulent diffusion causes sperm clouds to spread laterally and vertically at a much slower rate than on a rough intertidal bottom. In other words, where larger roughness elements cause more drag, larger turbulent eddies are formed, causing gametes to disperse more quickly by turbulent diffusion.

Without considering other aspects of the physical environment or of the biology of these animals, we might predict on the basis of the results in figure 9.4 that external fertilization should be more successful in the deep sea than in shallow water. If we also assume that fertilization is more successful among brooders than free spawners (this seems likely, as brooders either fertilize internally or free-spawn only the male gametes), we might further predict that brooding should be more common in the intertidal zone than in the deep sea!

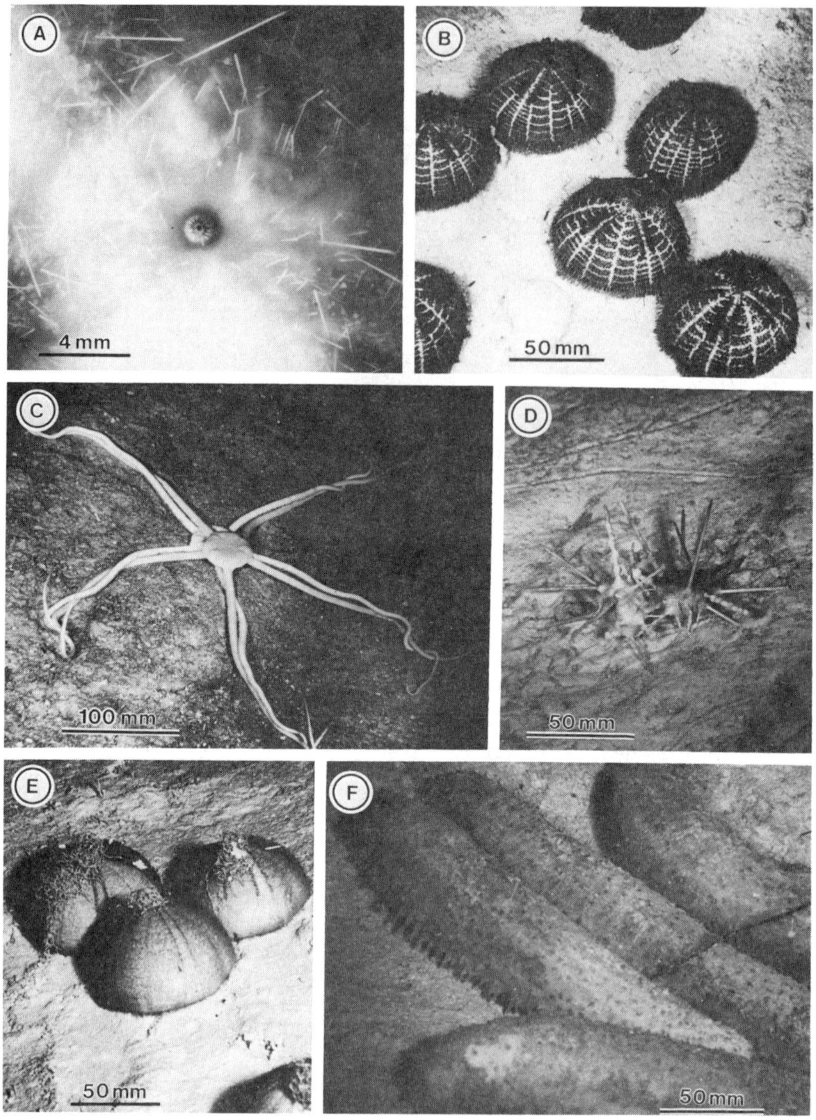

FIGURE 9.3. A: Masses of sperm resting on the aboral surface of a spawning echinothuriid sea urchin, *Araeosoma fenestratum*. B: A herd of the bathyal echinoid *Palaeopneustes cristatus* on soft substratum in the Bahamas. C: Reproductive pair of unidentified ophiuroids in the Bahamas. D: Pair of *Stylocidaris lineata* during the reproductive season at about 600 m depth in the Bahamas. E: Part of a herd of *Conolampas sigsbei* on the Bahamian slope. F: Four individuals of the hermaphroditic holothurian *Paroriza pallens* moving together across the bottom at 918 m depth in the Bahamas.

This pattern, which is the opposite of Thorson's Rule (1950), appears to hold for at least asteroids and holothuroids. Lawrence (1987) in reviewing the literature on brooding asteroids mentioned 8 asteriids, 2 asterinids, and 1 echinasterid, all from shallow water, which brood their young. Pearse (this volume) adds five additional species from shallow polar and temperate habitats. By contrast, no brooders are known from the Rockall Trough fauna, which comprises 13 asteroid species, and only a handful of deep-sea brooding asteroids are known from other regions (Hyman 1955). Only two abyssal holothurians, *Ocnus sacculus* (Pawson 1983) and *Oneirophanta mutabilis* are known brooders (Hansen 1968), but more than 40 species of shallow-water apodans and dendrochirotes have adopted a brooding strategy (Thompson 1878; Hyman 1955; Lawrence 1987; Smiley et al. 1991).

The two-dimensional sperm concentration calculated for a "typical" abyssal habitat is shown in figure 9.5A. If we place contour lines (isopleths of equal sperm concentration) on this diagram then rotate to view it from directly below or above, we can visualize the shape of the region where sperm concentration is high enough to fertilize an egg (figure 9.5B). This region (where sperm concentration exceeds 10^2 sperm per ml) takes the shape of an elongate ellipse ("flame-shaped," in Denny's 1988 terminology) having a length of about 5 meters. How likely is a sperm cloud of this size and shape to encounter eggs in an abyssal habitat?

One of the most comprehensive data sets on the distribution and abundance of deep-sea echinoids comes from the study of bathyal species in the northwest Atlantic Ocean by Grassle et al. (1975). I have reconstructed a randomly

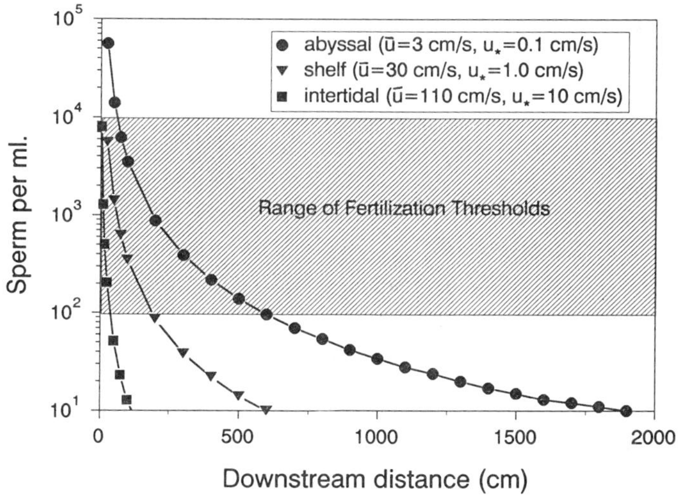

FIGURE 9.4. Sperm concentration as a function of distance from a single spawning male sea urchin in "typical" abyssal, shelf, and intertidal habitats, as predicted by Denny's (1988) model. The approximate range of sperm concentrations required for high fertilization in sea urchins is shaded.

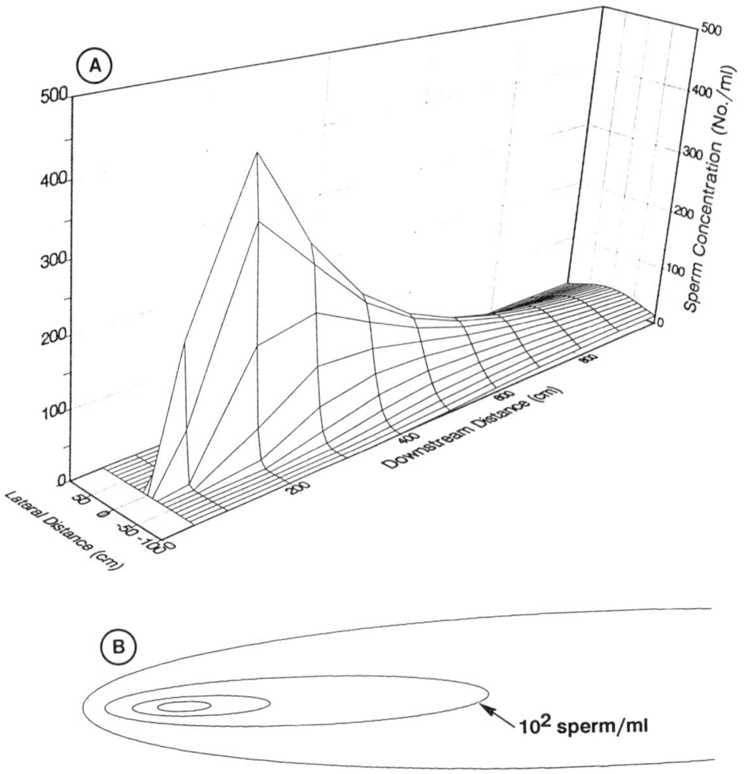

FIGURE 9.5. A: Two-dimensional plot of the sperm concentrations at a height 10 cm off the bottom downstream from a spawning sea urchin at abyssal depths. B: Isopleths of sperm concentration obtained by viewing "A" from below.

dispersed, intermediate density population of *Echinus affinus* from Alvin Dive 436 (1796–1831 m) by assigning the appropriate number of individuals to random Cartesian coordinates. Half of the individuals are randomly assigned to each sex. The positions of all individuals are shown in figure 9.6A. The elliptical portion of the sperm plume wherein more than 50 percent of eggs are likely to be fertilized (i.e., the threshold at which sperm concentration falls below 10^2 sperm per ml) is plotted as a 2-dimensional "cloud" extending downstream from each male, and an egg plume of similar dimensions, representing a high (but unknown) concentration of eggs, is plotted downstream from each female. If we assume synchronous spawning by the entire population the high concentration regions of the majority of individuals overlap one or more of those of other individuals in the population (figure 9.7). Wherever egg and sperm clouds overlap, fertilization is likely to be successful. Where sperm clouds overlap each other, the total sperm concentration increases, thereby increasing the effective sperm release rate and extending downstream the distance where concentration remains high enough for fertilization.

Although figure 9.6A is a reasonable representation of the regions where sperm will be effective, similar thresholds do not apply to eggs. A single,

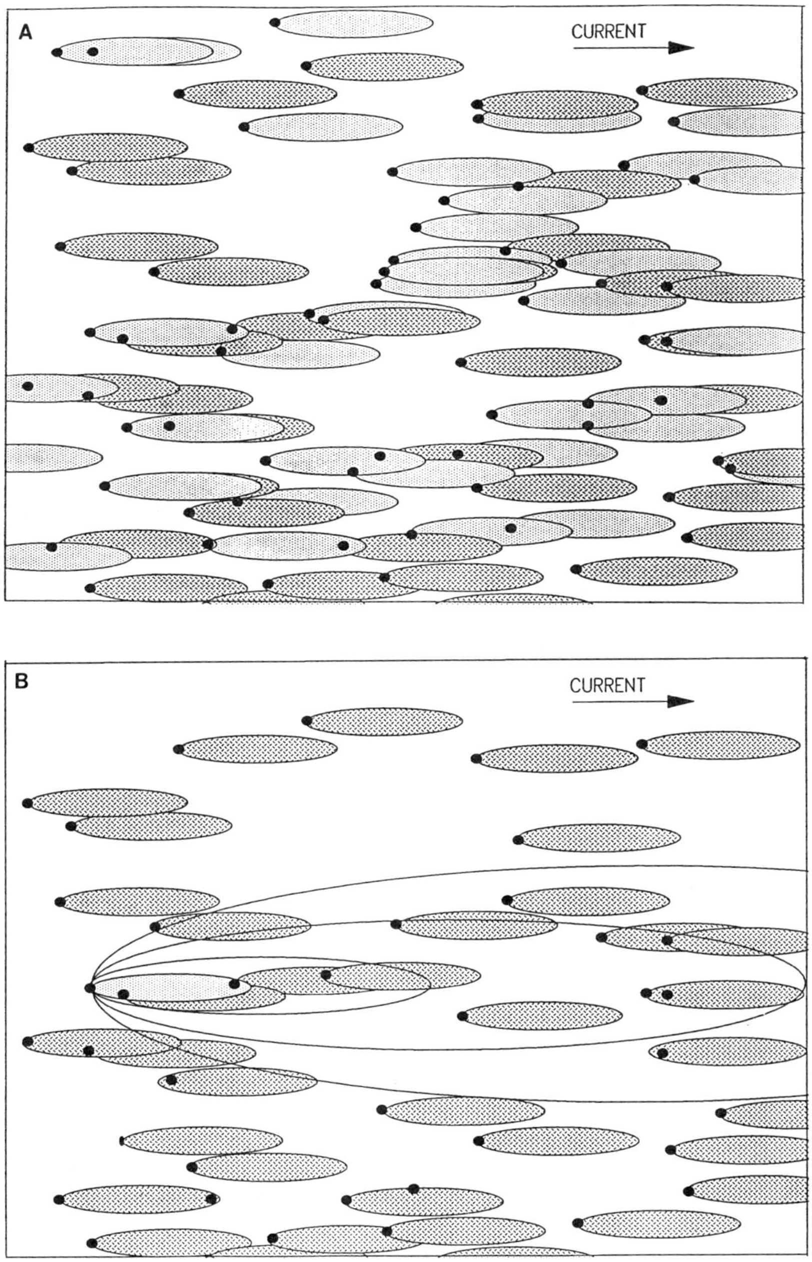

FIGURE 9.6. A: A reconstructed population of *Echinus affinus* from Alvin Dive 436 (Grassle et al. 1975) spawning synchronously and showing the extent of overlap in gamete clouds. The sperm clouds (hatched) represent the threshold sizes and shapes obtained from figure 9.5B. Egg clouds (stippled) are plotted the same shape for consistency, even though there is no threshold egg concentration required for fertilization. B: The gamete plume of a single representative female urchin superimposed on the sperm clouds of males from "A."

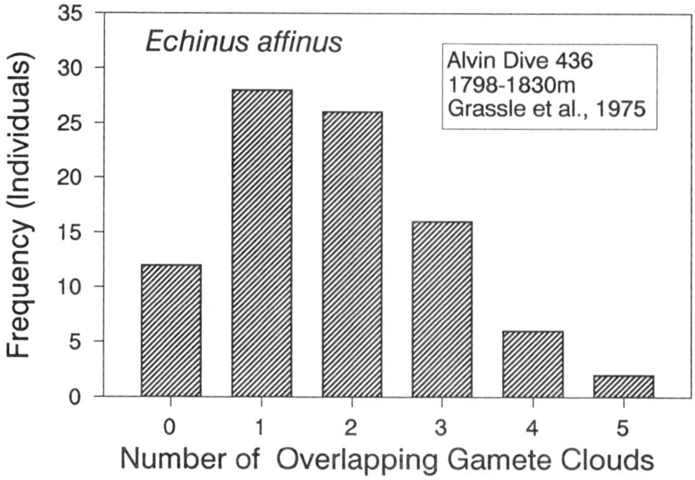

FIGURE 9.7. The frequency of individual gamete plumes in figure 9.6A that overlap various numbers of adjacent gamete clouds.

isolated egg can be fertilized as long as it occurs in a region of high enough sperm density; sperm concentration is the limiting factor, and the number of surrounding eggs is irrelevant. For this reason, the egg plume is more accurately represented as a series of concentric ellipses of increasing area, each representing a region of lower egg concentration. A single egg cloud is plotted in this manner in figure 9.6B. Even at only low current speeds, eggs should be able to drift for a considerable distance before losing viability. In the population illustrated here, most eggs are likely to encounter a high-density sperm plume before drifting very far downstream. Several field experiments on fertilization success (Pennington 1985; Yund 1990; Levitan 1991; Levitan et al. 1992) have examined the fertilization rate of eggs as a function of downstream distance from a spawning male. All show that fertilization is a very uncertain event for eggs held at any particular point in space. However, the few available estimates of population-wide fertilization rates following natural spawnings (Oliver and Willis 1987; Sewell and Levitan 1992) indicate that eggs allowed to drift freely have a much higher chance of becoming inseminated. In the recent study of natural fertilization rates in *Cucumaria miniata* by Sewell and Levitan (1992), more than 90 percent of naturally spawned eggs were found to be fertilized. The simple model with *Echinus affinus* presented here suggests that fertilization should be easily accomplished in a moderately dense deep-sea population as long as spawning is synchronous.

Dealing with Isolation

Not all deep-sea populations occur at densities as high as those of *Echinus affinus,* nor are their spatial patterns always random. Indeed, as in animal populations everywhere, individuals in the deep sea are virtually always aggregated on some spatial scale. Even where the overall population density is low, aggregation may facilitate encounters between males and females, as long as aggregation occurs on a sufficiently small scale.

Some deep-sea organisms occur at such low densities that fertilization must be a very rare event indeed. For example, I have often encountered only one or two individuals of certain starfish species during submersible dives in which several kilometers of bottom were surveyed. Even if we assume good temporal synchrony of spawning, it seems unlikely that some of these populations will produce any offspring unless there are mechanisms for locally enhancing the population density at the time of breeding.

Some shallow-water echinoderms aggregate during the spawning season, presumably to maximize zygote production (Orton 1914; Boschma 1924; Dix 1969; Komatsu 1983; Run et al. 1988). Numerous bathyal and abyssal echinoderms have been observed in dense aggregations (Grassle et al. 1975; Billett and Hansen 1982; Rice et al. 1982) or small, discrete groups (Pawson 1976, Fujioti et al. 1987). Although some workers have suggested that such aggregations form to facilitate breeding (Pawson 1976), others have interpreted them as feeding aggregations (Grassle et al. 1975). These hypotheses are not mutually exclusive; we have shown experimentally that at least one species of sea urchin, *Stylocidaris lineata,* aggregates at macrophyte detrital falls throughout the year (Young et al. 1993), and also aggregates for breeding during the reproductive season (Young et al. 1992).

Aggregating behavior varies substantially among species of echinoids. At bathyal depths in the Bahamas, some seasonally reproducing species, including the spatangoids *Paleobrissus cristatus* (figure 9.3B) and *Archaeopneustes histrix* (Young 1992), the cassiduloid *Conolampas sigsbei* (figure 9.3E) and the arbaciid *Coelopleurus floridanus* (Young 1992) move about in aggregations during all seasons of the year. A typical aggregation of *A. hystrix* may contain several dozen individuals spread over a distance of tens of meters. Individual urchins are spaced a few centimeters to several meters apart, but the distance to the nearest adjacent aggregation may be on the order of hundreds of meters to kilometers.

Other seasonally reproducing species are dispersed more or less randomly during most of the year, but aggregate just prior to the breeding season. The best studied example is the cidaroid *Stylocidaris lineata,* which lives at bathyal depths on the Bahamian slope (Young et al. 1992). In this species, discrete aggregations (often pairs, figure 9.3D) begin to form three to four months before spawning. During the spawning season, individuals in pairs were significantly more ripe than isolated individuals, most of which had spent gonads (Young et al. 1992). In a typical spawning aggregation, spines of adjacent individuals are actually interlocked. Heterosexual and homosexual pairings occur at the frequency that would be expected by chance, indicating that individuals are incapable of discriminating gender. Another Bahamian cidaroid,

Cidaris blakei, demonstrates the same kind of seasonal aggregating behavior (Young and Tyler unpublished). A major advantage of these very tight aggregations is that gametes released by individuals in the clump will disperse along identical axes, thereby maximizing the chances of fertilization. Where two males spawn at the same time, the sperm release rate is effectively doubled, thereby increasing the downstream distance that concentrations remain high enough to successfully fertilize eggs (Young et al. 1992). Apparent reproductive pairing has also been observed in bathyal ophiuroids (figure 9.3C). It is not known if this species broods or broadcasts its eggs. However, in the pair illustrated, both individuals were very ripe and the male was resting on the female with arms intertwined.

Perhaps the most surprising reproductive aggregations have been observed in hermaphroditic holothurians of the genus *Paroriza* (Tyler et al. 1992). Pairs to quadruplets (figure 9.3F) of these animals have been observed in the North Atlantic (Mauviel and Sibuet 1985; Tyler et al. 1992), in the Bahamas (Tyler et al. 1992, figure 9.1F) and in the Philippine Trench (Fujioti et al. 1987). Mauviel and Sibuet (1985) present a remarkable photograph showing tracks of two individuals that pass over a substantial area of the bottom. This and other observations suggest that individuals may remain together for extended periods, perhaps spawning together over an extended reproductive season (Tyler et al. 1992).

The physiological and behavioral mechanisms by which deep-sea echinoderms form aggregations remain uninvestigated. Some shallow-water starfish are known to release pheromones that synchronize spawning (Beach et al. 1975; Miller 1989). Such pheromones could also be used to attract individuals. However, since at least one species, *Stylocidaris lineata*, is apparently incapable of differentiating gender, it seems equally plausible that individuals encounter each other randomly, remaining together if spawning is imminent and going their separate ways if it is not.

Ecological Importance of Gamete Modifications

An unexpectedly high proportion of deep-sea echinoderms display sperm modifications, including gigantism, pigmentation, elongate nuclei, lipid stores in the midpiece, and dimorphism (Eckelbarger et al. 1989a, b, c; Eckelbarger, this volume). These have sometimes been interpreted as adaptations for fertilization in the deep sea, but some may be attributed to ancestry and none are apparently associated with internal fertilization.

The sperm of soft-bodied echinothuriid sea urchins have elongate nuclei and are endowed with large globules of lipid attached to the posterior end of the midpiece (Eckelbarger et al. 1989b). Phospholipids are known to fuel swimming in the sperm of other echinoderms (Mita and Yasumasu 1983) and are metabolized during swimming to the extent that ultrastructural differences between newly spawned and older sperm can be seen (Mita and Nakamura 1992). The sperm of most echinoderms remain motile and viable for only a few hours after they are diluted with seawater (Gemmill 1900; Cohn 1918). However, we have found that diluted echinothuriid sperm remain motile much longer than would be expected for the sperm of any shallow-water species investigated to date (figure 9.8). More than 50 percent of the sperm of *Araeosoma fenestratum* remained active 15 hours after initial dilution and some remained active for up to 25 hours. Even more remarkably, more than 60 percent of the sperm of *Sperosoma antillense* were still swimming actively 30 hours after dilution, at which time observations were stopped. Unfortunately, we have not been successful at fertilizing the large lecithotrophic eggs of either of these species, so the relationship between viability and motility is not specifically known for either species. However, motility is generally considered to be a good indicator of sperm viability. Sperm of the shallow-water echinoid *Arbacia punctulata,* which lack large stores of lipid on the midpiece, lose their motility and their ability to fertilize eggs within 8 hours after being diluted in seawater (figure 9.8) at both 10° and 24°C.

The gametes of three species of the deep-water echinoid genus *Aspidodiadema, A. jacobyi* from Bahamian waters (Eckelbarger et al. 1989a; Young et al. 1989), *Aspidodiadema arcitum* from the Hawaiian slope (Young et al. unpublished), and a species from Japanese waters (Emlet personal communication) have been examined. The sperm of all three have extremely long heads (26 μm in *A. jacobyi*) containing elongate, curved, and pointed nuclei. The function of these elongate sperm remains unknown. However, as noted previously, these urchins spawn both eggs and sperm in masses of mucus which are retained on

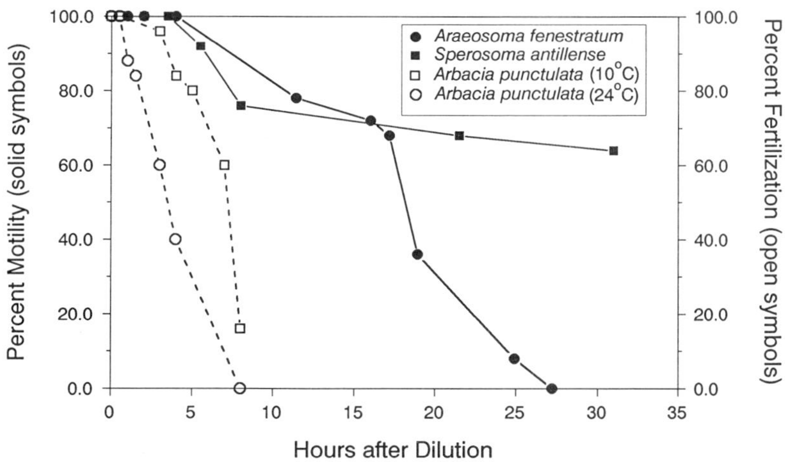

FIGURE 9.8. Percentage of echinothuriid sperm still swimming as a function of time after dilution, and percentage of eggs fertilized by sperm of *Arbacia punctulata* incubated at two temperatures, as a function of time after dilution.

the spines of the adults. In seawater, the sperm swim in tight circles in the direction of the head curvature, presumably because the head acts as an anterior rudder. It seems likely that these very long and unusual sperm may be adapted for locomotion in the viscous mucus in which the eggs are initially embedded.

Males of the abyssal echinoid *Phrissocystis multispina* produce dimorphic spermatozoa including an euspermatozoon typical of echinoids and a paraspermatozoon which is bipolar-tailed. Superficially, the paraspermatozoon resemble the sperm of *Aspidodiadema*. However, in this case, the curved anterior process is not formed of nuclear material, but of the anteriorly directed portion of the sperm tail. This is the only species of echinoderm known to produce dimorphic sperm, and the function of two sperm types remains unknown. We have speculated elsewhere (Eckelbarger et al. 1989c) that the bipolar-tailed sperm may bind the functional sperm into packets, thereby increasing the distances which sperm can disperse without having their concentration reduced by turbulent diffusion. However, recent observations (Young, Tyler, and Emson, unpublished) of the living biflagellate sperm indicate that they do not form sperm aggregations and that the bipolar-tailed sperm swim by beating the posterior flagellum in much the same way as typical sperm. The anterior portion of the flagellum does not beat; it is held stationary, pointing forward, and causes the sperm to swim in circles, like the sperm of *Aspidodiadema*. Perhaps *Phrissocystis* also spawns gametes in mucus and retains them on the spines. The spines have spinelets similar to those of the three species known to have mucus-bound gametes (figure 9.1D). Unfortunately, gametes of *Phrissocystis* have only been obtained by dissection, so the nature of the spawning process is not known. The function of these unusual dimorphic sperm will remain a mystery until the fertilization process is observed in spawned, living gametes.

Acknowledgments

Many of the results and ideas in this paper originated in discussions with colleagues on cruises, especially P. Tyler, D. Levitan, B. Bingham, W. Jaeckle, L. Cameron, and K. Eckelbarger. I thank the crews of RRS Challenger, RV Seward Johnson, RV Edwin Link, and RV Kila as well as the pilots and technicians associated with the Johnson-Sea-Link and Pisces V submersibles for their professional assistance. This research was supported by NSF grants OCE-8717922 and OCE-8916264, by grants from the NOAA National Undersea Research Program, and by NATO Collaborative Research Grant CRG-900628. Harbor Branch Contribution No. 1010.

References

Beach, D. H., N. J. Hanscomb, and R. F. G. Ormond. 1975. Spawning pheromone in crown-of-thorns starfish. *Nature, London*. 254:135–136.

Billett, D. S. M. and B. Hansen. 1982. Abyssal aggregations of *Kolga hyalina* (Echino-

Boschma, H. 1924. Uber einen Fall von Kopulation bei einer Asteridae (*Archaster typicus*). *Zool. Anz.* 58:283–285.
Brazeau, D. A. and H. R. Lasker. 1990. Sexual reproduction and external brooding by the Caribbean gorgonian *Briareum asbestinum. Mar. Biol.* 104:465–474.
Chia, F. S. 1966. Development of a deep-sea cushion star, *Pteraster tesselatus. Proc. California Acad. Sci.* Ser. 4 34:505–510.
Cohn, E. J. 1918. Studies in the physiology of spermatozoa. *Biol. Bull.* 34:167–218.
David, B. and R. Mooi. 1990. An echinoid that "gives birth": Morphology and systematics of a new Antarctic species, *Urechinus mortenseni* (Echinodermata, Holasteroida). *Zoomorphology* 110:75–89.
Denny, M. W. 1988. Biology and the mechanics of the wave-swept environment. Princeton: Princeton University Press.
Denny, M. W. and M. F. Shibata. 1989. Consequences of surf-zone turbulence for settlement and external fertilization. *Am. Nat.* 134:859–889.
Dix, T. G. 1969. Aggregating in the echinoid *Evechinus chloroticus. Pac. Sci.* 23:123–124.
Eckelbarger, K. J., C. M. Young, and J. L. Cameron. 1989a. Modified sperm in echinoderms from the bathyal and abyssal zones of the deep sea. In J. S. Ryland and P. A. Tyler, eds., *Reproduction, Genetics, and Distributions of Marine Organisms,* pp. 67–74. Fredensborg, Denmark: Olsen, and Olsen.
Eckelbarger, K. J., C. M. Young, and J. L. Cameron. 1989b. Modified sperm ultrastructure in four species of soft-bodied echinoids (Echinodermata: Echinothuriidae) from the bathyal zone of the deep sea. *Biol. Bull.* 177:230–236.
Eckelbarger, K. J., C. M. Young, and J. L. Cameron. 1989c. Ultrastructure and development of dimorphic sperm in the abyssal echinoid *Phrissocystis multispina* (Echinodermata: Echinoidea): Implications for deep sea reproductive biology. *Biol. Bull.* 176:257–271.
Elliott, A. J. and S. A. Thorpe. 1983. Benthic observations on the Madeira abyssal plain. *Oceanologica Acta* 6:463–466.
Emlet, R. B., L. R. McEdward, and R. R. Strathmann. 1987. Echinoderm larval ecology viewed from the egg. In M. Jangoux and J. M. Lawrence, eds., *Echinoderm Studies 2,* pp. 55–136. Rotterdam: A. A. Balkema.
Fisher, W. K. 1940. Asteroidea. *Discovery Reports* 20:69–306.
Fujioti, K., A. Taira, K. Kobayeshi, K. Nakamura, T. Iiyama, J.-P. Cadet, S. Lallemand, and D. Girard, eds. 1987. *6000 m deep: A Trip to the Japanese Trenches.* Tokyo: University of Tokyo Press.
Gage, J. D. and P. A. Tyler. 1991. *Deep-Sea Biology: A Natural History of Organisms at the Deep-Sea Floor.* Cambridge: Cambridge University Press.
Gemmill, J. F. 1900. On the vitality of ova and spermatozoa of certain animals. *J. Anat. Physiol.* 34:163–181.
Genin, A., P. K. Dayton, P. F. Lonsdale, and F. N. Spiess. 1986. Corals on seamount peaks provide evidence of current acceleration over deep-sea topography. *Nature* 322:59–61.
Giese, A. C. and H. Kanatani. 1987. Maturation and spawning. In A. C. Giese, J. S. Pearse, and V. B. Pearse, eds., *Reproduction of Marine Invertebrates.* Vol. 9: *General Aspects: Seeking Unity in Diversity,* pp. 251–329. Pacific Grove: Blackwell; Palo Alto: Boxwood.
Grassle, J. F., H. L. Sanders, R. R. Hessler, G. T. Rowe, and T. McLellan. 1975. Pattern and zonation: A study of the bathyal megafauna using the research submersible Alvin. *Deep-Sea Res.* 22:457–481.
Hansen, B. 1968. Brood-protection in a deep-sea holothurian, *Oneirophanta mutabilis* Theel. *Nature* 217:1062–1063.
Hayes, S. P. 1980. The bottom boundary layer in the Eastern Tropical Pacific. *J. Phys. Oceanogr.* 10:315–329.
Healy, J. M., F. W. E. Rowe, and H. E. S. Clark. 1988. Spermatozoa and spermiogenesis in

Xyloplax (Class Concentricycloidea): A new type of spermatozoon in the Echinodermata. *Zool. Scripta* 17:297–310.

Hyman, L. H. 1955. *The Invertebrates.* Vol. 4: *Echinodermata.* New York: McGraw-Hill.

Klein, H. and E. Mittelstaedt. 1992. Currents and dispersion in the abyssal northeast Atlantic: Results from the NOAMP field program. *Deep-Sea Res.* 39:1727–1745.

Knauss, J. A. 1965. A technique for measuring deep ocean currents close to the bottom with an unattached current meter, and some preliminary results. *J. Mar. Res.* 23:237–245.

Komatsu, M. 1983. Development of the sea star, *Archaster typicus,* with a note on male-on-female superposition. *Annot. Zool. Japon.* 56:187–195.

Lampitt, R. S., D. S. M. Billett, and A. L. Rice. 1986. Biomass of the invertebrate megabenthos from 500 to 4100 m in the northeast Atlantic Ocean. *Mar. Biol.* 93:69–81.

Lawrence, J. M. 1987. *A Functional Biology of Echinoderms.* Baltimore: Johns Hopkins University Press.

Levitan, D. R. 1991. Influence of body size and population density on fertilization success and reproductive output in a free-spawning invertebrate. *Biol. Bull.* 181:261–268.

Levitan, D. R., M. A. Sewell, and F. S. Chia. 1992. How distribution and abundance influence fertilization success in the sea urchin *Strongylocentrotus franciscanus. Ecology* 73:248–254.

Mauviel, A. and M. Sibuet. 1985. Répartition des traces animales et importance de la bioturbation. In L. Laubier and C. Monniot, eds., *Peuplements profunds du Golfe de Gascogne: Campagnes BIOGAS,* pp. 157–173. IFREMER, Brest.

McEdward, L. R. 1992. Morphology and development of a unique type of pelagic larva in the starfish *Pteraster tesselatus* (Echinodermata: Asteroidea). *Biol. Bull.* 182:177–187.

Miller, R. L. 1989. Evidence for the presence of sexual pheromones in free-spawning starfish. *J. Exp. Mar. Biol. Ecol.* 130:205–221.

Mita, M. and M. Nakamura. 1992. Ultrastructural study of an endogenous energy substrate in spermatozoa of the sea urchin *Hemicentrotus pulcherrimus. Biol. Bull.* 182:298–304.

Mita, M. and I. Yasumasu. 1983. Metabolism of lipid and carbohydrate in sea urchin spermatozoa. *Gamete Res.* 7:133–144.

Mortensen, T. 1914. Development of some Japanese echinoderms. *Annot. Zool. Jap.* 8:543–552.

Mullineaux, L. S. and C. A. Butman. 1990. Recruitment of encrusting benthic invertebrates in boundary-layer flows: A deep-water experiment on Cross Seamount. *Limnol. Oceanogr.* 35:409–423.

Nance, J. M. and L. F. Braithwaite. 1979. The function of mucous secretions in the cushion star *Pteraster tesselatus* Ives. *J. Exp. Mar. Biol. Ecol.* 40:259–266.

Oliver, J. K. and B. L. Willis. 1987. Coral-spawn slicks in the Great Barrier Reef: Preliminary observations. *Mar. Biol.* 94:521–529.

Orton, J. H. 1914. On the breeding habits of *Echinus miliaris,* with a note on the feeding habits of *Patella vulgata. J. Mar. Biol. Assoc. U. K.* 10:254–257.

Pawson, D. L. 1976. Some aspects of the biology of deep-sea echinoderms. *Thal. Jugosl.* 12:287–293.

Pawson, D. L. 1983. *Ocnus sacculus* new species (Echinodermata: Holothuroidea), a brood-protecting holothurian from southeastern New Zealand. *New Zeal. J. Mar. Freshwater Res.* 17:227–230.

Pawson, D. L. and J. E. Miller. 1979. Secondary sex characters in *Coelopleurus floridanus* A. Agassiz, 1872 (Echinodermata: Echinoidea). *Bull. Mar. Sci.* 29:581–586.

Pennington, J. T. 1985. The ecology of fertilization of echinoid eggs: the consequences of sperm dilution, adult aggregation, and synchronous spawning. *Biol. Bull.* 169:417–430.

Pratt, R. M. 1963. Bottom currents on the Blake Plateau. *Deep-Sea Res.* 10:245–249.

Rice, A. L., R. G. Alred, E. Darlington, and R. A. Wild. 1982. The quantitative estimation of the deep-sea megabenthos: A new approach to an old problem. *Oceanol. Acta* 5:63–72.

Richardson, M. J., M. Wimbush, and L. Mayer. 1981. Exceptionally strong near-bottom flows on the continental rise of Nova Scotia. *Science* 213:887–888.

Rowe, G. T. 1971. Benthic biomass and surface productivity. In J. D. Costlow, ed., *Fertility of the Sea,* vol. 2. New York: Gordon and Breach.

Rowe, G. T. 1983. Biomass and production of the deep-sea macrobenthos. In G. T. Rowe, ed., *The Sea,* 8:97–121. New York: Wiley Interscience.

Run, J.-Q., C.-P. Chen, K.-H. Chang, and F.-S. Chia. 1988. Mating behavior and reproductive cycle of *Archaster typicus* (Echinodermata: Asteroidea). *Mar. Biol.* 99:247–253.

Saunders, P. M. 1988. Bottom currents near a small hill on the Madeira Abyssal Plain. *J. Phys. Oceanogr.* 18:868–879.

Schmitz, W. J. Jr. 1987. Observations of new, large, and stable abyssal currents at midlatitudes along 165°E. *J. Phys. Oceanogr.* 17:1309–1315.

Sewell, M. A. and D. R. Levitan. 1992. Fertilization success during a natural spawning of the dendrochirote sea cucumber *Cucumaria miniata. Bull. Mar. Sci.* 5:161–166.

Smiley, S., F. S. McEuen, C. Chaffee, and S. Krishnan. 1991. Echinodermata: Holothuroidea. In A. C. Giese, J. S. Pearse, and V. B. Pearse, eds., *Reproduction of Marine Invertebrates,* vol. 6: *Echinoderms and Lophophorates,* pp. 663–750. Pacific Grove, Calif.: Boxwood Press.

Smith, C. R. and S. C. Hamilton. 1983. Epibenthic megafauna of a bathyal basin off southern California: Patterns of abundance, biomass and dispersion. *Deep-Sea Res.* 30A:907–928.

Thistle, D. and J. E. Eckman. 1990. The effect of a biologically produced structure on the benthic copepods of a deep-sea site. *Deep-Sea Res.* 37:541–554.

Thomson, C. W. 1878. Notice of some peculiarities in the mode of propagation of certain echinoderms of the Southern Sea. *J. Linn. Soc. Zool. (London)* 13:55–79.

Thorson, G. 1950. Reproductive and larval biology of marine bottom invertebrates. *Biol. Rev.* 25:1–45.

Thorson, G. 1966. Some factors influencing the recruitment and establishment of marine benthic communities. *Neth. J. Sea Res.* 3:267–293.

Tyler, P. A. 1988. Seasonality in the deep sea. *Oceanogr. Mar. Biol. Ann. Rev.* 26:227–258.

Tyler, P. A., D. S. M. Billett, C. M. Young, and L. A. Giles. 1992. Reproductive biology and behaviour in the deep-sea holothurian genus *Paroriza* (Holothuroidea: Synallactidae. *J. Mar. Biol. Assoc. U. K.* 72:447–462.

Van Dover, C. L. In press. *In situ* spawning of hydrothermal vent tubeworms *Riftia. Biol. Bull.*

Vogel, H., G. Czihak, P. Chang, and W. Wolf. 1982. Fertilization kinetics of sea urchin eggs. *Math. Biosci.* 189–216.

Warren, B. A. and W. B. Owens. 1988. Deep currents in the central subarctic Pacific Ocean. *J. Phys. Oceanogr.* 18:529–551.

Weatherly, G. L. 1972. A study of the bottom boundary layer of the Florida Current. *J. Phys. Oceanogr.* 2:54–72.

Wimbush, M. 1976. The physics of the benthic boundary layer. In I. N. McCare, ed., *The Benthic Boundary Layer,* pp. 3–10. New York: Plenum.

Wimbush, M. and W. Munk. 1970. The benthic boundary layer. In A. E. Maxwell, ed., *The Sea,* pp. 731–758. New York: Wiley-Interscience.

Young, C. M. 1992. Episodic recruitment and cohort dominance in echinoid populations at bathyal depths. In G. Colombo, I. Ferrari, V. U. Ceccherelli, and R. Rossi, eds., *Marine Eutrophication and Population Dynamics,* pp. 239–246. Fredensborg, Denmark: Olsen and Olsen.

Young, C. M., J. L. Cameron, and K. J. Eckelbarger. 1989. Extended pre-feeding period in a planktotrophic echinoid larva from the bathyal zone of the deep sea. *J. Mar. Biol. Ass. U. K.* 69:695–702.

Young, C. M., P. A. Tyler, J. L. Cameron, and S. G. Rumrill. 1992. Seasonal breeding aggregations in low-density populations of the bathyal echinoid *Stylocidaris lineata*. *Mar. Biol.* 113:603–612.

Young, C. M., P. A. Tyler, R. H. Emson, and J. D. Gage. 1993. Perception and selection of macrophyte detrital falls by the bathyal echinoid *Stylocidaris lineata*. *Deep-Sea Res.* 40:1475–1486.

Yund, P. O. 1990. An *in situ* measurement of sperm dispersal in a colonial marine hydroid. *J. Exp. Zool.* 253:102–106.

CHAPTER TEN

Implications of Mesoscale Flows for Dispersal of Deep-Sea Larvae

Lauren S. Mullineaux

Abstract

Dispersal of larvae between isolated benthic habitats in the deep sea may be strongly influenced by mesoscale flows. For instance, mesoscale flow features near seamounts, such as eddies and Taylor caps, are capable of retaining water parcels for periods of weeks to months. These flow features are likely to retain and accumulate larvae spawned by the seamount benthos, increasing the probability that they recolonize their source habitats. When topographic flow features become unstable and advect away from the seamount, they may transport the larvae to remote habitats as discrete concentrated patches. At hydrothermal vents hot water released at the seafloor forms buoyant plumes that entrain large volumes of near-bottom water. Zooplankton, including larvae of benthic invertebrates, may thereby be transported 10s of kilometers away from the source vent communities. If the vertical velocity of the buoyant plume is sufficiently strong, it may form vortices, retaining vent fluids and their associated fauna in the vicinity of the vent. These vortices, and the benthic larvae entrained into them, may then advect away as discrete patches. Thus, dispersal of larvae from vents and seamounts may occur infrequently, but in concentrated patches, causing episodic recruitment into remote habitats.

Sediment covers most of the deep seafloor, but rocky outcrops are common on mid-ocean ridges and seamounts in all ocean basins. These rocky habitats often support communities dominated by large, suspension-feeding organisms quite different than those found on the sediment-covered seafloor. Many of these organisms cannot disperse long distances as adults, and so populations living on individual ridge segments or on seamounts are effectively isolated from each other by expanses of uninhabitable seafloor. The isolation of rocky habitats in the deep sea requires that dispersal between benthic populations occur by dispersal of larval or juvenile forms through the water column.

Dispersal between isolated habitats in the deep sea has received much less attention than dispersal between isolated islands or isolated terrestrial habitats. Nevertheless, similar ecological questions and approaches, such as those derived from theory of island biogeography (MacArthur and Wilson 1967) can be applied in all these habitats. For instance, habitat area, distance from source populations and immigration rates are likely to play important roles in determining the diversity and species composition of communities living on deep-water seamounts or ridges. In addition, the genetic makeup of populations in isolated deep-water habitats are likely to be strongly affected by larval dispersal rates, and consequent gene flow. While recent studies of the isolated endemic communities at hydrothermal vents have begun to explore these genetic approaches (e.g., Vrijenhoek et al. 1991), very little is known about how benthic invertebrates disperse in the deep sea.

The purpose of this paper is to examine mesoscale hydrodynamic processes (on scales of kilometers to 100s of kilometers) that could potentially control larval dispersal between isolated habitats in the deep sea. Although reproductive characteristics of adults and behaviors of larvae are unquestionably important in dispersal, many fundamental questions about dispersal mechanisms can be addressed by investigating the dynamics of water parcels carrying the dispersing larvae. This approach, using readily available oceanographic information and models of flow dynamics to predict dispersal trajectories and episodicities, allows us to address large-scale processes that are not presently tractable from studies of larval morphology or behavior.

An assumption of this approach is that a larva is advected with its surrounding water parcel. In order to completely satisfy this assumption, larvae must be neutrally buoyant, swim nondirectionally, and swim slowly relative to oceanic flows. Although these assumptions may not be suitable for large crustacean larvae or for many shallow-water larvae with known tactic responses to light, gravity, and chemical gradients, they are likely to be reasonable for small, weakly swimming larvae, particularly those living in deep water where light is not available as an orientation cue and chemical gradients are generally weak. Mesoscale flow features are large relative to individual larvae and are likely to have a strong influence on dispersal between isolated habitats. Note that not all deep-sea larvae are necessarily expected to disperse passively, and observed deviations of individual species' larval dispersal patterns from those predicted from mesoscale flow dynamics will provide a focus for future species-specific studies of larval behavior. Hydrodynamic approaches have been used to investigate larval transport in small-scale boundary-layer flows (e.g., Mullineaux and Butman 1991) and in shallow-water environments (Sammarco and Andrews 1988), but the use of mesoscale flow dynamics to investigate dispersal between deep-ocean habitats is yet to be explored.

Only a few deep-sea larvae have been recorded from past examinations of plankton samples (e.g., Wishner 1980) and they appear to be very rare in the open ocean. Some have suggested that this rarity is due, in part, to low abundances and fecundities of benthic adults, and propose that a large proportion of deep-sea taxa exhibit direct development or release demersal larvae (Wilson and Hessler 1987); conversely, others have explained rarity by showing that some species rise to the surface for a portion of their life cycle (Bouchet and Warén 1979; Killingley and Rex 1985). The ideas presented in this paper pose an alternative mechanism for dispersal that could also explain the observed rarity of larvae in deep-ocean waters. Near-bottom mesoscale flow features may accumulate larvae and transport them, as concentrated patches, to distant habitats. Sampling schemes that do not sample these mesoscale patches effectively would likely result in the conclusion that larvae do not disperse in the water column. Thus, one consequence of dispersal occurring in concentrated mesoscale patches is that most deep ocean waters would appear devoid of larvae. A second consequence is that dispersal and colonization would occur as an episodic, but possibly predictable, process rather than a steady, advection-diffusion process.

Seamounts and hydrothermal vents are similar in that they are isolated by expanses of sediment-covered seafloor, but differ greatly in the taxonomic and functional characteristics of the associated faunas. Most seamount habitats remain relatively unchanged on ecological time scales, so dispersal may be less critical for maintaining populations on ecologic time scales than on evolutionary ones. Hydrothermal vent habitats, however, may be ephemeral on time scales as short as 10s to 100s of years, requiring that endemic vent populations disperse and recolonize active vents frequently. The consequences of dispersal are very different in these two habitats. Populations on seamounts may derive short-term benefits from larval retention and restricted dispersal because fewer larvae are lost to unsuitable habitats, and more larvae recolonize the source population. The short-term benefits of larval retention near vents, however, may be outweighed by the long-term benefits of larval emigration because the habitat of the source population may become unsuitable within only a few generations.

The following two sections present descriptions of mesoscale flows near seamounts and hydrothermal vents. They describe mechanisms potentially retaining or dispersing plankton, in order to provide a theoretical framework for future field investigations of larval dispersal. These dispersal scenarios are then used to further speculate on the possible roles of dispersal in mesoscale flows in population maintenance and gene flow on both ecological and evolutionary time scales.

Seamounts

Seamount habitats

Large seamounts (those extending > 1 km above the seafloor) are abundant throughout the ocean basins (figure 10.1), with densities ranging from 2 to 55 seamounts per 10^6 km^2 (Smith 1991). Biological communities have been found on all seamounts sampled. Although seamounts are abundant, they tend to be separated by at least 5 to 10 base diameters, which is on the order of 10s to 100s of kilometers. Thus, benthic populations living on seamounts may be widespread, but are often separated by very large distances.

The dominant megafaunal organisms living on seamounts tend to be suspension-feeding invertebrates (Grigg et al. 1987; Levin et al. 1991), making the community structure very different from that found in sediment-dwelling communities dominated by small deposit feeders. Little is known about the ecology of these organisms, although biogeographic studies (reviewed in Wilson and Kaufmann 1987) suggest that the deeper species tend to be more cosmopolitan than the shallower ones. On many North Pacific seamounts, the most prominent taxa are gorgonians, corals, anemones, and sponges. The reproductive and larval ecology of these deep-water species is not well known, but comparisons with related shallow-water species suggest that larvae from the majority of these organisms are liberated into the water column. The lengths of the larval lifespans are unknown, but larvae of at least 32 benthic taxa have been captured in the water column over Fieberling Guyot, a seamount in the North Pacific (unpub. data).

Unusually high abundances of benthic invertebrates (Genin et al. 1986; Wishner et al. 1990) and fish (Uchida and Tagumi 1986) have been observed on seamounts, relative to other communities from similar depths. Several theories have been put forth to explain this abundance, most of them focusing on enhanced food abundance (e.g., Uda and Ishino 1958). An alternate possibility is that the community of benthic suspension feeders is neither food nor space limited, and abundant populations are maintained by enhanced recolonization of the source seamount by larvae retained in mesoscale flows.

Larvae from many of these seamount species, especially the small planulae larvae of the coelenterates, are likely to be mixed away from source adults in turbulent near-bottom flows. Near-bottom flow velocities at most seamounts are much stronger than open-ocean currents at comparable depths, and flows at seamounts are often forced by diurnal and semidiurnal tides (Noble and

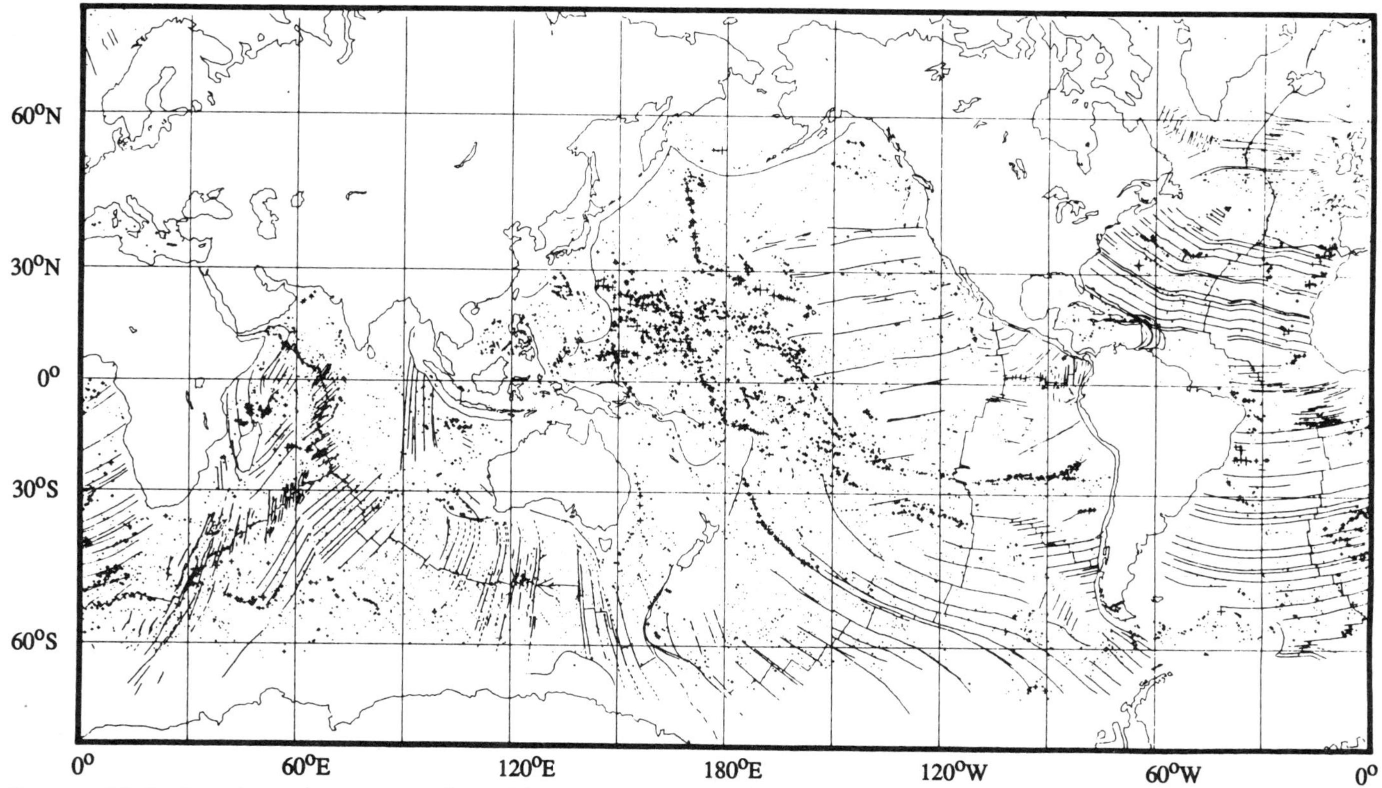

FIGURE 10.1. Locations of seamounts (derived from Seasat altimetry data) and ridges on the seafloor. Seamounts are plotted with symbol size proportional to estimated height. Seamounts larger than 15 km in diameter are included. Mid-ocean ridges are represented by straight lines broken into ridge segments, offset by transform faults (curving, normal lines). Hydrothermal vents have been found along many of these ridges. (Modified from Craig and Sandwell 1988.)

Mullineaux 1989; Eriksen 1991). Given realistic mean near-bottom flow velocities of approximately 10 cm s^{-1}, a substantial number of larvae would be swept off a moderately sized seamount (10 km diameter at summit) in a single 12-hour tidal excursion. In the absence of mesoscale retention processes, these larvae would then drift in the plankton at dilute concentrations. Once a larva is transported away from a seamount, it is unlikely to colonize another seamount because seamounts cover only a small fraction of the seafloor (e.g., 6–13 percent in the Pacific; Smith 1991) and only a portion of each seamount provides suitable habitat. Since colonization from remote seamount habitats is likely to be infrequent, most recruitment on a seamount is probably from local populations. Thus, retention of larvae may enhance the probability of their recolonizing the local seamount, and help ensure the availability of sufficient numbers of colonists to maintain the observed abundant populations. The mechanisms by which mesoscale flows might retain or disperse larvae are explored in the following section.

DISPERSAL IN TOPOGRAPHIC FLOWS

Mesoscale flow features, such as eddies and Taylor caps (Roden 1987), can retain particles near a seamount or other habitat if they increase the residence times of local water parcels. These processes should be effective for retaining larvae spawned at the seamount, since the larval source is located within the flow feature, but they are not effective for concentrating larvae from remote sources. If retention occurs during spawning periods, then larvae will accumulate in the mesoscale flow features. Observations of enhanced larval abundances in known mesoscale flow features would provide strong evidence that retention is occurring. Furthermore, if larvae are retained near the seamount for a sufficiently long period, they will have a higher probability of recolonizing the local seamount. The term "sufficiently long" must be defined in terms of larval competency periods—retention must occur long enough to extend past any precompetent period in the larval life span. Since larval precompetency periods are unknown for deep-water seamount species, this interval cannot be easily defined. Nevertheless, knowledge of precompetency periods of shallow-water species suggests that they are not likely to extend past several weeks to months. Rapid recruitment of seamount species has been observed (Mullineaux 1989; Mullineaux and Butman 1990) and the precompetency periods for some species may be very short or nonexistent.

One prominent mesoscale feature thought to occur near many seamounts is a Taylor cap. This feature forms as oceanic flows accelerate over a seamount or other topographic feature, and are deflected as a result of the earth's rotation. As velocity of the oncoming flow (U) increases over the seamount, so does the Coriolis acceleration $Uf = 2\, U\, \omega \sin \phi$, where ω is the angular velocity imparted by the earth's rotation, and ϕ is the geographic latitude. The Coriolis effect causes the currents to veer toward the right in the northern hemisphere (and to the left in the southern hemisphere), generating an anticyclonic gyre over the seamount (figure 10.2). Early studies assumed the gyre reached the surface, forming a column-like structure called a Taylor column. Recent studies have indicated that the anticyclonic gyre decreases in intensity with distance above the seamount, forming a conelike structure renamed a Taylor cap (Chapman and Haidvogel 1992).

The anticyclonic gyre may remain over the seamount for a period of weeks

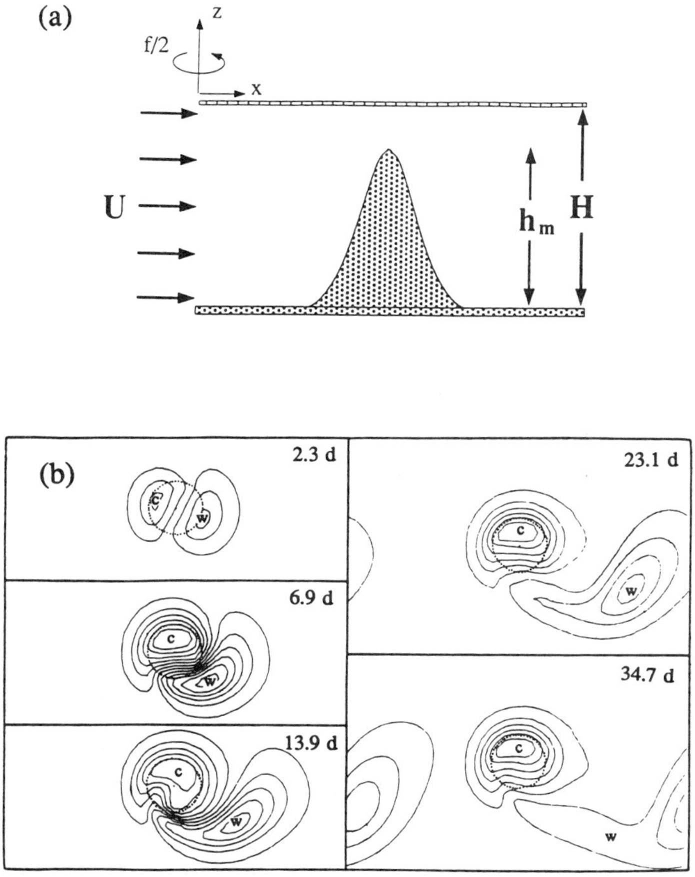

FIGURE **10.2.** Model of Taylor column (or cap) formation.

A. A sketch depicting the isolated topographic feature and general parameters used in Taylor column/cap models. *H* is the total water depth, h_m the height of the seamount, *U* the incident flow velocity, and *f* the Coriolis parameter. (Modified from Chapman and Haidvogel 1992.)

B. Isopycnals (lines of equal density) contoured with solid lines at a depth of 3720 m over a 200 m tall seamount in an incident flow of 5.1 cm s^{-1}. The cold anomaly (c) implies anticyclonic rotation, and the warm anomaly (w) implies cyclonic rotation. Dotted circle shows the seamount base. Successive boxes show a time sequence over 34.7 days. (Modified from Huppert and Bryan 1976.)

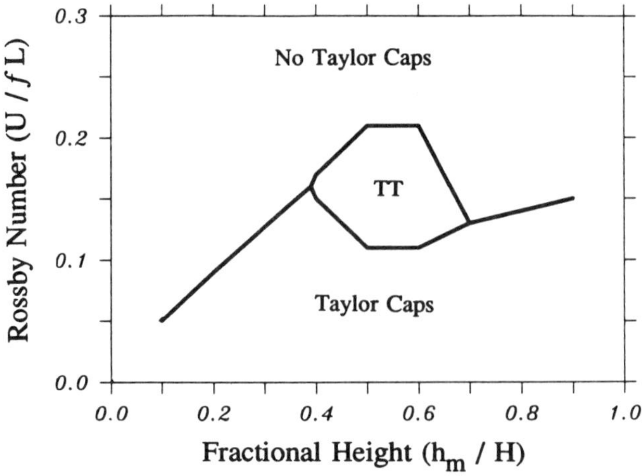

FIGURE 10.3. Flow regimes over a Gaussian-shaped seamount in steady, stratified flow, estimated from a numerical model. Taylor caps occur only below the lower heavy line, while flows above the upper heavy line move directly over the seamount without forming a Taylor cap. Within an intermediate region (TT), part of the flow is temporarily trapped over the seamount before being swept downstream. (Adapted from Chapman and Haidvogel 1992.)

or months, and is eventually advected off the seamount. The seamounts and oceanographic conditions that will generate a Taylor cap can be predicted by the Chapman and Haidvogel (1992) numerical model. This model indicates that Taylor caps should form over seamounts with low Rossby Numbers $Ro = U/fL$ (where L is the diameter of the seamount base, U the incident flow velocity, and f the Coriolis parameter) and high fractional height $\delta = h_m/H$, where h_m is the seamount height, and H the ocean depth at the base of the seamount (figure 10.3). In ocean basins, this means that most tall seamounts (> 1000 m above the seafloor) in moderate oceanic currents (< 10 cm s^{-1}) will form Taylor caps. Alternatively, Taylor caps are not likely to form over short seamounts, or those in swift currents such as the Gulf Stream. Temporary trapping, where particles remain in the vicinity of the seamount for only a short period, occurs only over mid-sized seamounts.

The structure of a Taylor cap and its capacity to retain water parcels and trap larvae can be visualized by "releasing" neutrally buoyant floats (figure 10.4) into the numerical model described in Chapman and Haidvogel (1992). These floats define the approximate limit of the Taylor cap, as floats only slightly further away are swept downstream. The shape of the cap tends to be more conical than columnar, and can be asymmetric in naturally occurring stratified flows. The flow is enhanced on the left side of the seamount (looking downstream), because the parallel orientation of the eddy streamlines and incident streamlines make the velocities additive.

The capability of predicting Taylor cap formation using easily measured oceanographic parameters of seamount height, water depth, and mean incident

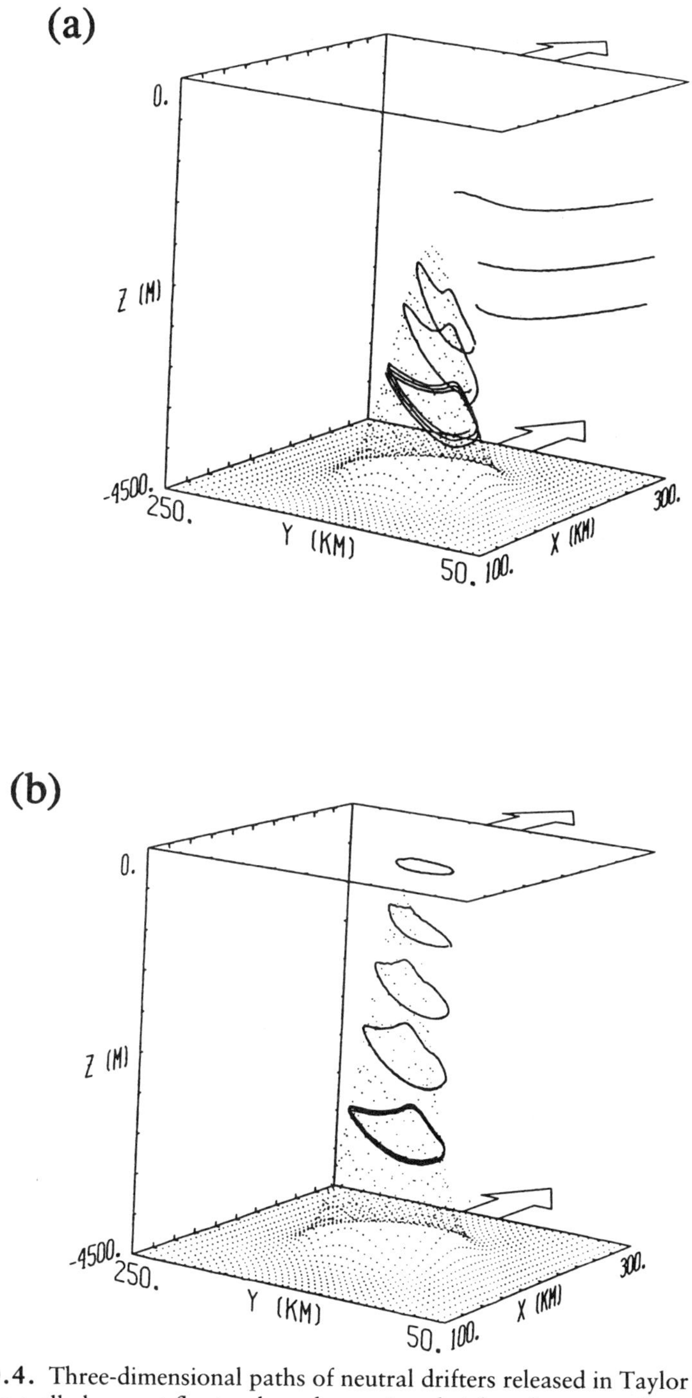

FIGURE 10.4. Three-dimensional paths of neutral drifters released in Taylor cap. Trajectories of six neutrally buoyant floats released at various heights above a moderate-sized (A) and tall (B) seamount in stratified flow with Rossby number $Ro = 0.1$. Arrows show direction of incident flow. The seamount topography is shown by a dot at each grid point. (Modified from Chapman and Haidvogel 1992.)

flow velocity, is an invaluable tool for dispersal studies. Predictions of larval dispersal trajectories and retention times can be made, a priori, for large numbers of seamounts on scales of ocean basins. These dispersal predictions can, in turn, be combined with plankton collections and other biological measurements to determine whether topographically induced flows do indeed retain and accumulate larvae. Once the link between larval retention and mesoscale flow features is established, oceanographic parameters become useful in predicting such biological parameters as larval emigration and immigration rates, as well as genetic exchange rates for use in population genetic and biogeographic studies.

Other mesoscale flow features may also trap larvae through mechanisms similar to those occurring in Taylor caps. For instance, bottom-trapped waves may generate circulation patterns (Brink 1989) that cause particles to follow rotational paths around the seamount, rather than being swept off. These bottom-trapped waves are probably excited near seamounts by resonance of ambient waves that propagate through ocean basins. In some latitudes, these wave motions are a result of diurnal tides. Although the circulation patterns differ qualitatively from those in Taylor caps (figure 10.5) they are formed under similar oceanic conditions (tall seamounts in stratified, slow flows) and share the characteristic of being bottom-intensified.

Eddies occurring in the wakes of seamounts may also act to retain and accumulate larvae, and possibly to advect them away from the seamount as concentrated patches. Laboratory experiments over a seamount-shaped topographic feature in rotational flow have shown that wake eddies can form under realistic oceanic conditions, such as tidally varying currents in a stratified water column (Boyer and Zhang 1990). At moderate Rossby Numbers ($Ro = 0.07$), corresponding to flow speeds expected over moderate-to-tall seamounts in the North Pacific basin, fluid parcels can be trapped in the lee of, and below the top of, isolated topographic features (figure 10.6). If flow speeds are increased, these trapped "bubbles" may be advected downstream. Thus, the biological consequences are similar to those from Taylor caps and bottom trapped waves: retention of larvae at moderate flow speeds, and possible advection of larval aggregations in stronger flows.

Evidence for the existence of Taylor caps (e.g., Owens and Hogg 1980; Gould et al. 1981), and bottom-trapped waves (Hogg 1980), has been gathered in the field from a variety of seamounts. Field evidence for wake eddies is less convincing, although temperature and current records near seamounts do indicate increased variability downstream from seamounts (reviewed in Boyer and Zhang 1990). Field experiments on larval processes near seamounts have only just begun, but specific predictions of larval retention times, accumulation rates, and dispersal trajectories can now be made at individual seamounts, based on theoretical considerations combined with site-specific physical measurements. For instance, water-mass residence times can be combined with

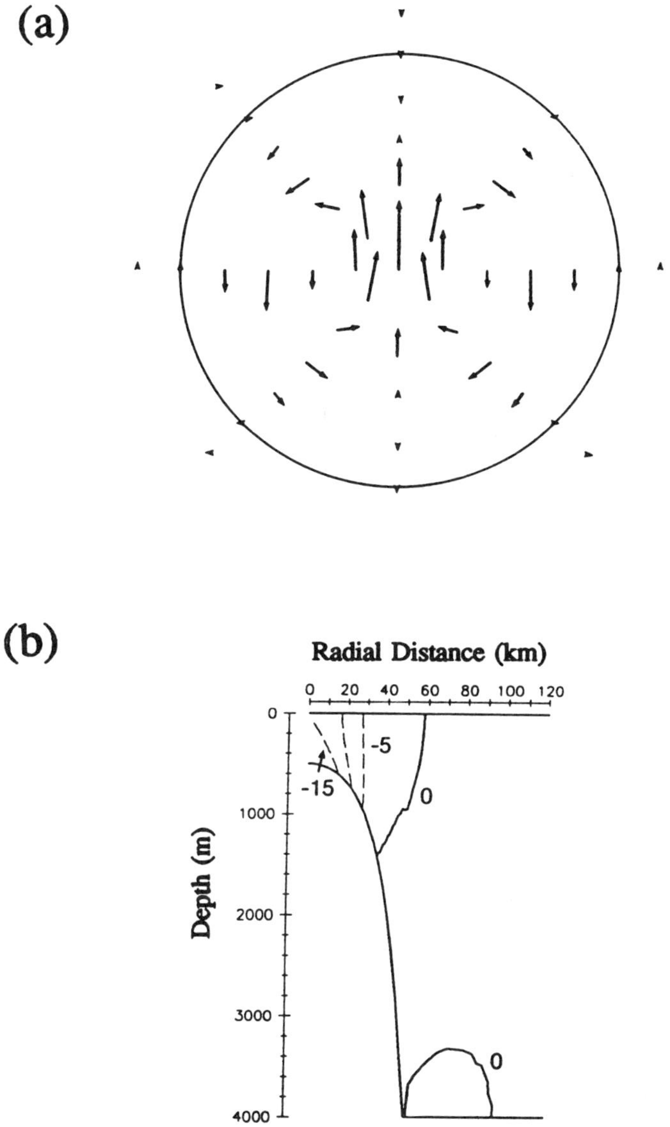

FIGURE 10.5. Circulation generated by bottom-trapped waves, from a model of waves over a tall seamount in deep water. Circulation patterns vary with wave mode, wave number and buoyancy frequency; only selected cases shown here. (Modified from Brink 1989.)

A. Plan-view plot of horizontal velocity vectors showing currents associated with bottom trapped waves in a weakly stratified regime. Vector orientation shows current direction, and vector length shows current speed. The circle outlines the seamount base.

B. Radial velocities (cm s^{-1}) contoured in side view over a seamount, showing bottom-trapping of wave properties in stratified regime.

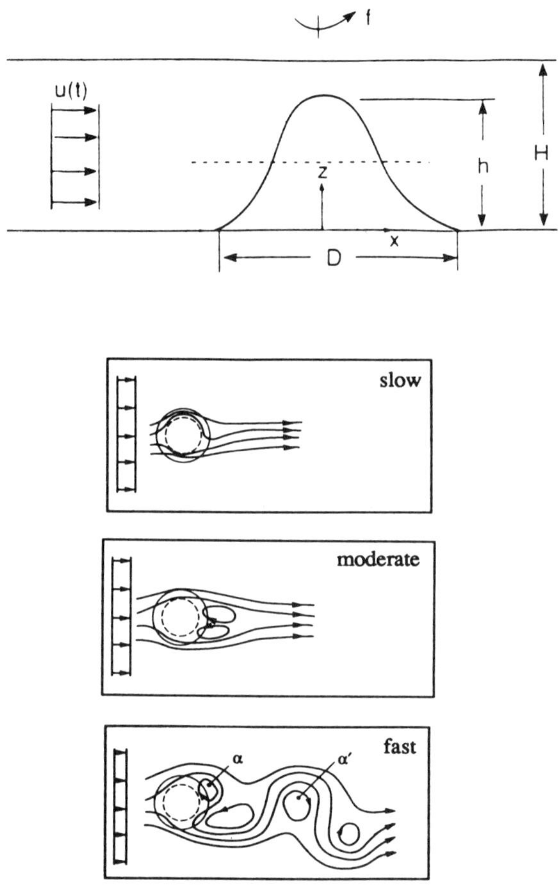

Figure 10.6. Interpretive sketches of characteristic flow regimes downstream from a Gaussian-shaped seamount in stratified water column with tidally varying currents *u(t)*, in a rotating system (Coriolis parameter f). Top views of laboratory system show seamount base (solid circle), seamount circumference at mid-height (dashed circle), and streamlines in relatively slow incident flow velocity ($Ro = 0.01$), moderate flow velocity ($Ro = 0.07$) and fast flow velocity ($Ro = 0.18$) where periodic shedding of anticyclonic (α) and cyclonic (α') eddies occurs.

near-bottom ("source") larval abundances to predict larval abundance and distribution patterns in the water column near a seamount. These predictions can be used to design plankton sampling schemes on a larger (meso) scale, and investigate the specific roles of mesoscale flow dynamics in larval dispersal between seamounts.

Implications for Seamount Populations

The processes of larval retention, aggregation, and episodic dispersal have several implications for population dynamics and genetic exchange of seamount benthos. If larvae are retained near seamounts, their probability of recolonizing the source seamount will be increased. In deep-sea communities of suspension feeders where space is not limiting and food supply is not likely to be controlled by density dependent processes, an increase in larval supply can result in an increase in population numbers. A larger population may, in turn, be more resistant to eradication from local catastrophic events, and persist longer on evolutionary times scales. Shields (1982; reviewed in Jackson 1986) has suggested that larval retention and inbreeding should be the best way to maintain genotypes over many generations. Alternatively, if larvae recolonize the source populations, genetic exchange with other populations will be reduced, possibly causing the source population to be more homogeneous and less resistant to environmental changes. The reduction in genetic exchange might also result in significant genetic differences between populations on separate seamounts.

The episodic instabilities of mesoscale flows near seamounts that cause eddies to detach and advect away also have intriguing consequences. Eddies of this size can retain distinctive properties, including zooplankton (or larval) assemblages, for periods of weeks or even months (Wiebe and Flierl 1985). The arrival of larval cohorts as concentrated patches, rather than as a diffuse trickle of individuals, may result in founder events (establishment of new populations) being very rare, but very successful when they do occur. New recruits are much more likely to establish self-maintaining populations if they attain reproductive maturity at the same time as a large number of conspecifics. These colonization events, although rare, may help avoid permanent reduction of genetic heterogeneity, and subsequent loss of adaptability on evolutionary time scales that could result from larval retention and restricted dispersal.

Hydrothermal Vents

Hydrothermal Vent Habitats

Hydrothermal vents are located on the deep seafloor in tectonically and volcanically active areas such as mid-ocean spreading centers, back-arc basins, and hot spots. Venting occurs when sea water percolates down through the seafloor, is heated by volcanic processes, and escapes upward through a network of conduits. Chemical transformations occurring during this process alter the vent fluid and enrich them in reduced compounds such as hydrogen sulfide (H_2S) and methane (CH_4). These and other reduced compounds provide the energy source for bacterial chemosynthesis and form the base of the food chain for a bizarre assemblage of organisms endemic to the vents (Jannasch and Wirsen 1985). Conspicuous members of the earliest discovered vent communities included enormous vestimentiferan tube worms, some over 3 m long and up to

4 cm in diameter, as well as large clams, mussels, and polychaetes. Subsequent discoveries of new vent fields have revealed a variety of vent communities, some dominated by shrimp and others by gastropods and anemones.

Vent habitats are scattered over the seafloor, so dispersal between vent communities is influenced by the spatial distribution of vent habitats, as well as their persistence through time. For instance, at mid-ocean ridges, the subsurface hydrology of the vent system causes vents to cluster in vent fields separated by 10s of kilometers along a ridge segment. Individual vents within an active hydrothermal field may last for periods of less than a year (pers. obs.), to 10s of years (MacDonald et al. 1980), while entire vent fields may cease to emit vent fluids on time scales of 100s of years (Lalou and Brichet 1982). Since the organisms living at individual vents cannot disperse long distances as adults, their larvae must be capable of dispersing to remote habitats in order to maintain their populations. While short-range dispersal may be sufficient to colonize new vents within a vent field, or even colonize a neighboring vent field a few kilometers away, it does not result in colonization of remote ridge segments, 100s of kilometers away. Since vents along an entire ridge segment may eventually become extinct, vent communities must be able to disperse 100s of kilometers on evolutionary time scales.

Hydrothermal vent habitats are more ephemeral than many other hard-substrate habitats, creating a unique challenge in population maintenance for the organisms living there. The explanation of how these endemic organisms colonize and maintain populations on both short and long time scales is one of the major ecological questions still unanswered for deep-sea vent habitats. Lutz et al. (1984) speculated that vent larvae should be highly dispersive and therefore planktotrophic. These investigators did find planktotrophic larvae for some vent invertebrates (e.g., the mussel, *Bathymodiolus,* the crab *Bythograea,* the shrimp *Alvinocaris,* and a few turrid gastropod species), but nonplanktotrophic larvae appear to be much more common, and have been identified from the vent clam *Calyptogena,* the galatheid *Munidopsis,* two genera of polychaete worms, and many species of trochoid gastropods and limpets (e.g., Gustafson and Lutz, this volume). In shallow water, lecithotrophic forms are generally thought to disperse less than planktotrophic forms. Lutz et al. (1986), however, suggest that a reduction in developmental rate at the low temperatures found in the deep sea would dramatically increase the dispersal capabilities of nonfeeding larvae. Further support for the suggestion that nonfeeding larvae may be widely dispersed comes from the discovery of nonplanktotrophic larvae and early juveniles collected 10s–100s of meters off the deep seafloor (Berg and Van Dover 1987; Wiebe et al. 1988; Mullineaux et al. 1991). The absence of swimming appendages in the juveniles and some of the larvae indicate they are being passively advected by deep-ocean currents.

Dispersal in Hydrothermal Vent Plumes

Although larvae of vent invertebrates appear to be capable of dispersing long distances even without swimming or feeding appendages, the mechanisms by which they disperse are still largely unknown. Here I explore the hypothesis that larvae are entrained in buoyant plumes emanating from hydrothermal vents and are advected away from the source vents in the plumes. When hydrothermal fluids are released from the seafloor, they are less dense than the surrounding seawater because of their elevated temperatures. As the fluids rise off the bottom, they form a buoyant plume, entraining surrounding, cooler water. Buoyant plumes can entrain very large volumes of near-bottom water, and are often diluted by volumes 10^4 times greater than the volume of the original effluent. Much of the entrainment occurs near the seafloor and is likely to include larvae and other benthopelagic zooplankton living near the vent (figure 10.7). As the plumes rise off the

FIGURE 10.7. Diagram showing hypothesized entrainment of near-bottom zooplankton into a buoyant hydrothermal vent plume.

seafloor they continue to entrain cooler, ambient water until they reach a level (usually about 200 m above the source) where they become the same density as the ambient seawater. The plumes then spread laterally (figure 10.8), generally in the direction of mean horizontal currents (e.g., Baker and Massoth 1987).

The amount of near-bottom water entrained into hydrothermal plumes can be estimated from models of turbulent, buoyancy-driven convective fluids. These models are derived from simple plume models that assume turbulent entrainment of ambient fluid is proportional to vertical plume velocity (e.g., Morton et al. 1956). Flow studies in vent plumes have recorded vertical velocities as large as 5–10 cm s^{-1} (Converse et al. 1984; Little et al. 1987). These vertical velocities result in substantial entrainment into the plume, with large amounts of ambient water transported vertically within the plume and distributed at the spreading level. Lupton et al. (1985) calculated that 70 percent of the water in the plume at the spreading level (the lateral plume) had been entrained below the equilibrium height, and vertical volume fluxes of 500 m^3 s^{-1} are possible (Speer and Rona 1989). Much of this water appears to be entrained very near the seafloor, where larvae and other zooplankton have been observed to be most abundant (Smith 1985; Berg and Van Dover 1987). Thus, it is reasonable to suppose that large numbers of particles in the bottom waters near a vent would become entrained and transported upwards in the buoyant plumes.

Larvae of benthic invertebrates and other zooplankton are likely to be transported vertically as passive particles in the plume if their swimming speeds are much lower than the flow velocities or if they are swimming in a nondirected fashion on spatial scales smaller than the length scale of the flow feature (in this case, the width of the plume). Swimming speeds, even of relatively large shallow water larvae, tend to be much slower (0.1–1.0 cm s^{-1}: Chia et al. 1984) than most oceanic flows and even highly motile zooplankton are advected within mesoscale oceanic water masses (Wiebe and Flierl 1983). Thus mero- and holoplankton whose settling velocities are smaller than plume vertical velocities are likely to be entrained in plumes, and models of plume dynamics can be used to predict the vertical and lateral transport of these organisms.

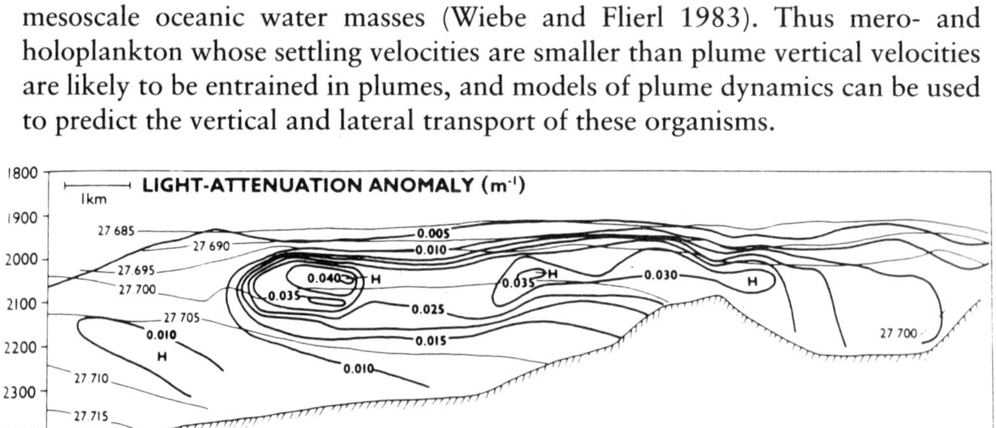

FIGURE 10.8. Contours of light-attenuation anomaly (related to particle abundance) showing structure of a lateral hydrothermal vent plume over the Cleft Segment of Juan de Fuca Ridge. (Adapted from Baker and Massoth 1987.)

Larvae entrained into the buoyant plume are likely to be carried up to the equilibrium plume level, and disperse horizontally. This horizontal transport is controlled by a combination of vertical motions such as swimming and passive sinking, and by lateral advection of the plume by currents. Most large (> 150 μm), dense particles, such as mineral precipitates, are not carried far by the lateral plume, and are either deposited near the plume, or recycled into the buoyant plume (e.g., Converse et al. 1984). Smaller or less dense particles, such as larvae, however, can be transported for long distances in the lateral plume, as evidenced by depositional patterns of small manganese particles in sediments (Klinkhammer and Hudson 1986), and by particle distributions in the plume itself (Cowen et al. 1986; Feely et al. 1987; Roth and Dymond 1989). Larvae collected from vent organisms range in diameter from 90 μm to 1 cm (Van Dover et al. 1985; Berg and Van Dover 1987; Wiebe et al. 1988; Jones and Gardiner 1989). Since most previous sampling efforts have used relatively coarse-mesh (> 300 μm) nets, many of the smaller larvae may have been missed. Settling velocities of vent larvae are unknown, but the larvae are, in general, small and are likely to have lower settling velocities than larger, shallow-water species. These low estimated settling velocities, coupled with observations of vent larvae 10s to 100s of meters above the bottom (Berg and Van Dover 1987; Wiebe et al. 1988) indicate that deep-sea larvae may disperse 10s to 100s of kilometers in the hydrothermal vent plumes.

The earth's rotation introduces an interesting twist to the scenario of larval dispersal in buoyant vent plumes. Recent developments in modeling of plume dynamics have indicated that the mesoscale evolution of hydrothermal plumes can be influenced by planetary rotation. Speer (1989) has shown with a simple model that a typical hydrothermal vent can generate a baroclinic vortex. The slow radial outflow of plume fluid at the neutral buoyancy level is inhibited by rotation leading to the formation of an anticyclonic eddy at the spreading level (figure 10.9). Similarly, as the ambient fluid around the rising plume is drawn inward by entrainment into the plume it will develop cyclonic circulation. Secondary (vertical) flows are set up to balance upward fluid transport in the buoyant plume with diffuse downward transport in the region surrounding it. These studies suggest that lateral plumes should develop rotational velocities of about 10 cm s^{-1}, and could remain above a vent field sufficiently long to trap and accumulate particles within the vortex.

Recent laboratory experiments (Helfrich and Battisti 1991) suggest that these plume-forced vortices are unstable, leading to the unsteady production of baroclinic vortex pairs from a steady buoyancy source. These vortex pairs, each consisting of a cyclonic and anticyclonic vortex, propagate away from the source in a direction perpendicular to the lateral offset in the vortex centers (i.e., as hetons, described in Hogg and Stommel 1985). These results suggest that plume vortices are able to propagate away from vents in directions other than those dictated by mean oceanic currents. Since isolated vortices form closed streamlines, they tend to retain any anomalous properties for long times.

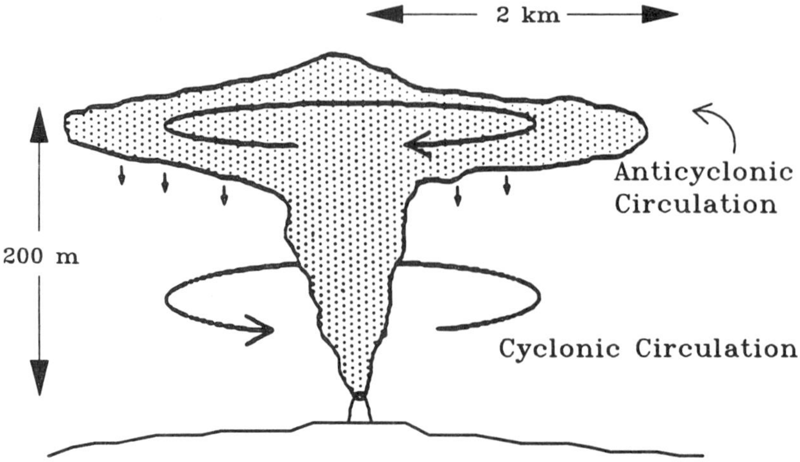

FIGURE 10.9. Diagram of vorticity in hydrothermal vent plume, scaled for conditions at Juan de Fuca Ridge, North Pacific.

Thus, plume-forced vortices could isolate and transport plume fluid (and associated larvae) away from vent regions, effectively isolating the plume fluid from oceanic waters. At the meso-scale, the merger of plume fluid with the background may be patchy rather than smooth. Plume instabilities should be more intense in plumes with large heat fluxes. Large plume features, or "megaplumes," reaching 700 m above the seafloor, and several kilometers in diameter, have been observed over Juan de Fuca Ridge in the northeast Pacific (Baker et al. 1987) and may be signatures of plume vortices. The dynamics of plume entrainment and vorticity suggest that larvae disperse from vent environments as discontinuous, episodic patches, rather than as a steady stream of individuals.

IMPLICATIONS FOR VENT POPULATIONS

Mesoscale flow processes are useful for modeling larval dispersal, but provide less information on specific mechanisms controlling recruitment into new, or existing, vent habitats. One possible hydrodynamic mechanism is the entrainment of dispersing larvae into diffuse convection cells that are generated near vents by buoyant plumes. Alternate biological mechanisms include ontogenetic changes in the larvae that cause them to sink, or behavioral responses to reduced chemicals that induce larvae to swim down toward vents. Mesoscale flow models do, however, predict that cohorts of recruits will arrive episodically in concentrated patches. Few recruitment studies have been conducted at vent habitats, but some evidence suggests that the recruitment process is episodic rather than continuous over the lifespan of the vent. Studies of several vent species (Hessler and Smithey 1983) revealed distinct size classes within populations, suggesting that these species colonize in discrete pulses. Further evidence for episodic colonization can be found in the genetic structure of vent populations. Grassle (1985) found little genetic variability within each of two size classes of the vent mussel *Bathymodiolus* at one site, but significant genetic differences between the two size classes. These results indicate that two distinct recruitment events occurred, rather than a steady stream of recruits, or local reproduction. Genetic studies of the vestimentiferan *Riftia pachyptila* (Bucklin 1988) showed surprisingly low variability in genetic structure, possibly as a result of population bottlenecks during repeated colonization and extinction events. Experiments with recruitment panels (e.g., Van Dover et al. 1988) are the most direct way to monitor recruitment, but are difficult to conduct on the temporal and spatial scales necessary to test hypotheses on mesoscale recruitment events.

Summary

Studies of mesoscale flow processes near seamounts and hydrothermal vents have demonstrated their potential to retain larvae of benthic invertebrates, to accumulate larvae near source populations, and to advect larvae as concentrated patches in discrete eddies. Quantitative models of these flows can be used to generate hypotheses about the trajectories and episodicities of larval dispersal and recruitment. For instance, seamounts may retain larvae near the source populations, enhancing local population abundances but reducing genetic exchange with other seamount populations. Studies of larval abundance and distribution in the water column, larval recruitment rates into the benthos, and genetic diversity within and between seamount populations can be used to address these hypotheses. Hydrothermal vent plumes, on the other hand, can potentially transport larvae long distances from endemic vent populations, allowing the species to persist despite the ephemeral nature of their habitat. Studies of larval abundances in plumes, as well as ongoing studies of recruitment and population genetics in vent communities will provide data to address the hypothesis that plume dynamics are controlling dispersal and recruitment of vent larvae.

Acknowledgments

Thought-provoking discussions with many colleagues, including Ken Brink, Dave Chapman, Karl Helfrich, Terry Joyce, and Peter Wiebe, have been the basis for this study. The manuscript was improved substantially by suggestions from Bob Hessler, Craig Young, and an anonymous reviewer. Support has been provided by ONR Grant N00041–89-J-1431 and NSF Grant OCE-9019575. This manuscript is contribution 7956 from Woods Hole Oceanographic Institution.

References

Baker, E. T. and G. J. Massoth. 1987. Characteristics of hydrothermal plumes from two vent fields on the Juan de Fuca Ridge, northeast Pacific Ocean. *Earth and Planetary Science Letters* 85:59–73.

Baker, E. T., G. J. Massoth, and R. A. Feely. 1987. Cataclysmic hydrothermal venting on the Juan de Fuca Ridge. *Nature* 316:342–344.

Berg, C. J. and C. L. Van Dover. 1987. Benthopelagic macrozooplankton communities at and near deep-sea hydrothermal vents in the eastern Pacific Ocean and the Gulf of California. *Deep-Sea Res.* 34:379–401.

Bouchet, P. and A. Warén. 1979. Planktotrophic larval development in deep-water gastropods. *Sarsia* 64:37–40.

Boyer, D. L. and X. Zhang. 1990. Motion of oscillatory currents past isolated topography. *J. Phys. Ocean.* 20(9):1425–1448.

Brink, K. H. 1989. The effect of stratification on seamount-trapped waves. *Deep-Sea Res.* 36:825–844.

Bucklin, A. 1988. Allozymic variability of *Riftia pachyptila* populations from the Galapagos Rift and 21°N hydrothermal vents. *Deep-Sea Res.* 35:1759–1768.

Chapman, D. C. and D. B. Haidvogel. 1992. Formation of Taylor caps over a tall isolated seamount in a stratified ocean. *Geophys. Astrophys. Fluid Dynamics* 64:31–65.

Chia, F. S., J. Buckland-Nicks, and C. M. Young. 1984. Locomotion of marine invertebrate larvae: a review. *Can. J. Zool.* 62:1205–1222.

Converse, D. R., H. D. Holland, and J. M. Edmond. 1984. Flow rates in the axial hot springs of the East Pacific Rise (21°N): Implications for the heat budget and the formation of massive sulfide deposits. *Earth and Planetary Science Letters* 69:159–175.

Cowen, J. P., G. J. Massoth, and E. T. Baker. 1986. Bacterial scavenging of Mn and Fe in a mid- to far-field hydrothermal particle plume. *Nature* 322:169–171.

Craig, C. H. and D. T. Sandwell. 1988. Global distribution of seamounts from seasat profiles. *J. Geophys. Res.* 93:10408–10420.

Eriksen, C. C. 1991. Observations of amplified flows atop a large seamount. *J. Geophys. Res.* 96:15227–15236.

Feely, R. A., M. Lewison, G. J. Massoth, G. Robert-Baldo, J. W. Lavelle, R. H. Byrne, K. L. Von Damm, and H. C. Curl, Jr. 1987. Composition and dissolution of black smoker particulates from active vents on the Juan de Fuca Ridge. *J. Geophys. Res.* 92:11347–11363.

Genin, A., P. K. Dayton, P. F. Lonsdale, and F. Spiess. 1986. Corals on seamount peaks provide evidence of current acceleration over deep sea topography. *Nature* 322:59–61.

Gould, W. J., R. Hendry, and H. E. Huppert. 1981. An abyssal topographic experiment. *Deep-Sea Res.* 28:409–440.

Grassle, J. P. 1985. Genetic differentiation in populations of hydrothermal vent mussels (*Bathymodiolus thermophilus*) from the Galapagos Rift and 13°N on the East Pacific Rise. *Bull. Biol. Soc. Wash.* 6:429–442.

Grigg, R. W., A. Malahoff, E. H. Chave, and J. Landahl. 1987. Seamount benthic ecology and potential environmental impact from manganese crust mining in Hawaii. In B. H. Keating et al., eds., *Seamounts, Islands, and Atolls.* AGU Geophysical Monograph 43:355–378.

Helfrich, K. R. and T. M. Battisti. 1991. Experiments on baroclinic vortex shedding from hydrothermal plumes. *J. Geophys. Res.* 96:12511–12518.

Hessler, R. R. and W. M. Smithey. 1983. The distribution and community structure of megafauna at the Galapagos Rift hydrothermal vents. In P. S. Rona, K. Bostrom, L.

Laubier, and K. L. Smith, Jr., eds., *Hydrothermal Processes at Seafloor Spreading Centers,* NATO Conference Series 4:735–770. New York: Plenum Press.

Hogg, N. G. 1980. Observations of internal Kelvin waves trapped round Bermuda. *J. Mar. Res.* 44:51–69.

Hogg, N. G. and H. M. Stommel. 1985. The heton, an elementary interaction between discrete baroclinic geostrophic vortices, and its implications concerning eddy heat-flow. *Proc. Roy. Soc.* Lond. A397:1–20.

Huppert, H. E. and K. Bryan. 1976. Topographically generated eddies. *Deep-Sea Res.* 23:655–679.

Jackson, J. B. C. 1986. Modes of dispersal of clonal benthic invertebrates: consequences for species' distributions and genetic structure of local populations. *Bull. Mar. Sci.* 39:588–606.

Jannasch, H. W. and C. O. Wirsen. 1985. The biochemical versatility of chemosynthetic bacteria at deep-sea hydrothermal vents. *Bull. Biol. Soc. Wash.* 6:325–334.

Jones, M. L. and S. L. Gardiner. 1989. On the early development of the vestimentiferan tube worm *Ridgeia* sp. and observations on the nervous system and trophosome of *Ridgeia* sp. and *Riftia pachyptila. Biol. Bull.* 177:254–276.

Killingley, J. S. and M. A. Rex. 1985. Mode of larval development in some deep-sea gastropods indicated by oxygen-18 values of their carbonate shells. *Deep-Sea Res.* 32:809–818.

Klinkhammer, G. and A. Hudson. 1986. Dispersal patterns for hydrothermal plumes in the South Pacific using manganese as a tracer. *Earth and Planetary Science Letters* 79:241–249.

Lalou, C. and E. Brichet. 1982. Ages and implications of East Pacific Rise sulfide deposits at 21°N. *Nature* 300:169–171.

Levin, L. A., C. L. Huggett, and K. F. Wishner. 1991. Control of deep-sea benthic community structure by oxygen and organic-matter gradients in the eastern Pacific Ocean. *J. Mar. Res.* 49:763–800.

Little, S. A., K. D. Stolzenbach, and R. P. von Herzen. 1987. Measurements of plume flow from a hydrothermal vent field. *J. Geophys. Res.* 92:2587–2496.

Lupton, J. E., J. R. Delaney, H. P. Johnson, and M. K. Tivey. 1985. Entrainment and vertical transport of deep-ocean water by buoyant hydrothermal plumes. *Nature* 316:621–623.

Lutz, R. A., D. Jablonski, and R. D. Turner. 1984. Larval development and dispersal at deep- sea hydrothermal vents. *Science* 226:1451–1454.

Lutz, R. A., P. Bouchet, D. Jablonski, R. D. Turner, and A. Warén. 1986. Larval ecology of mollusks at deep-sea hydrothermal vents. *Am. Malacol. Bull.* 4:49–54.

MacArthur, R. H. and E. O. Wilson. 1967. *The Theory of Island Biogeography.* Princeton: Princeton University Press.

MacDonald, K. C., K. Becker, F. N. Spiess, and R. D. Ballard. 1980. Hydrothermal heat flux of the "black smoker" vents on the East Pacific Rise. *Earth and Planetary Science Letters* 48:1–7.

Morton, B. R., G. I. Taylor, and J. S. Turner. 1956. Turbulent gravitational convection from maintained and instantaneous sources. *Proc. R. Soc. London,* Ser. A 234:1–23.

Mullineaux, L. S. 1989. Vertical transport of benthic invertebrate propagules in flows over Volcano 7 seamount. *Trans. Am. Geophys. Un.* 71:155.

Mullineaux, L. S. and C. A. Butman. 1990. Recruitment of benthic invertebrates in boundary-layer flows: A deep water experiment on Cross Seamount. *Limnol. Oceanogr.* 35:409–423.

Mullineaux, L. S. and C. A. Butman. 1991. Initial contact, exploration, and attachment of barnacle (*Balanus amphitrite*) cyprids settling in flow. *Mar. Biol.* 110:93–103.

Mullineaux, L. S., P. H. Wiebe, and E. T. Baker. 1991. Hydrothermal vent plumes: Larval highways in the deep sea? *Oceanus* 34:64–68.

Noble, M. and L. S. Mullineaux. 1989. Internal tidal currents over the summit of Cross Seamount. *Deep-Sea Res.* 36:1791–1802.

Owens, W. B. and N. G. Hogg. 1980. Oceanic observation of stratified Taylor column near a bump. *Deep-Sea Res.* 27:1029–1045.

Roden, G. I. 1987. Effect of seamounts and seamount chains on ocean circulation and thermohaline structure. In B. H. Keating et al., *Seamounts, Islands, and Atolls. AGU Geophysical Monograph* 43:335–354.

Roth, S. E. and J. Dymond 1989. Transport and settling of organic material in a deep-sea hydrothermal plume: Evidence from particle flux measurements. *Deep-Sea Res.* 36:1237–1254.

Sammarco, P. W. and J. C. Andrews. 1988. Localized dispersal and recruitment in Great Barrier Reef Corals: The Helix experiment. *Science* 239:1422–1424.

Shields, W. M. 1982. *Philopatry, Inbreeding, and the Evolution of Sex.* Albany: State University of New York Press.

Smith, D. K. 1991. Seamount abundances and size distributions, and their geographic variations. *Rev. Aq. Sci.* 5(3–4):197–210.

Smith, K. L. Jr. 1985. Macrozooplankton of a deep sea hydrothermal vent: *In situ* rates of oxygen consumption. *Limnol. Oceanogr.* 30:102–110.

Speer, K. G. 1989. A forced baroclinic vortex around a hydrothermal plume. *Geophys. Res. Lett.* 16:461–464.

Speer, K. G. and P. A. Rona. 1989. A model of an Atlantic and Pacific hydrothermal plume. *J. Geophys. Res.* 94:6213–6220.

Uchida, R. N. and D. T. Tagami. 1986. Groundfish fisheries and research in the vicinity of seamounts in the North Pacific Ocean. *Marine Fisheries Review* 46:1–17.

Uda, M. and M. Ishino. 1958. Enrichment pattern resulting from eddy systems in relation to fishing grounds. *J. Tokyo Univ. Fish.* 44:105–119.

Van Dover, C. L., C. J. Berg, and R. D. Turner. 1988. Recruitment of marine invertebrates to hard substrates at deep-sea hydrothermal vents on the East Pacific Rise and Galapagos spreading center. *Deep-Sea Res.* 35:1833–1849.

Van Dover, C. L., J. R. Factor, A. B. Williams, and C. J. Berg, Jr. 1985. Reproductive patterns of decapod crustaceans from hydrothermal vents. *Biol. Soc. Wash. Bull.* 6:223–227.

Vrijenhoek, R. C., C. Craddock, S. Schultz, and R. A. Lutz. 1991. Genetic studies at deep-sea hydrothermal vents. *Deep-Sea Biology Symposium Programme* 6:69–70.

Wiebe, P. H. and G. R. Flierl. 1983. Euphausiid invasion/dispersal in Gulf Stream cold core rings. *Aust. J. Mar. Freshwat. Res.* 34:625–652.

Wiebe, P. H., N. Copley, C. L. Van Dover, A. Tamse, and F. Manrique. 1988. Deep-water zooplankton of the Guaymas Basin hydrothermal vent field. *Deep-Sea Res.* 35:985–1013.

Wilson, G. D. F. and R. R. Hessler. 1987. Speciation in the deep sea. *Ann. Rev. Ecol. Syst.* 18:185–207.

Wilson, R. R. and R. S. Kaufmann. 1987. Seamount biota and biogeography. In B. H. Keating et al., eds., *Seamounts, Islands, and Atolls.* AGU Geophysical Monograph 43:355–378.

Wishner, K. F. 1980. Aspects of the community ecology of deep-sea, benthopelagic plankton with special attention to gymnopleid copepods. *Mar. Bio.* 60:179–187.

Wishner, K., L. Levin, M. Gowing, and L. Mullineaux. 1990. Involvement of the oxygen minimum in benthic zonation of a deep seamount. *Nature* 346:57–59.

CHAPTER ELEVEN

Recruitment Ecology and Age Structure of Deep-Sea Invertebrate Populations

John D. Gage

Abstract

This paper briefly reviews recent literature on recruitment and age structure of bottom-dwelling invertebrates in the context of results from a long-running sample time series at a 2900 m permanent station in the Rockall Trough (NE Atlantic). Overall, results suggest that deep-sea populations encompass a much wider range in life history traits than previously perceived: Some deep-sea species show seasonal reproductive periodicities, relatively high longevity, low rates of adult mortality, and marked recruitment variability, while other opportunist forms, such as wood borers, appear capable of high rates of recruitment and individual growth, and short life spans associated with relatively high fecundity. Several macrofaunal species from the Rockall Trough sediment community demonstrate characteristics of annually pulsed recruitment and relatively rapid growth rates, and an age structure where fast-growing juveniles may be numerically conspicuous in size frequencies. Although data are sparse, there are clear indications that an annual peak in recruitment to subadult sizes occurs in populations of many species regardless of whether gametogenesis is seasonal. Hence, seasonal recruitment may not necessarily be a corollary of an annual cycle in reproduction. The size and composition of the postlarval population for the majority of nonseasonally breeding species remains unclear. However, demographic models suggest differing population dynamics apply to the postlarvae compared to adults of deposit-feeding bivalves and brittle stars. This is consistent with the postlarval "bottleneck" theory where body-size related constraints on nutrition of meiofaunal postlarvae, and competition with (and vulnerability to grazing by) macro- and megafaunal deposit feeding adults for high quality labile food particles, creates a restrictive transition between the postlarval and subadult/adult stages.

The recruitment and age structure of populations in the deep sea is of interest for several reasons. First, it is important in understanding processes structuring the highly diverse communities of small-bodied invertebrates associated with deep-ocean sediments. Second, knowledge of rates of recolonization is crucial for reliable predictions of recovery times from perturbations associated with deep-ocean mining and the dumping of nontoxic organic wastes.

In the past, the deep-sea bed has been perceived to support populations characterized not only by an astonishingly rich diversity in species, but also by small brood size, an age structure dominated by older stages, and slow individual growth rates (Grassle and Sanders 1973). This view has been difficult to test because of the difficulty in collecting rate data on population processes from the deep sea. Such research topics require expensive and logistically difficult time series sampling and/or experiments using submersibles and free vehicles. As a consequence, data on rates of individual growth and colonization and species succession of populations are sparse.

In situ studies have employed sediment trays placed on the bottom at various sites in the North Atlantic (Desbruyères et al. 1980, 1985; Grassle and Morse-Porteous 1987). The results seem surprisingly constant with regard to water depth and sediment type, but do indicate low rates of colonization when compared to similar experiments in shallow water (Zajac and Whitlach 1982). When extrapolated to background levels of faunal abundance, recolonization data suggest that recovery of deep-sea communities from severe disturbance may take more than two to five years (Smith and Hessler 1988). Another difference from shallow water is that colonizing species show great variability in successional patterns; while some of the most abundant recolonizers are also dominants of the background community, other abundant colonists seem to be opportunists that are rare or absent in samples from background sediment. However, inshore experiments (Smith and Brumsickle 1989) suggest that the sediment trays may imperfectly mimic normal sediment colonization patterns on the deep-sea floor as a result of unnatural flow conditions over the experiments and owing to sediment isolation.

The alternative approach to the experimental study of recruitment and succession has involved emplacement of a variety of parcels of organic material, ranging from wood blocks to fish carcasses and macroalgae, on the sea bed. Experiments with wood date back to classic experiments in the early 1970s when blocks exposed for 104 days became heavily colonized by wood-boring molluscs (Turner 1973). This and subsequent experiments (Turner 1977) were the first to show in deep-sea taxa, the high population density, high reproductive rate, early maturity and rapid growth typical of ecological opportunists in shallow water. However, the highly specific and rapid response of wood borers differs from the more varied colonization pattern found associated with parcels of macroalgae such as *Sargassum.* In the northwestern Atlantic, this material attracted high densities of apparently opportunistic polychaetes similar to those found in sediment trays and around wood (Grassle and Morse-Porteous 1987). The physical disturbance to the sediment immediately adjacent to a carcass reduces both population density and species richness. However, such disturbance may elicit a variable response, from rapid recolonization within three to six weeks by both dominant and rare species in the background community, to a much slower community response at a deeper site (Smith 1986). In contrast to the slow rates of recruitment to sediment trays, manipulative experiments, using DSV Alvin, on artificial mounds as hydrodynamic analogs of the large faecal mounds produced by large burrowing echiurans indicated that the sediment macrofauna may recover to background levels within eleven months after this disturbance. But controlled experiments with floored "mounds" indicates most of this rapid response results from burrowing immigrants, and complete community recovery following this burial disturbance may require more than three years (Smith et al. 1986; Kukert and Smith 1992).

The reason for the slow rate of recovery to background abundance in sediment tray experiments in the deep sea remains problematic. Some species, such as the small aplacophoran mollusc *Prochaetoderma yongei,* had clearly grown very rapidly, reaching adult size within 1 month and sexual maturity within one year (A. Scheltema 1987). Furthermore, the results provided evidence that individuals of other molluscs, such as the protobranch bivalve *Deminucula atacellana,* may have grown to the nearly mature size of 2 mm shell length within 26 months. This suggests that individual growth rates may be comparable to those of shallow-water relatives (e.g., Rachor 1976; Ansell and Parulekar 1978; Lewis et al. 1982). Overall, however, populations of nonopportunists, such as small, sediment-dwelling macrofauna, seem to expand more slowly than would be expected in shallow water. But there may be inhibiting effects of the unnatural flow regime over the tray to water-borne colonizing larvae (Smith 1985; Smith and Brumsickle 1989), and from isolation by the sides and bottom of the tray to potential postlarval recruits from the adjacent sediment (Kukert and Smith 1992). Such factors need to be considered in relation to the complex hydrodynamics of the benthic boundary layer (Richards 1990).

Knowledge on rates of recruitment to the exotic populations associated with hydrothermal vents remains fragmentary, but is sufficient to indicate marked contrasts with these processes elsewhere in the deep sea. Clearly, because of their ephemeral habitat in an environment of abundant food, there has been selection for large size, at least in part, to maximize fecundity in a limited time (Grassle 1986; Tunnicliffe 1991). The isolation of vent fields along the spreading ridge crest implies the need to develop efficient means for recolonization at a rate commensurate with the generation time of vent taxa. However, because hydrothermal activity along a segment of ridge crest is likely to last over several generations, an ability for within-generation dispersal of larvae over large distances may not be as important a requirement for survival as might be expected (Grassle 1986). Collection at 100–200 m above a vent field of postlarvae of various forms of bivalves, gastropods, polychaetes, and crustaceans normally associated with hydrothermal vents indicate that vent organisms also may be dispersed as postlarvae by the vent plume (Berg and Van Dover 1987; Mullineaux, this volume).

Recruitment experiments using slate substrata set out at vent sites show broad size ranges in early life history stages of various attached species. This indicates that intermittent or continuous recruitment occurs (Van Dover et al. 1988), and this proposition is supported from polymodal size frequencies in large samples of alvineliid worms (McHugh 1989). The best evidence for the rapid rate of population expansion and succession in the hydrothermal vent community comes from revisiting vent sites in the eastern Pacific (Laubier and Desbruyères 1984; Fustec et al. 1987; Hessler et al. 1988). This is reflected in sparse data on individual growth rates, using radiometric dating, shell dissolution data, and by measuring the shell-growth increment of individually file-marked specimens of hydrothermal bivalves (Turekian et al. 1981; Rhoads et al. 1982; Turner and Lutz 1984; Lutz et al. 1983, 1988). Overall, it has become clear that the dynamic metabolism of vent faunas is reflected in the rapid rates of colonization and successional change during the growth and decline of individual vent fields.

The results of the influential study by Turekian et al. (1975), employing radiometric dating, is one of the few data on individual growth rate available for the nonvent, sediment-dwelling species found only at abyssal depths. Turekian and his colleagues measured the activity of the naturally occurring radioisotope ^{228}Ra (half-life 5.75 years) that occurs along with the calcite in different growth increments of the shells of specimens trawled from the northwestern Atlantic. Although showing considerable scatter, the results suggested that these tiny bivalves reach reproductive maturity at about 50 years. The largest specimen was estimated to be around 100 years old, albeit with a confidence interval of plus or minus 76 years! The assumptions underlying this interpretation are discussed by Gage (1991), who points out that the function describing change in body mass from the pronounced sculptural rings on the shell was flawed because these rings probably are not laid down at the regular time interval assumed. Other data have come from study of shell-growth rings of bivalves from upper bathyal depths (Hutchings and Haedrich 1984; Gilkinson et al. 1986). These data suggest both rapid and seasonally intermittent growth similar to that of specimens from shallow depths (Skalkin and Tabunkov 1969). Other data on the probable age of deep-sea brachiopods and crinoids obtained from growth marks (Zezina 1975; Roux 1977; Duco and Roux 1981) and radiometric dating (Cochran 1982) are reviewed by Gage (1991). These data also suggest growth and age structure comparable to shallow-water populations.

The Rockall Study

Study of the large samples of echinoderms and bivalves species from a 2.9 km depth permanent station in the Rockall Trough (figure 11.1) has provided the first opportunity to study population dynamics, including recruitment and age structure, from a long time series in a deep-sea basin (Gage et al. 1980). The station is accessible from western ports in Scotland and England yet provides sufficient depth to render the results generally applicable to the abyssal zone in the deep sea. However, it is recognized that the Rockall Trough is a relatively productive area compared to more southerly latitudes in the North Atlantic. The implications of this regarding the population dynamics of deposit-feeding populations of the sediment, beyond allowing a greater biomass to be supported on a unit area basis, are not known. In practical terms, the relatively high deep-sea benthic standing crop of the Rockall Trough means that single epibenthic sledge tows usually provide adequate sample sizes for the more common species.

What species, then, should one study in order best to characterize recruitment and life-history patterns in the deep sea? Because of the high species richness and relatively high evenness in species abundance relations characteristic of the smaller-sized sediment-dwelling fauna of the deep sea, there is a relatively large suite of species whose numerical representation justifies them as community dominants, and hence as important structural components of the community. Of these species, polychaetes, peracarid crustaceans, molluscs, and echinoderms are the most numerous. In the Rockall study, the echinoderms and molluscs initially have been chosen for study because taxonomic knowledge of these groups is more advanced than in others. Taxonomic knowledge is important in population studies encompassing juvenile stages, whose identification may be difficult at best. The echinoderms and molluscs also provide skeletal hard parts showing growth marks which are useful in constraining size/age relations within certain limits in analyzing size structure and developing demographic models.

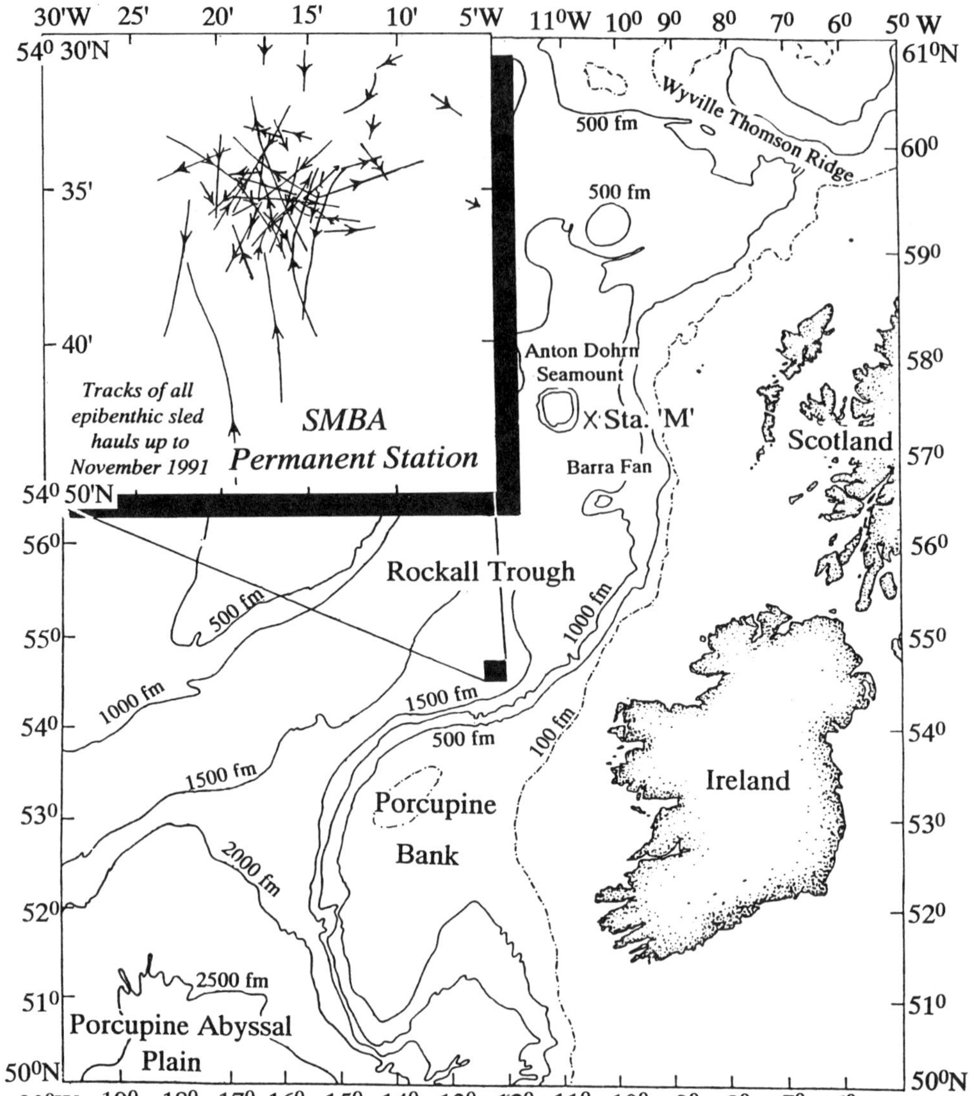

FIGURE 11.1. Bathymetric chart of Rockall Trough showing the location of the SMBA permanent station and Station "M." Inset shows the plotted tracks of the epibenthic sled hauls on the bottom.

Sample Variability

Size structure in samples from the deep sea are particularly influenced by factors such as mesh size and excessive sample washing during trawl recovery in bad weather (Gage et al. 1980). Weather-related factors affecting the degree of epibenthic-sled winnowing during recovery could bias against representation of small sizes in some samples. However, analysis of weather conditions experienced during sampling provide no support for this proposition; samples obtained during unusually calm conditions in winter contain no greater proportion of small individuals than those taken when sea conditions were rougher. However, on the few occasions when the mesh size of the main sampling bag of the epibenthic sled was reduced from 1 mm to 0.5 mm, a considerable increase was evident in the representation of smaller sizes. Another consequence of the smaller mesh size is a considerably smaller catch of animals; this and the limited seasonal coverage of 0.5 mm samples has, until recently, prevented us from evaluating size-related bias in the sample time series, particularly for fauna such as bivalves whose smooth, rounded shape makes them more easily lost through the net than the spiky small echinoderms.

Seasonal Recruitment

Populations of the numerically dominating species of echinoderm and bivalve mollusc are not necessarily dominated by adults. On the contrary, in the summer months the epibenthic sled samples yield numerous postlarvae, particularly of brittle stars, which presented a stark contrast to the samples taken in winter and spring (Lightfoot et al. 1979). It was found that many of these brittle star postlarvae belonged to a species, *Ophiocten gracilis,* normally found as an adult breeding population not at the upper abyssal permanent station, but at about 1000 m depth on the adjacent continental slope (Gage and Tyler 1981a). This species is a seasonal breeder known to release a planktotrophic ophiopluteus (formerly known as *Ophiopluteus ramosus*) whose occurrence in surface plankton far off the edge of the continental slope suggests dispersion prior to metamorphosis and settlement (Geiger 1963; Semenova et al. 1964; Tyler and Gage 1982). An apparently depth-related survivorship pattern could be tracked in the gradient of size structure, where the representation of adult sizes became rapidly reduced in samples along a bathymetric transect from the upper bathyal off the Hebrides to upper abyssal depths at the permanent station off western Ireland. The implication of this unexpected and apparently wasteful dispersal of larvae far off the continental slope is that in the deep sea as well as coastal areas planktotrophy requires high fecundity. Reasons for the depth-related differences in juvenile survival remain obscure.

A similar pattern, but with the gradient running in the opposite direction, occurs in the brittle star *Ophiura ljungmani* (Gage and Tyler 1981a). This upper abyssal species is thought on the basis of egg size, fecundity, and the large numbers of postlarvae appearing in summer, to have a planktotrophic ophiopluteus (Schoener 1972). At the permanent station, size structure extended up to the maximum known disc diameter, and adults formed a progressively smaller proportion of total summertime size distribution in shallower samples, with the postlarvae remaining conspicuous up to about 1600 m on the adjacent slope. As first demonstrated by Schoener (1968), this brittle star shows very clear seasonal recruitment, with postlarval frequencies appearing as a conspicuous "spike" in summer samples (Tyler and Gage 1980; Gage and Tyler 1981b).

The large epifaunal brittle star *Ophiomusium lymani* has a cosmopolitan distribution on the lower continental slope around ocean basins and, perhaps related to this wide latitudinal distribution, shows no evidence from study of gametogenic histology of seasonal breeding in the Rockall Trough. This species has been studied from large samples taken with both the epibenthic sled and Agassiz trawl at Station "M", a repeat station located north of the permanent station in soundings of about 2200 m. Samples of this species taken with the epibenthic sled, although smaller than those obtained with the Agassiz trawl, do include postlarval sizes. The size structure is characteristically bimodal, consisting of a peak in postlarval frequencies less than 3 mm disc diameter and a right-hand mode of adult frequencies (Gage 1985). Although the number of sorted epibenthic sled samples from Station "M" presently available is small, there is little indication from these that the number of the characteristically spiny postlarvae (see Schoener 1967) varies significantly with time of year. However, comparison of adult frequencies suggests a seasonal recruitment to adult sizes does occur. This results in a widely dispersed, low mode of presumably fast-growing subadults averaging about 8 mm disc diameter in samples taken during the following winter/spring which occur between the conspicuous postlarval and adult modes in winter/spring hauls (Gage and Tyler 1982a). A further low mode at about 18 mm is thought to represent individuals recruited during the previous summer. This suggests that recruitment of young of this species may nevertheless show a seasonal periodicity.

The Bivalvia include some macrofaunal species showing seasonal periodicities in gametogenesis, and others that possess small numbers of mature gametes in the gonad year round. The seasonally pulsed recruitment expected from seasonally breeding lecithotropic species, such as the protobranch bivalves *Ledella pustulosa* and *Yoldiella jeffreysi,* is less well defined than the pronounced postlarval spike in the smallest size frequencies seen in summertime samples of the brittle star *Ophiura ljungmani.* This probably is a direct result of differences in fecundity associated with the two contrasting strategies in early life history; maybe one or two orders of magnitude fewer propagules are released by lecithotrophic species than by planktotrophic species. However, unlike *Ophiomusium lymani.* if the size distributions of these two species are arranged according to collection date in a "composite" year (figure 11.2) then smaller sizes (corresponding to late postlarvae and subadults) clearly dominate summer samples.

The appearance of small sizes in the summer samples would be expected in the two seasonally breeding species, both of which release ripe eggs and sperm in winter (Lightfoot et al. 1979) and have lecithotrophic development (see Ockelmann 1965). Although the durations of a nonfeeding pelagic larval stages remain unknown in the deep sea, we might guess that the low temperature will depress the rate of development compared to shallow water. If slow develop-

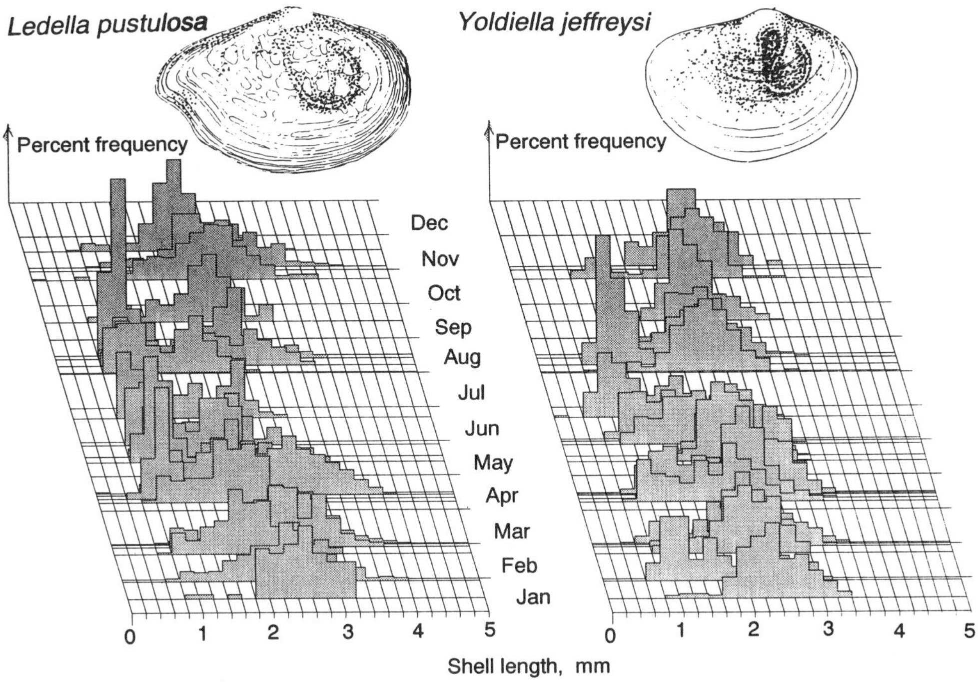

FIGURE 11.2. Shell-length frequencies for *Ledella pustulosa* (left) and *Yoldiella jeffreysi* (right) plotted as percentages with the distributions arranged as a "composite" single year in order to show the differing representation of juvenile frequencies at different times of the year. Sample sizes range from 42 to 426 (*L. pustulosa*) and 74 to 688 (*Y. jeffreysi*).

ment results in several weeks of dispersal, settlement would occur some time in spring or early summer, coinciding with the peak arrival of phytodetritus sinking from the surface at a rate of 100–150 m^{-1} (Billett et al. 1983). This would provide sufficient extra resources to allow for growth of these recruits so that they first appear in numbers in summertime samples. Of course, such a scenario could not apply to the nonseasonal breeders, *Malletia cuneata* and *Deminucula atacellana,* which also occur sufficiently abundantly in the samples to provide meaningful size frequency distributions. These two species appear to remain reproductively active throughout the year, and hence might be expected to produce otherwise similar lecithotrophic larvae all the year round. Yet, at least for *D. atacellana* size frequency distributions from the Rockall time series suggest a clear pattern of seasonally pulsed recruitment similar to those of *Ledella pustulosa* and *Yoldiella jeffreysi.*

The apparent pattern of seasonal recruitment, which can be inferred from the fluctuating representation of small individuals in the samples, might be caused by varying survivorship of postlarvae through the year. Certainly, the gametogenic cycle of *Ledella pustulosa* and *Yoldiella jeffreysi,* along with those of other seasonally breeding invertebrates within the Rockall Trough (Tyler et al. 1982), is roughly synchronized to produce young in the early part of the year. These benefit from the peak in detrital fallout from the surface that occurs in May/June in this part of the northeastern Atlantic (Billett et al. 1983; Lampitt 1985; Rice et al. 1986; Rice et al. 1991). Although rates of recruitment from these data would appear to be quite high, it cannot necessarily be assumed that the recruits that subsequently form a discrete cohort of subadults are all derived from settlement during the same year. It seems likely that growth of postlarvae may be quite slow, as described for inshore fauna (Muus 1973, 1981; Bosselmann 1991). Furthermore, some larval bivalves may have a protracted meiobenthic phase allowing them to await suitable conditions for growth in order to join the breeding population (e.g., Muus 1973; Webb 1984; Bachelet 1988).

Analysis of Age Structure

The discovery of seasonal breeding and recruitment from studies of gametogenic histology and population size structure at different times of the year (Schoener 1968; Lightfoot et al. 1979; Tyler and Gage 1980; Gage and Tyler 1981b) prompted application of classical techniques for analysis of population structure to the brittle star *Ophiura ljungmani* based on identifying and tracking the growth in modes in size frequencies (Gage and Tyler 1981b). Use of a conspicuous recruitment spike as a time marker for tracking the growth of cohorts usually is practicable for only a limited period after recruitment. This is because juvenile mortality quickly reduces numbers, and because of growth is increasingly variable with age. This means that size frequency distributions, even of large samples, often show little or no resolvable modal structure. Despite this difficulty, it has been possible to estimate age structure by means of growth curves fitted to the movement of the earliest recognizable modes in the Rockall time series (Gage and Tyler 1981b). These data suggest an annual periodicity in rate of growth that is supported by the presence of skeletal growth marks. The regularly spaced growth bands are similar to the seasonal growth zones in ossicles of shallow-water species. Their number and spacing, considered in relation to the size of the animal, provide independent support for the growth model developed for this species (Gage 1990a).

A growth banding pattern has also been found in the ossicles of the large, cosmopolitan brittle star *Opiomusium lymani*. However, for this species the large size of the postlarval "spike" in size frequencies compared to the low frequencies in subadult size classes suggests that only a small percentage of the small individuals survive to adulthood. By tracking the growth of the low juvenile mode in frequencies of consecutive samples it can be concluded that growth of subadults is rapid. In contrast, that of adults appears to be much slower, the demographic model developed to describe the underlying age structure of the marked unimodal adult peak being composed of a large stack of slowly growing year classes (Gage and Tyler 1982b). The heavily calcified adult brittle stars appear, unlike postlarvae, to be subject to only low mortality loss.

Ledella pustulosa and *Yoldiella jeffreysi* are examples of apparently seasonal

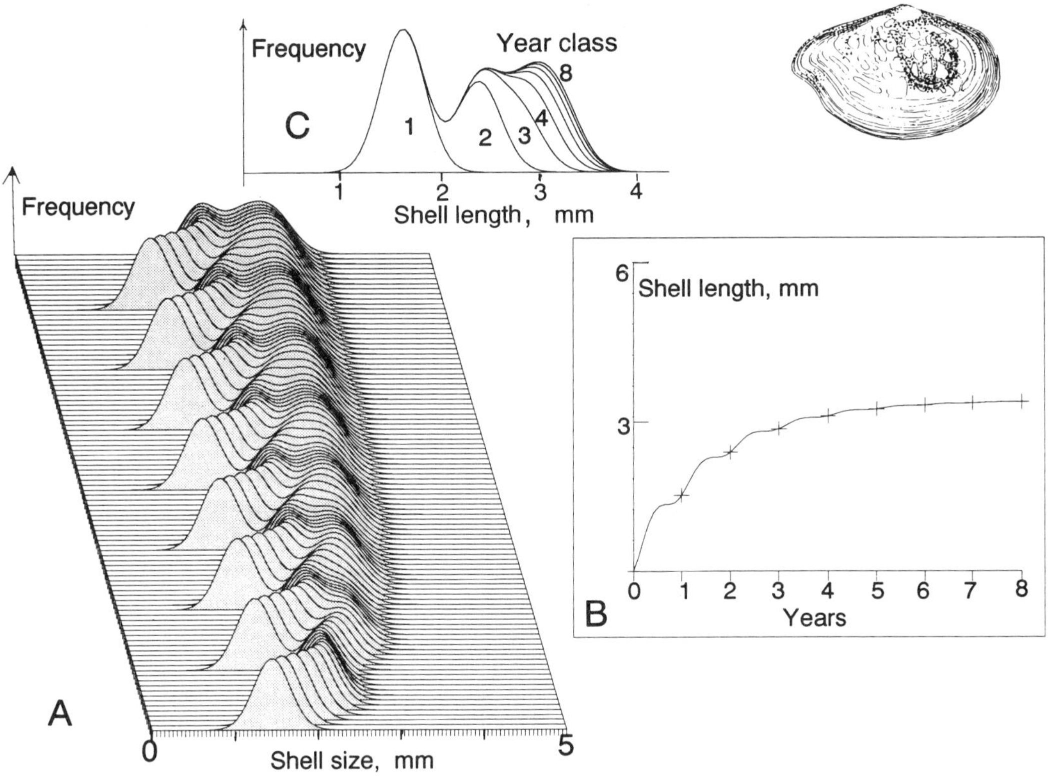

FIGURE 11.3. Demographic model for *Ledella pustulosa*. A: Size frequency curves plotted at monthly intervals from an initial recruitment of one-year old subadults at bottom, applying a seasonally oscillating von Bertalanffy growth function, giving rise to the "growth rings." The model size structure is shown being built up over 8 years, with regular and equal annual recruitments. B: Seasonally oscillating growth curve used by the model. C: Age-class composition of the final model distribution generated by A. Seasonally oscillating von Bertalanffy parameters, $t_o = -0.126$, $L_{oo} = 3.315$, K = 0.601, the constant C (expressing the intensity of oscillation) = 0.1 and t_s (expressing the fraction of a year between recruitment and the onset of first oscillation) = 0.5 where:

$$\text{shell length at time } t = L_{oo}\{1 - e^{-K(t-t^{O})} \pm C.K/2\pi \sin 2\pi(t-t^{S})\}$$

and the Mortality Coefficient $Z = 0.46$ where:

$$Z = -(\log_e N_{t+1} - \log_e N_t)$$

and N is the number of survivors from an initial recruitment of 1000 individuals.

breeders that show a seasonally varying size structure dominated in varying degree by immature sizes. Like other species that are sufficiently abundant to allow shell length measurements of large samples, the frequencies appear to show little resolvable modal structure that might be used for tracking growth-marked cohorts in consecutive samples. A ringlike growth pattern observed on the shells of *Ledella pustulosa* and *Yoldiella jeffreysi* (see photographs in Gage and Tyler 1991; Gage 1991) is thought from SEM examination to reflect annual growth checks, rather than shell sculpture (unpublished data). They are reminiscent of growth banding on the shells of inshore protobranch bivalves. However, it remains uncertain whether seasonal intermittency in shell growth is the result of metabolic stress of seasonal breeding, or reflects growth impetus provided by a seasonal pulse in labile detrital particles. Certainly, a regular pattern in shell growth banding seems to be absent in the nonseasonally breeding species *Malletia cuneata* and *Deminucula atacellana,* suggesting such growth marks are related to gametogenic periodicities.

The rapidly changing pattern in shell-size frequencies seen in small size classes of *Ledella pustulosa, Yoldiella jeffreysi* and *Deminucula atacellana* suggests initial growth after recruitment is rapid. Because recruitment of these species appears to be annually pulsed, growth models can be developed based on discrete annual recruitment waves passing into and incrementing an adult size distribution such as that shown in figure 11.3. The merging and stacking of frequencies shown in this model relating to size classes > ca 2 mm (when the gonad has started to develop a significant mass) clearly would make analysis by normal mixture analysis (see Macdonald and Pitcher 1979), or by tracking growth of cohort modes, unrewarding.

MODELING DEMOGRAPHIC STRUCTURE OF DEEP-SEA POPULATIONS

This approach extends the methods of Macdonald and Pitcher (1979) and Schnute and Fournier (1980) based on calculation of frequencies expected from a normal mixture model. A demographic model may be based on the parameters of a specified growth function, a mortality coefficient, and a suitable function describing any necessary changes in dispersion (standard deviation) of the normal frequencies relating to each age class. Such models are easily calculated on a desktop microcomputer by using a suitable normal-distribution approximation. Varying growth/survivorship scenarios can be plotted as frequency distributions in order to appreciate what options may give rise to observed frequencies (e.g., Gage 1985; Barry and Tegner 1989). Model parameters, constrained by a choice of tractable options and by data from growth lines or other information (such as results from sediment tray recolonization experiments), can be fitted to observed frequencies by means of nonlinear optimization. Models applied to the Rockall data considerably extend the approach of previous computer-based methods for size-frequency analysis by (1) including the option for fitting model frequencies to a time series of observations in a manner similar to the MULTIFAN likelihood-based method of Fournier et al. (1990); (2) by parameterization of a seasonally oscillating growth rate based on that described by Cloern and Nichols (1978); and (3) by allowing the search to find the optimum number of age classes present within certain limits. Such an approach employing several sets of frequency distributions taken

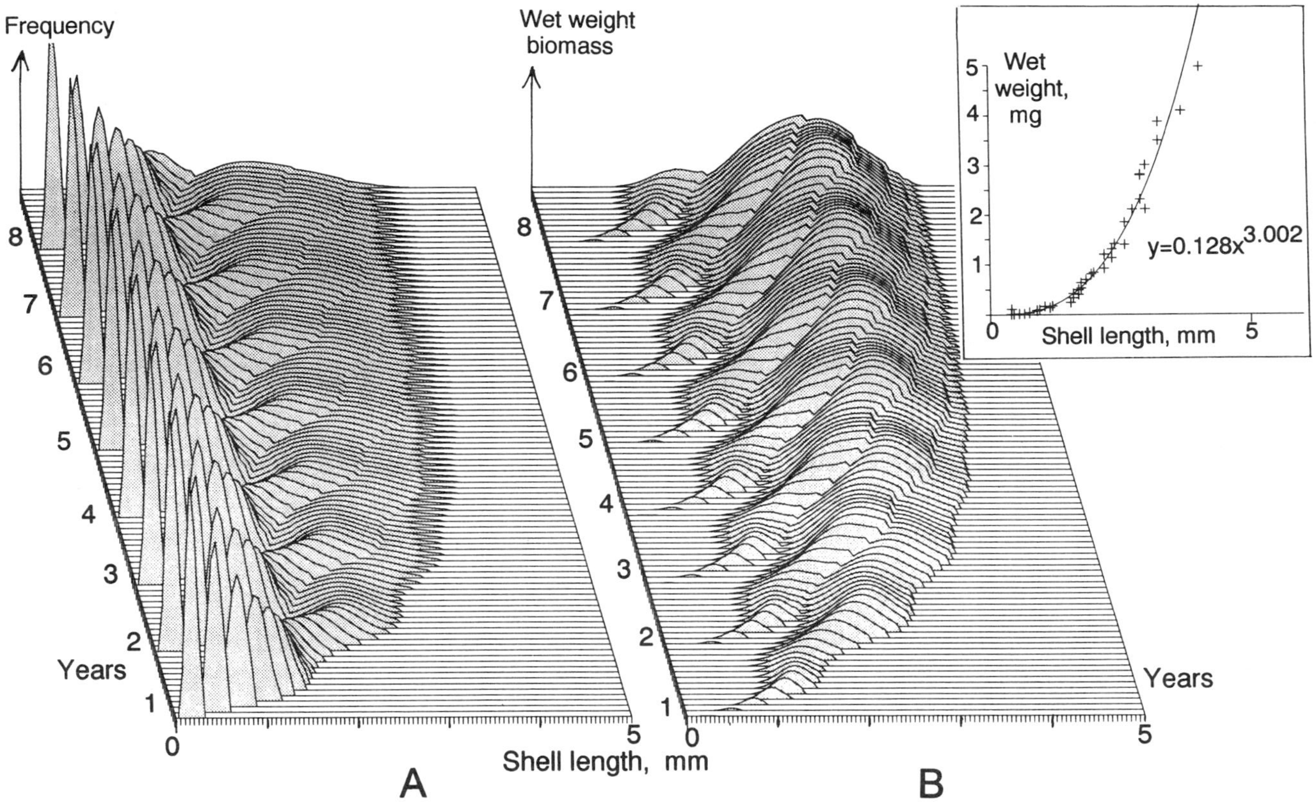

FIGURE 11.4. Demographic models for *Ledella pustulosa* including undersampled postlarval sizes. A: This model employs growth parameters identical to those for the model shown in figure 11.3, but differs in attempting to provide a simulation of postlarval growth from time soon after settlement, applying a 90% mortality during the first year and thereafter as for figure 11.3. B: Wet-weight production model for *Ledella pustulosa* based on frequencies in (A) transformed to wet-weight biomass by applying the power curve relationship shown fitted to body-size measurements in inset.

at different times of the year has been shown by Rosenberg and Beddington (1987) to yield less-biased estimates of number of significant age classes present than when single frequency distributions are analyzed separately. I will describe details of the present model and the accompanying software package separately.

The frequencies generated by demographic models fitted to the data on the bivalve *Ledella pustulosa,* are presented (figures 11.3 and 11.4) in a form comparable to that employed by Astheimer (1986) for Antarctic krill. That shown in figure 11.3 is based on fit to observed frequencies in a "composite" year. The size-frequency model shown in figure 11.4A is similar to that in figure 11.3, but includes postlarval sizes thought to be undersampled in the time series samples, and therefore is somewhat speculative regarding the size of the postlarval frequencies in relation to those appearing in the samples. Transformation of these model frequencies to biomass (figure 11.4B) suggests an instructive approach for assessing the relative importance of juveniles and adults as consumers competing for limited food resources.

Discussion

Recruitment and the Postlarval "Bottleneck" Theory

Available evidence (see Tyler 1988) suggests that larvae of many seasonally breeding species seem to be released early in the year well before the time when a spring bloom at the surface might be expected. However, the timing of the bloom varies from year to year, and several weeks may be required both for larvae to complete development at low temperature to a stage ready to settle on the deep-sea bed, and for the material to sink to the bottom as phytodetritus and provide a labile food source for postlarvae.

Jumars et al. (1990) reasoned, on inferences based on constraints on size-related gut volumes and residence times, that juvenile and adult food resources may differ for deposit feeders. In order to achieve digestive efficiency, gut residence times of ingested material must match digestive and absorptive kinetics. Because ingestion rate and gut volume influence residence time, a small gut volume results in selective pressure toward specialization on food material that can return high digestive dividends rapidly. Jumars et al. (1990) point out that this may lead to development needing to bridge a selective dichotomy between macrophagy by meiofauna (including small postlarvae) and microphagy by macrofaunal and smaller megafaunal deposit feeders, which of course include the adults of the species considered in this paper. The transition corresponds to the marked trough in equivalent spherical diameter of about 1 mm in plots of size-related biomass and species abundance (Schwinghamer 1985). Jumars et al. (1990) suggest that bridging this gap may well depend on windfalls of excess labile food material, such as the seasonal phytodetrital layers on the bed in the northeastern Atlantic (Billett et al. 1983).

In this scenario, a pool of slow- or nongrowing postlarvae exist as components of the meiofauna. Postlarval growth may be stimulated by seasonal availability of nutritious food particles in labile planktonic detritus. However, they will be subject to intense competition with other meiofauna, and with macro- and megafaunal deposit-feeding microphages, possibly feeding preferentially on aggregates of these nutritious particles (Billett et al. 1983, 1988; Lampitt 1985; Graf 1989). The postlarval pool will also decline in numbers as a result of the nonparticle selective sediment grazing by these macro- and megafauna, as well as from selective predation by smaller predators. The potential vulnerability of even large numbers of larvae, in competition with the microphagous deposit feeding of adult sizes, is indicated in simulations of population dynamics where the relatively large peaks in size frequencies, when transformed to biomass, become insignificant (figure 11.4). However, their numbers are augmented by irregular but year-round settlement of lecithotrophic larvae.

The apparently inexplicable summertime maxima in numbers of young of nonseasonal breeders will reflect growth stimulated by this temporary increase in nutritional resources. In seasonal breeders, the size of the recruitment pool will vary within wider limits, with a substantial additional input in summer (coinciding with maximum availability of phytodetritus) after breeding in winter/spring. In planktotrophic species of echinoderms and bivalves, fecundity is higher (Schoener 1972; Lightfoot et al. 1979) and the seasonal appearance of postlarval sizes may be more numerically obvious, as with *Ophiura ljungmani*, but these high densities are subject to heavier mortality in the early post-settlement stages. Such species seem to be in the minority in the deep sea, and seasonal breeding is thought as a whole to apply to only a small percentage of deep-sea invertebrate species investigated (Tyler 1988).

Overall, the postlarval bottleneck means that nutritional constraints and probably competition with, and vulnerability to grazing by, larger deposit-feeding adult stages, might determine why population dynamics for this postlarval pool will be different from that of adults. However, we are far from a complete understanding of the dynamics of the postlarval pool. A first step toward a better understanding of this important period in the life history of benthic animals will be an improvement in the quantitative sampling of small individuals.

Is there a Characteristic Life-History Strategy among Deep-Sea Invertebrates?

The data reviewed in this paper, though still sparse and incomplete, have come from the efforts of several deep-sea biologists, and from a wide variety of deep-sea taxa. These make clear, as shown by Gage (1991), that the prediction of the deep sea as a nonseasonal environment supporting a low activity ecosystem, with its component populations showing unusually slow rates of recruitment and growth and an age structure dominated by old individuals, has no universal validity. Furthermore, we now have tantalizing indications not only of seasonal periodicities in populations, but also of interannual changes (Gage 1990b) that place the deep-sea ecosystem firmly in the context of comparable events known in shallow water. Despite some adaptation to a low energy environment, such as generalist lifestyles in deposit feeders, rates of other processes with important survival value, such as the rapid population growth of xylophaginous bivalves, appear broadly similar to those in shallow water environments. However, there are indications that, where comparisons with closely related taxa are possible, those found deepest show the lowest fecundity (e.g., R. S. Scheltema 1972) and slowest rate of growth, and therefore greatest longevity (e.g., Gage et al. 1986). The important conclusion remains reinforced, particularly from results of studies from hydrothermal vents and from some of the more "normal" small deposit-feeding macrofauna from Rockall, that the life history characteristics of low reproductive production (and hence low recruitment to the adult population), slow growth to maturity, and great age are traits that are by no means the general rule in the deep sea.

ACKNOWLEDGMENTS

The Rockall time series work would not have been possible without continuing support of Academic Research Project 2 of the Scottish Marine Biological Association through grant aid from the U.K. Natural Environment Research Council. I also acknowledge the help and interest of many friends and colleagues, especially Dr. Paul A. Tyler of the Department of Oceanography at Southampton University, and the support received from the officers and crew of R.S.S. *Challenger* and other vessels of the N.E.R.C. research vessel fleet since 1973. I also acknowledge the support received since 1990 from Contract 037 in the Commission of European Communities MAST I program.

REFERENCES

Ansell, A. D. and A. H. Parulekar. 1978. On the rate of growth of *Nuculana minuta* (Muller) (Bivalvia: Nuculanidae). *J. Mollusc. Stud.* 44:71–82.

Astheimer, H. 1986. A length class model of the population dynamics of the Antarctic krill *Euphasia superba* Dana. *Mar. Biol.* 6:227–232.

Bachelet, G. 1989. Recruitment in *Abra tenuis* (Montagu) (Bivalvia, Semelidae), a species with direct development and a protracted meiobenthic phase. In J. S. Ryland and P. A. Tyler, eds., *Reproduction, Genetics, and Distributions of Marine Organisms,* pp. 23–30. Copenhagen: Olsen and Olsen.

Barry, J. P. and M. J. Tegner. 1989. Inferring demographic processes from size-frequency distributions: Simple models indicate specific patterns of growth and mortality. *Fish. Sci. U.S.* 88:13–19.

Berg, C. J. and C. L. Van Dover. 1987. Benthopelagic macrozooplankton communities at and near deep-sea hydrothermal vents in the eastern Pacific Ocean and Gulf of California. *Deep-Sea Res.* 34A:379–401.

Billett, D. S. M., R. S. Lampitt, A. L. Rice, and R. F. C. Mantoura. 1983. Seasonal sedimentation of phytoplankton to the deep-sea benthos. *Nature (Lond.)* 302:520–522.

Billett, D. S. M., C. Llewellyn, and J. Watson. 1988. Are deep-sea holothurians selective deposit feeders? In R. D. Burke, P. V. Mladenov, P. Lambert, and R. L. Parsley, eds., *Echinoderm Biology,* pp. 421–429. Rotterdam: Balkema.

Bossellmann, A. 1991. Recruitment and postlarval growth of some macrozoobenthos species in the German Bight. *Meeresforschung* 33:141–158.

Cloern, J. E. and F. H. Nichols. 1978. A von Bertalanffy growth model with seasonally varying coefficient. *J. Fish. Res. Bd. Can.* 35:1479–1482.

Cochran, J. K. 1982. The use of naturally occurring radionuclides as tracers for biologically related processes in deep-sea sediments. In W. G. Ernst and J. G. Morin, eds., *The Environment of the Deep Sea,* pp. 55–72. Englewood Cliffs, N.J.: Prentice-Hall.

Desbruyères, D., J. Bevas, and A. Khripounoff. 1980. Un cas de colonisation rapide d'une sediment profond. *Oceanol. Acta* 3:285–291.

Desbruyères, D., J. Deming, A. Dinet, and A. Khripounoff. 1985. Réactions de l'écosysteme benthique profond aux perturbations: nouveaux résultats expérimentaux. In L. Laubier and C. Monniot, eds., *Peuplements Profonds du Golfe de Gascogne,* pp. 121–142. Brest: Institut Français de Recherche pour l'Exploitation de la Mer.

Duco, A. and M. Roux. 1981. Modalités particulières de croissance liées au milieu abyssal chez les Bathycrinidae (Échinodermes, Crinoides pédoncules). *Oceanol. Acta* 4:389–393.

Fournier, D. A., J. R. Sibert, J. Majowski, and J. Hampton. 1990. MULTIFAN a likelihood-based method for estimating growth parameters and age composition from multiple length-frequency data sets illustrated using data for Southern Bluefin Tuna (*Thynnus maccoyii*). *Can. J. Fish. Aquat. Sci.* 47:301–317.

Fustec, A., D. Desbruyères, and S. K. Juniper. 1987. Deep-sea hydrothermal vent communities at 13°N on the East Pacific Rise: Microdistribution and temporal variations. *Biol. Oceanogr.* 4:99–164.

Gage, J. D. 1977. Structure of the abyssal macrobenthic community in the Rockall Trough. In B. F. Keegan, P. O'Ceidigh, and P. J. S. Boaden, eds., *Biology of Benthic Organisms,* pp. 247–260. Oxford: Pergamon.

Gage, J. D. 1985. The analysis of population dynamics in deep-sea benthos. In P. E. Gibbs, ed., *Proceedings of the Nineteenth European Marine Biology Symposium,* pp. 201–212. Cambridge: Cambridge University Press.

Gage, J. D. 1990a. Skeletal growth markers in the deep-sea brittle stars *Ophiura ljungmani* and *Ophiomusium lymani. Mar. Biol.* 104:427–435.

Gage, J. D. 1990b. Annual and interannual changes in benthic populations in the Rockall Trough. *J. Mar. Biol. Ass. U. K.* 70, 671 (abstract).

Gage, J. D. 1991. Biological rates in the deep sea: A perspective from studies on processes in the benthic boundary layer. *Rev. Aquat. Sci.* 5:49–100.

Gage, J. D. and P. A. Tyler. 1981a. Non-viable seasonal settlement of larvae of the upper bathyal brittlestar *Ophiocten gracilis* in the Rockall Trough abyssal. *Mar. Biol.* 64:153–161.

Gage, J. D. and P. A. Tyler. 1981b. Reappraisal of age composition, growth, and survivorship of the deep-sea brittle star *Ophiura ljungmani* from size structure in a time series from the Rockall Trough. *Mar. Biol.* 64:163–174.

Gage, J. D. and P. A. Tyler. 1982a. Growth and reproduction in the deep-sea brittlestar *Ophiomusium lymani* Wyville Thomson. *Oceanol. Acta* 5:73–83.

Gage, J. D. and P. A. Tyler. 1982b. Growth strategies in deep-sea ophiuroids. In J. M. Lawrence, ed., *Echinoderms: Proceedings of the International Conference, Tampa Bay,* pp. 305–311. Rotterdam: Balkema.

Gage, J. D. and P. A. Tyler. 1991. *Deep-Sea Biology: A Natural History of Organisms at the Deep-Sea Floor.* Cambridge: Cambridge University Press.

Gage, J. D., R. H. Lightfoot, M. Pearson, and P. A. Tyler. 1980. An introduction to a sample time series of abyssal macrobenthos: Methods and principle sources of variability. *Oceanol. Acta* 3:169–176.

Gage, J. D., P. A. Tyler, and D. Nichols. 1986. Reproduction and growth of *Echinus acutus* var. *norvegicus* and *E. elegans* on the continental slope off Scotland. *J. Exp. Mar. Biol. Ecol.* 101:61–83.

Geiger, S. R. 1963. *Ophiopluteus ramosus* between Iceland and Newfoundland. *Nature (Lond.)* 198:908–909.

Gilkinson, K. D., J. A. Hutchings, P. E. Oshel, and R. A. Haedrich. 1986. Shell microstructure and observations on internal banding patterns in the bivalves *Yoldia thraciaeformis* Storer, 1838, and *Nuculana pernula* Muller, 1779 (Nuculanidae), from a deep-sea environment. *Veliger* 29:70–77.

Graf, G. 1989. Benthic-pelagic coupling in a deep-sea community. *Nature (Lond.)* 342:437–439.

Grassle, J. F. 1977. Slow recolonization of deep-sea sediment. *Nature (Lond.)* 265:618–619.

Grassle, J. F. 1986. The ecology of deep-sea hydrothermal vent communities. *Adv. Mar. Biol.* 23:301–362.

Grassle, J. F. and L. S. Morse-Porteous. 1987. Macrofaunal colonization of disturbed deep-sea environments and the structure of deep-sea benthic communities. *Deep-Sea Res.* 34:1911–1950.

Grassle, J. F. and H. L. Sanders. 1973. Life histories and the role of disturbance. *Deep-Sea Res.* 20:643–659.

Hessler, R. R., W. M. Smithey, M. A. Boudrias, C. H. Keller, R. A. Lutz, and J. J. Childress. 1988. Temporal change in the megafauna at the Rose Garden hydrothermal vent (Galapagos Rift: eastern tropical Pacific). *Deep-Sea Res.* 35A:1681–1709.

Hutchings, J. A. and R. A. Haedrich. 1984. Growth and population structure in two species of bivalves (Nuculanidae) from the deep sea. *Mar. Ecol. Prog. Ser.* 17:135–142.

Jumars, P. A., L. M. Mayer, J. W. Deming, J. A. Baross and R. A. Wheatcroft. 1990. Deep-sea deposit feeding strategies suggested by environmental and feeding constraints. *Phil. Trans. R. Soc. Lond.* Ser. A 331:85–101.

Kukert, H. and C. R. Smith. 1992. Disturbance, colonization, and succession in a deep-sea sediment community: Artificial-mound experiments. *Deep-Sea Res.* 39A:1349–1371.

Lampitt, R. S. 1985. Evidence for seasonal deposition of detritus to the deep-sea floor and its subsequent resuspension. *Deep-Sea Res.* 32A:885–897.

Laubier, L. and D. Desbruyères. 1984. *Des Oasis au Fond des Mers.* Paris: Le Rocher.

Lewis, J. B., S. Saleh, H. M. Reiswig, and C. M. Lalli. 1982. Growth, production, and biomass of the burrowing protobranch mollusc *Yoldia limatula* in the Bideford River, Prince Edward Island, Canada. *Mar. Biol.* 70:173–179.

Lightfoot, R., P. A. Tyler, and J. D. Gage. 1979. Seasonal reproduction in deep-sea bivalves and brittlestars. *Deep-Sea Res.* 26A:967–973.

Lutz, R. A., L. W. Fritz, and D. C. Rhoads. 1983. Molluscan growth at deep-sea hydrothermal vents. *Bull. Biol. Soc. Wash.* 6:199–210.

Lutz, R. A., L. W. Fritz, and R. M. Cerrato. 1988. A comparison of bivalve (*Calyptogena magnifica*) growth at two deep-sea hydrothermal vents in the eastern Pacific. *Deep-Sea Res.* 35A:1793–1810.

Macdonald, P. D. M. and T. J. Pitcher. 1979. Age groups from size-frequency data: A versatile and efficient method of analyzing distribution mixtures. *Can. J. Fish. Aquat. Sci.* 36:987–1001.

McHugh, D. 1989. Population structure and reproductive biology of two sympatric hydrothermal vent polychaetes, *Paralvinella pandorae* and *P. palmiformes*. *Mar. Biol.* 103:95–106.

Muus, K. 1973. Settling, growth, and mortality of young bivalves in the Oresund. *Ophelia* 12:79–116.

Muus, K. 1981. Density and growth of juvenile *Amphiura filiformis* (Ophiuroidea) in the Oresund. *Ophelia* 20:153–168.

Ockelmann, K. W. 1965. Developmental types in marine bivalves and their distribution along the Atlantic coast of Europe. In L. R. Cox and J. Peake, eds., *Proceedings of the First European Malacological Congress, London, 1962,* pp. 25–35. London: Conchological Society of Great Britain and Ireland and the Malacological Society of London.

Rachor, E. 1976. Structure, dynamics, and productivity of a population of *Nucula nitidosa* (Bivalvia, Protobranchiata) in the German Bight. *Ber. Dtsch. Wiss. Komm. Meeresforsch.* 24:296–331.

Rhoads, D. C., R. A. Lutz, R. M. Cerrato, and E. C. Revelas. 1982. Growth and predation activity at deep-sea hydrothermal vents along the Galapagos Rift. *J. Mar. Sci.* 40:503–516.

Rice, A. L., D. S. M. Billett, J. Fry, A. W. G. John, R. S. Lampitt, R. F. C. Mantoura, and R. J. Morris. 1986. Seasonal deposition of phytodetritus to the deep-sea floor. *Proc. R. Soc. Edinb.* 88B:265–279.

Rice, A. L., D. S. M. Billett, M. H. Thurston, and R. S. Lampitt. 1991. The Institute of Oceanographic Sciences biology programme in the Porcupine Seabight: background and general introduction. *J. Mar. Biol. Ass. U. K.* 71:281–310.

Richards, K. J. 1990. Physical processes in the benthic boundary layer. *Phil. Trans. R. Soc. Lond.* Ser. A 331:3–13.

Rosenberg, A. A. and J. R. Beddington. 1987. Monte-Carlo testing of two methods for estimating growth from length-frequency data with general conditions for their applicability. In D. Pauly and G. R. Morgan, eds., *Length-Based Methods in Fisheries Research, ICLARM Conference Proceedings* 13:283–298. Manila, Philippines and Safat, Kuwait: International Center for Living and Aquatic Resources and Kuwait Institute for Scientific Research.

Roux, M. 1977. Les Bourguetcrinina (Crinoidea) recuillis par la "Thalassa" dans le golfe de Gascogne: Anatomie comparee des pedoncles et systematique. *Bull. Mus. Nat. Hist. Nat. (Zool.), Paris* Ser. 3, 426, 425–484.

Scheltema, A. H. 1987. Reproduction and rapid growth in a deep-sea aplacophoran mollusc, *Prochaetoderma yongei. Mar. Ecol. Prog. Ser.* 37:171–180.

Scheltema, R. S. 1972. Reproduction and dispersal of bottom-dwelling deep-sea invertebrates: a speculative summary. In R. W. Brauer, ed., *Barobiology and Experimental Biology of the Deep Sea,* pp. 58–68. Chapel Hill: University of North Carolina Press.

Schnute J. and D. Fournier. 1980. A new approach to length-frequency analysis: Growth structure. *Can. J. Fish. Aquat. Sci.* 37:1337–1351.

Schoener, A. 1967. Post-larval development in five deep-sea ophiuroids. *Deep-Sea Res.* 14:645–660.

Schoener, A. 1968. Evidence for reproductive periodicity in the deep-sea. *Ecology* 49:81–87.

Schoener, A. 1972. Fecundity and possible mode of development of some deep-sea ophiuroids. *Limnol. Oceanogr.* 17:193–199.

Schwinghamer, P. 1985. Observations on size structure and pelagic coupling of some shelf and abyssal benthic communities. In P. E. Gibbs, ed., *Proceedings of the Nineteenth European Marine Biology Symposium,* pp. 347–359. Cambridge: Cambridge University Press.

Semenova, T. N., S. A. Mileikovsky, and K. N. Nesis. 1964. The morphology, distribution, and seasonal incidence of the ophiuroid larva of *Ophiocten sericeum* (Forbes) in the northwest Atlantic, Norwegian Sea, and Barents Sea. *Okeanologiya (Moscow)* 4:669–683 (in Russian).

Skalkin, V. A. and V. D. Tabunkov. 1969. Biology, distribution, and reserves of *Nuculana pernula* (Taxodonta, Nuculanidae) of south-eastern littoral of Sakhalin. *Zool. Zhur.* 48:1147–1155 (in Russian).

Smith, C. R. 1985. Colonization studies in the deep sea: Are results biased by experimental design? In P. E. Gibbs, ed., *Proceedings of the Nineteenth European Marine Biology Symposium,* pp. 183–190. Cambridge: Cambridge University Press.

Smith, C. R. 1986. Nekton falls, low intensity disturbance, and community structure of infaunal benthos in the deep sea. *J. Mar. Res.* 44:567–600.

Smith, C. R. and S. J. Brumsickle. 1989. The effects of patch size and substrate isolation on colonization modes and rates in an intertidal sediment. *Limnol. Oceanogr.* 34:1263–1277.

Smith, C. R. and R. R. Hessler. 1988. Colonization and succession in deep-sea ecosystems. *Trends Ecol. Evol.* 2:359–363.

Smith, C. R., P. A. Jumars, and D. J. DeMaster. 1986. *In situ* studies of megafaunal mounds indicate rapid sediment turnover, and community response at the deep-sea floor. *Nature (Lond.)* 323:251–253.

Tunnicliffe, V. 1991. The biology of hydrothermal vents: Ecology and evolution. *Oceanogr. Mar. Biol. Ann. Rev.* 29, 319–407.

Turekian, K. K., D. P. Cochran, D. P. Kharkar, R. M. Cerrato, J. R. Vaisnys, H. L. Sanders, J. F. Grassle, and J. A. Allen. 1975. Slow growth rate of deep-sea clam determined by ^{228}Ra chronology. *Proc. Natl. Acad. Sci. U.S.A.* 180:2829–2832.

Turekian, K. K., J. K. Cochran, and Y. Nozaki. 1981. Growth rate of a clam from the Galapagos hot spring field using natura; radionuclide ratios. *Nature (Lond.)* 280:385–387.

Turner, R. D. 1973. Wood-boring bivalves, opportunistic species in the deep sea. *Science (Wash.)* 180:1377–1379.

Turner, R. D. 1977. Wood, molluscs, and deep-sea food chains. *Bull. Am. Malacol. Union* 1977:13–19.

Turner, R. D. and R. A. Lutz. 1984. Growth and distribution of molluscs at deep-sea vents and seeps. *Oceanus* 27:54–62.

Tyler, P. A. 1988. Seasonality in the deep sea. *Oceanogr. Mar. Biol. Ann. Rev.* 26:227–258.

Tyler, P. A. and J. D. Gage. 1980. Reproduction and growth of the deep-sea brittlestar *Ophiura ljungmni* (Lyman). *Oceanol. Acta* 3:177–185.

Tyler, P. A. and J. D. Gage. 1982. *Ophiopluteus ramosus,* the larval form of *Ophiocten gracilis* (Echinodermata: Ophiuroidea) from the Rockall Trough. *J. Mar. Biol. Ass. U. K.* 62:485–486.

Tyler, P. A., A. Grant, S. L. Pain, and J. D. Gage. 1982. Is annual reproduction in deep-sea echinoderms a response to variability in their environment? *Nature (Lond.)* 300:747–749.

Van Dover, C. L., C. J. Berg, and R. D. Turner. 1988. Recruitment of marine invertebrates to hard substrates at deep-sea hydrothermal vents on the east Pacific Rise and Galapagos spreading center. *Deep-Sea Res.* 35A:1833–1849.

Webb, C. M. 1984. Seasonal Variations and Settlement of Some Marine Benthic Invertebrate Larvae on Sediments. Ph.D. thesis, University of Wales.

Zajac, R. N. and R. B. Whitlach. 1982. Responses of estuarine infauna to disturbance. I. Spatial and temporal variation of initial recolonization. *Mar. Ecol. Prog. Ser.* 10:12–37.

Zezina, O. N. 1965. On some deep-sea brachiopods from the Gay Head-Bermuda transect. *Deep-Sea Res.* 22:903–912.

CHAPTER TWELVE

Life History Studies of Deep-Sea Benthic Infauna: Polychaeta, Aplacophora, and Cumacea from the Continental Slope off Massachusetts

James A. Blake and Les Watling

ABSTRACT

We studied the reproductive biology and life history of three species of benthic invertebrates from the continental slope off Massachusetts. *Aurospio dibranchiata,* a widely distributed deep-sea spionid polychaete from depths of 1250–2100 m exhibited seasonality in egg production, but no evidence of seasonal recruitment. Individuals at one 1350 m station on the Canadian border were consistently larger than individuals from 2100 m. *Prochaetoderma yongei,* an aplacophoran mollusc from four stations in middle slope depths of 1250–1350 m, appears to reproduce continuously, with no evidence of seasonality. *Diastylis stygia,* a cumacean, from the lower continental slope in depths of 2065–2115 m was, aside from one pulse of juveniles in April 1985, characterized by variable, but constant recruitment. It is likely that small-scale variability in recruitment is due to local responses to fluxes of phytal detritus that have been observed on the seafloor at the study site.

There have been few studies on the reproduction and life history of deep-sea benthic invertebrates because of the lack of seasonal collections and the lack of specimens in the quantity or condition necessary to adequately assess reproductive processes. Further, apart from the long time series on the larger megafaunal taxa in the Rockall Trough off Ireland and Scotland, there have been few opportunities to obtain seasonal samples to assess life history parameters. From 1983 to 1987, the U.S. Department of the Interior, Minerals Management Service (MMS), conducted the U.S. Atlantic Continental Slope and Rise Program (ACSAR) that was designed to assess biological processes in a deep-sea environment. The infaunal collections available from the replicated box cores taken during the ACSAR program have provided a unique resource for assessing reproductive and life history processes in the smaller benthic infaunal taxa. The ACSAR program included intensive studies of benthic infaunal community structure from the Canadian boundary off Massachusetts to off South Carolina (Blake et al. 1987; Maciolek et al. 1987a, b). The U.S. North ACSAR program included 15 stations ranging in depth from 255 m to 2180 m.

In order to develop additional data on seasonality of reproduction and recruitment in deep-sea benthic communities and to learn more about the reproduction and life history characteristics of the benthic infauna of the U.S. North ACSAR, we analyzed life-history parameters on three dominant species: *Aurospio dibranchiata,* a spionid polychaete; *Prochaetoderma yongei,* an aplacophoran; and *Diastylis stygia,* a cumacean. These species were chosen on the basis of their abundance in the benthic communities at particular depth ranges.

Methods

Three replicate samples were collected at each station with a 0.25 m^2 Hessler-Sandia box core. A subsample consisting of 0.09 m^2 was used for infaunal analysis. Samples were sieved through a screen having a mesh diameter of 300 μm. Samples were taken in the following months: November 1984 (Cruise 1), April 1985 (Cruise 2), July 1985 (Cruise 3), November 1985 (Cruise 4), April 1986 (Cruise 5), and July 1986 (Cruise 6). Station locations are shown in figure 12.1.

Aurospio dibranchiata was selected for analysis from two stations (3 and 13) at 1250 m and 1350 m, respectively, and five 2100-m stations (2, 5, 6, 8, 14, and 15). *Prochaetoderma yongei* was analyzed from four stations (3, 9, 10, and 13) representing the 1250-m to 1350-m depth range. *Diastylis stygia* was analyzed from one station (5) at 2065 m, where it occurred at particularly high densities, and from station 6 (2115 m). All individuals present in the samples were measured. Samples were not taken on Cruise 3 at stations 6, 13, and 15.

For *Aurospio dibranchiata,* measurements were made of thoracic width,

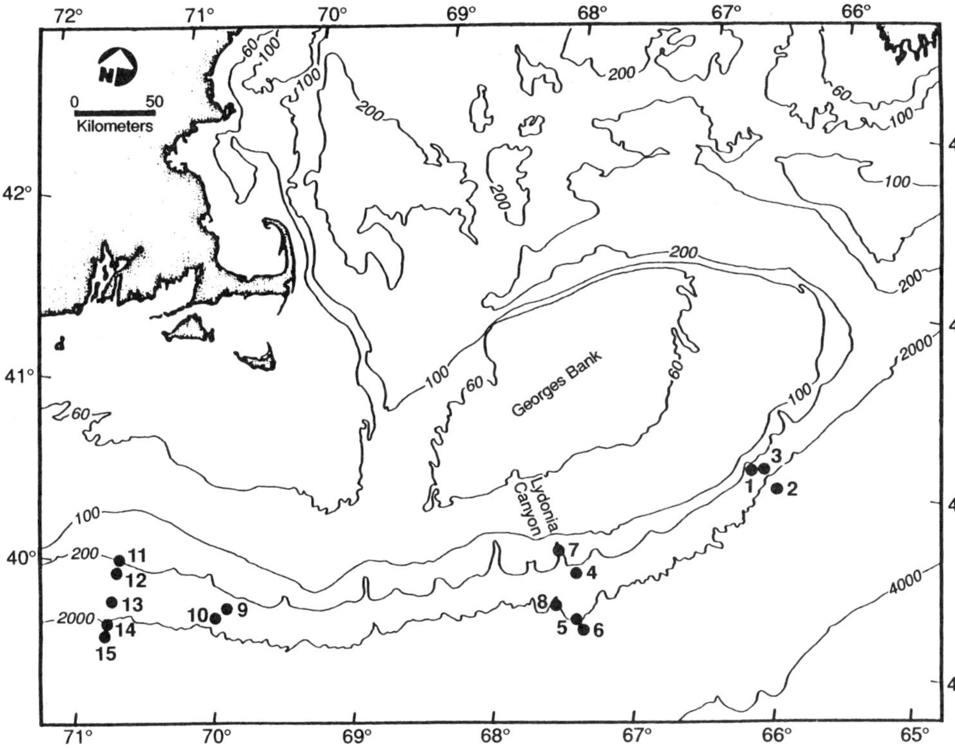

FIGURE 12.1. Map of the of the North Atlantic study area off Massachusetts.

prostomial length, length of the branchial region, and number of branchial pairs. A regression analysis of these measurements against the total number of setigers was made in order to determine the best fit. This method was necessary because *A. dibranchiata* and other soft-bodied polychaetes are frequently missing posterior body parts (Blake and Baptiste 1985; Blake 1993). In *Prochaetoderma yongei,* total length proved to be the best measurement; width was highly variable because of preservation artifacts. For *Diastylis stygia,* measurements were taken from the rostrum to the posterior end of pereonite 5; lengths were recorded to the nearest 0.1 mm. Each of the *D. stygia* specimens measured was also assigned to one of the following life history stages: manca, juvenile female, juvenile male, ovigerous female, or immature male with pleopods. At least two size modes were contained in each of the manca and juvenile categories; this was expected because the developmental sequence for this species includes two manca stages and at least two juvenile stages. For the purposes of this study, these stages were not differentiated.

All specimens of *Aurospio dibranchiata* and *Prochaetoderma yongei* were examined under the compound microscope for reproductive characteristics. The presence of eggs (or oocytes), their size and location in or on the body, the presence, location, and arrangement of sperm or spermatophores, the presence and degree of brooded young, and the presence of any obvious or unusual reproductive structures were noted.

Results

AUROSPIO DIBRANCHIATA MACIOLEK (FIGURES 12.2–12.6)

Aurospio dibranchiata is widely distributed in the North and South Atlantic Ocean (Maciolek 1981a) and has been reported from the Pacific Ocean (Maciolek 1981b). These reports have indicated that the species ranges from 300 to 3600 m in depth but is most common below 1000 m. This species is one of the dominants along the 2000-m to 2100-m isobath on the U.S. South, Mid-, and North Atlantic continental slope and rise. It is the most abundant species at many of the stations (Maciolek et al. 1987a, b). The only previous data on reproduction and life history is from off North Carolina (Blake 1993).

As part of the original description of *Aurospio dibranchiata,* Maciolek (1981a) noted gravid females with eggs measuring 50–96 μm across their widest dimension. Blake (1993) reported eggs measuring up to 112 μm from a 2000-m site off Cape Lookout, North Carolina. He noted that the largest eggs were found in November 1985, and no eggs were observed at all in spring collections, suggesting a seasonal pattern. In the present study, most of the large eggs were also observed in November 1985, with mean diameters of 100 μm and 106 μm at stations 5 and 6 respectively (table 12.1). The largest eggs recorded were found at station 6 in the November 1985 collection (126 μm) and at station 5 in the July 1986 collection (up to 134 μm); the smallest egg diameters in April occurred at station 2, the deepest station on the most northerly transect. All stations except station 13 exhibited higher mean egg diameters in the November 1985 collection. Egg diameters from station 13 increased in size in the following spring and summer collections. These data indicate

TABLE 12.1 *Mean Egg Diameters (μm) for* Aurospio dibranchiata *at Eight U.S. North Atlantic Continental Slope Stations on Six Sampling Dates*

Date/Stations	2	3	5	6	8	13	14	15
Nov 1984	NE	80	74	72	80	91	91	69
Apr 1985	21	53	62	NE	67	64	72	70
Jul 1985	NE	61	73	NS	65	NS	NE	NS
Nov 1985	47	83	100	106	98	64	NE	76
Apr 1986	20	70	63	48	70	77	NE	62
Jul 1986	86	70	72	87	67	80	NE	69

NE = No eggs seen in animals examined.
NS = No samples taken.

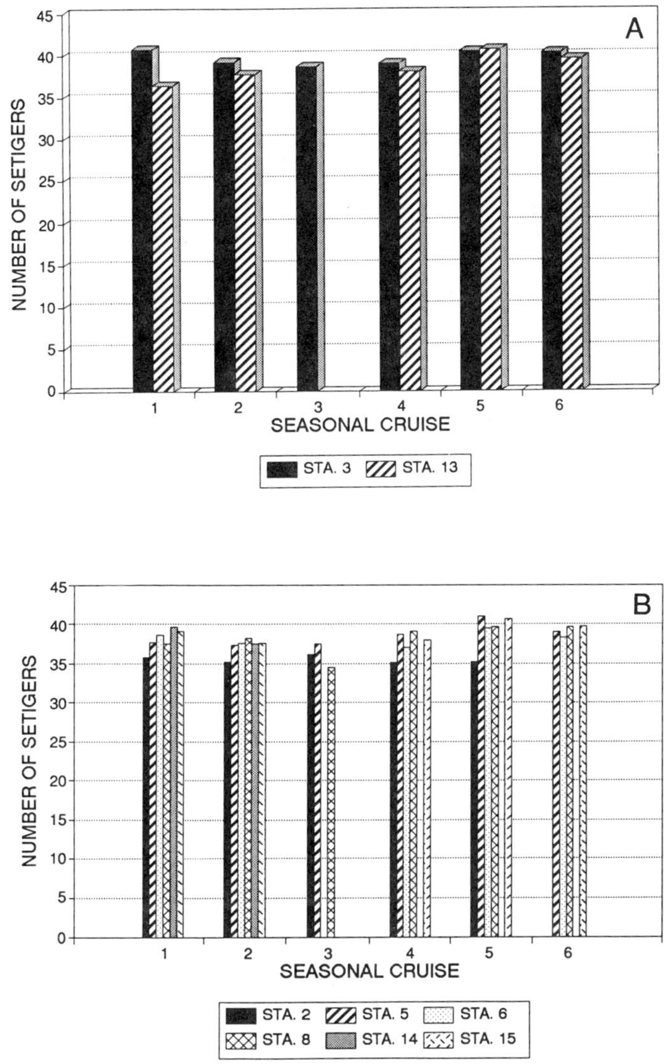

FIGURE 12.2. A: Histogram showing mean numbers of setigers for *Aurospio dibranchiata* from U.S. North Atlantic Station 3 (1350 m) and Station 13 (1250 m). B: Histogram showing mean numbers of setigers for *Aurospio dibranchiata* from six 2100-m U.S. North Atlantic stations. Some stations were not sampled on Cruise 3.

that eggs are present in some individuals in all seasons, but that larger eggs are present in November collections. This result suggests seasonality in egg maturation and supports the results from the North Carolina station (Blake 1993).

The size-frequency data do not provide any strong evidence of seasonality in recruitment of juveniles. When the mean number of setigers of individuals at

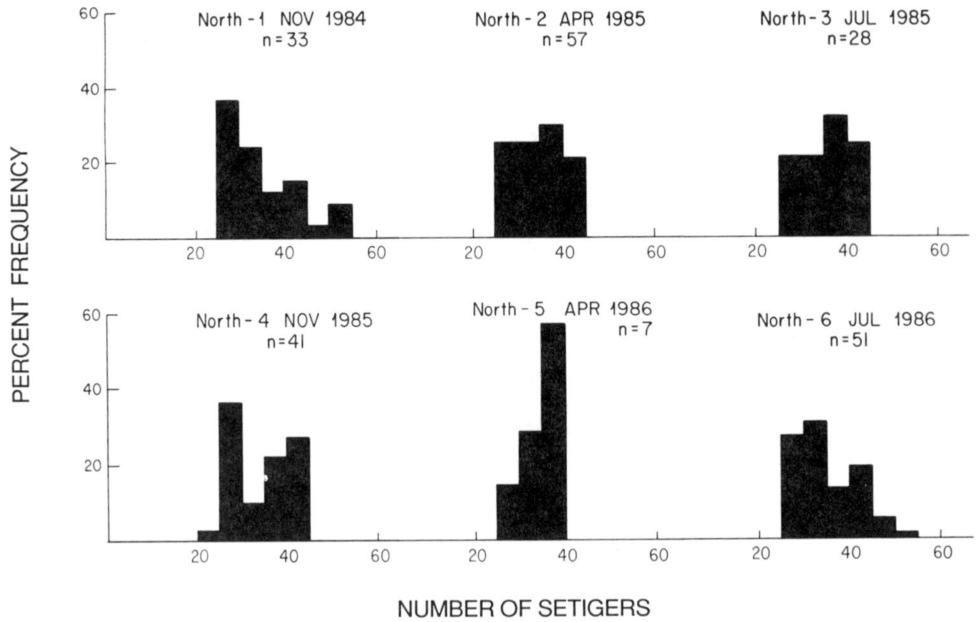

FIGURE 12.3. Frequency distribution of size classes of *Aurospio dibranchiata* at U.S. North Atlantic Station 2 (2100 m).

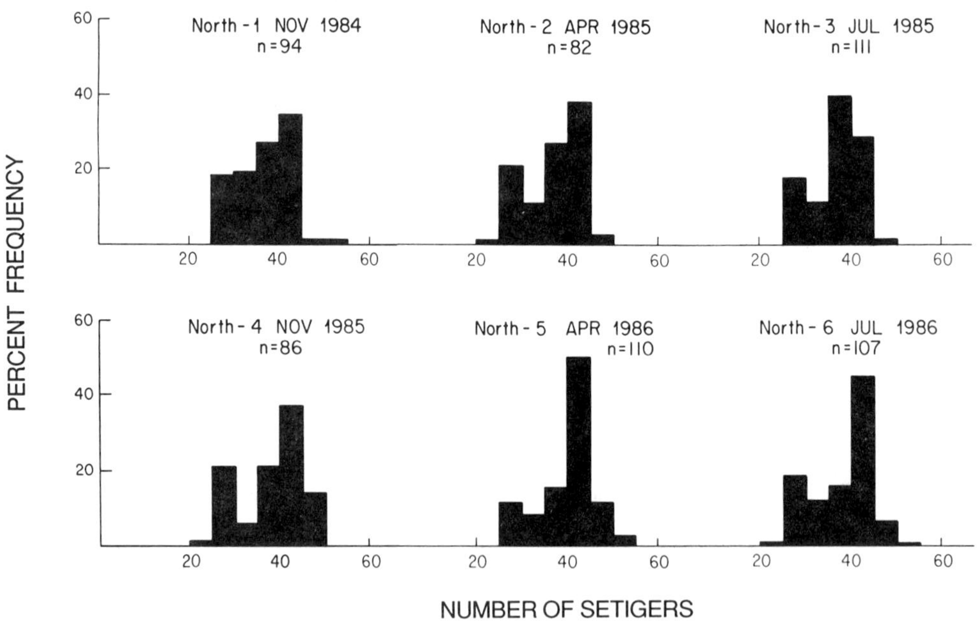

FIGURE 12.4. Frequency distribution of size classes of *Aurospio dibranchiata* at U.S. North Atlantic Station 5 (2065 m).

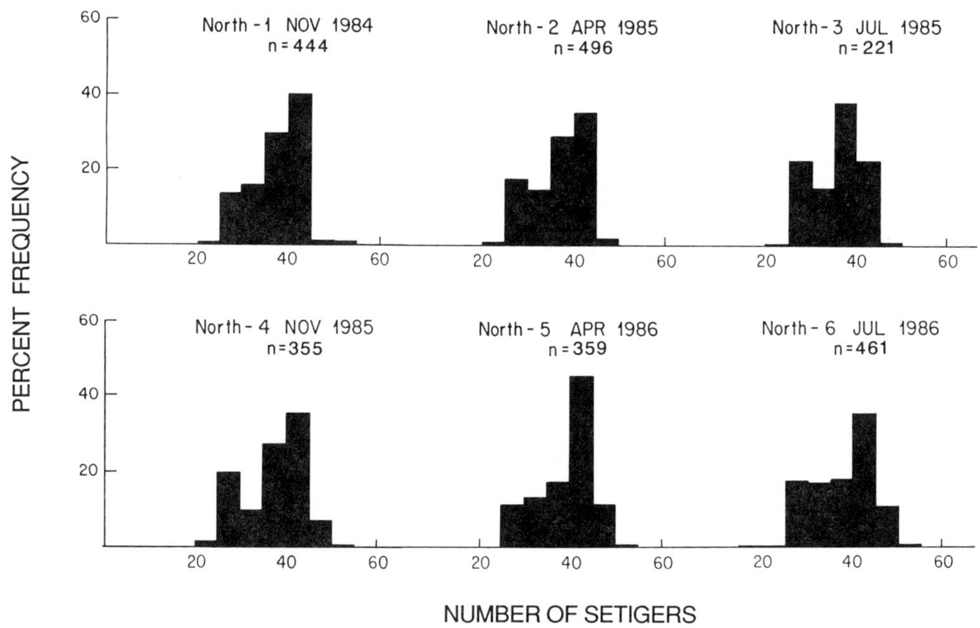

FIGURE 12.5. Frequency distribution of size classes of *Aurospio dibranchiata* combined for five 2000-m to 2100-m U.S. North Atlantic stations (2, 5, 6, 8, 15).

each of the stations is plotted, there is a consistency among all six sampling dates (figure 12.2A-B). For each station, there are apparent differences in population structure from one sampling date to the next. For example, at station 2 (2100 m on the U.S./Canadian boundary transect), the smallest size classes (26–30 setigers, figure 12.3) occurred in November 1984 and 1985. This pattern is repeated to some extent at most of the stations, but some stations also exhibit peaks in this small size class in the spring and summer collections (e.g., station 5, figure 12.4). A composite of the data collected from all of the 2000–2100 m stations is presented in figure 12.5.

In an effort to determine if population structure differs with depth, the mean number of setigers was compared between stations 13 (1250 m) and 15 (2155 m) off southern New England and between stations 3 (1350 m) and 2 (2100 m) on the Canadian boundary transect (figure 12.6). The specimens from the shallower station 3 are always larger than those from station 2 (figure 12.6A) on the Canadian boundary, but specimens from the two depths are essentially identical in the more southerly transect (figure 12.6B).

Although the egg diameter data suggest some seasonality in the occurrence of larger egg sizes in November, the size frequency data results are mixed. It appears that *Aurospio dibranchiata* reproduces year round, with representatives of the smallest size classes present in some populations in each of the three seasons sampled. Reproduction in this species appears to be continuous, but not asynchronous.

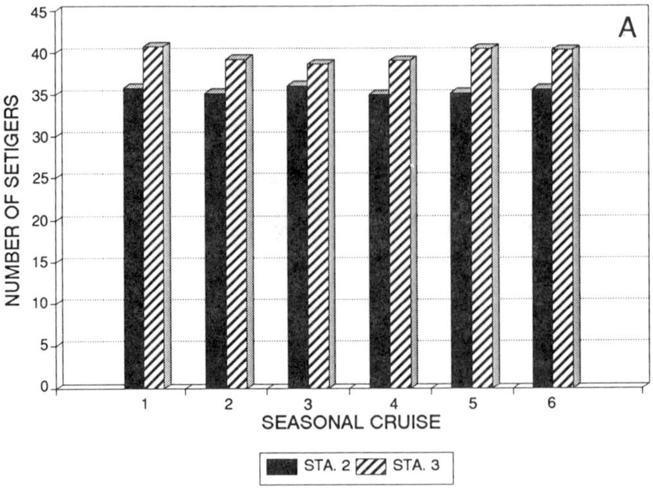

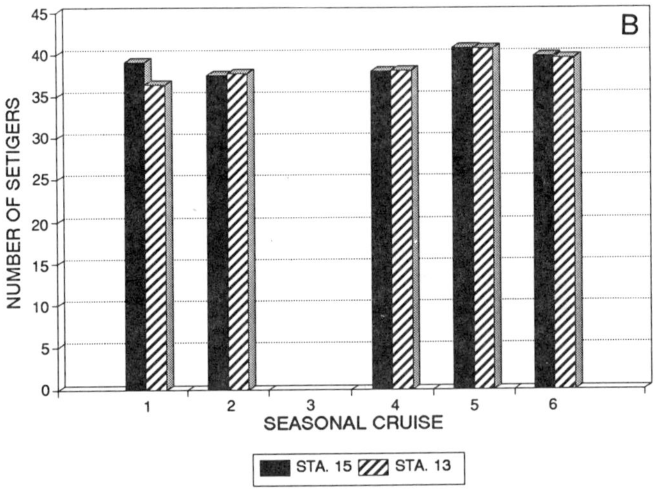

FIGURE 12.6. A: Histogram showing mean number of setigers of *Aurospio dibranchiata* at stations 2 (2000 m) and 3 (1300 m) on the U.S. Canadian Boundary Transect. B: Histogram showing mean number of setigers at Stations 15 (2000 m) and 13 (1250 m) south of Cape Cod (No samples taken on Cruise 3).

Prochaetoderma yongei Scheltema (Figures 12.7–12.8)

Prochaetoderma yongei is an aplacophoran mollusc with a wide distribution throughout the northwestern and eastern Atlantic continental slope and rise (Scheltema 1985). The species is abundant in the North Atlantic basin at depths of 800–2000 m but is not present south of Cape Hatteras (Scheltema 1985; Blake et al. 1987). In this study, *P. yongei* was among the dominant species at stations between 1250 m and 1350 m depths.

Prochaetoderma yongei is a dioecious species that grows extremely fast, achieving mature size in just two months, and reaches sexual maturity in one year (Scheltema 1987). Scheltema determined that individuals less than 1.0 mm long are juveniles, but that individuals between 1.0 and 1.8 mm may be either juveniles or sexually mature adults. This inconsistency causes age classes based on body length to overlap. *P. yongei* is reproductive throughout the year, although not at a constant rate (Scheltema 1987). Scheltema (1987) has determined that females from summer collections have a higher average number of eggs than females from winter collections. Reproductive females from all seasons contain eggs at various stages of development. No reproductive characteristics were observed in the present study.

The size frequency data reveal little evidence of cyclic growth patterns in *Prochaetoderma yongei.* However, the overlap of size classes makes the data difficult to interpret. Individuals in the smallest size class (0.5 to 1.0 mm) are present in all collections, with the highest percentage occurring in April 1985 (figure 12.7). The four-month gap between collections is probably too great to observe recruitment based on length frequency data. Differences in length between newly recruited individuals and older individuals are probably obscured within two months (Scheltema 1987).

The larger size classes dominated during the summer months, especially in July 1986 when the average length was 2.1 mm (figure 12.8). There was a decrease in the average length from November 1984 to April 1985 followed by an increase at all stations except station 13 (figure 12.8). Average length of the specimens at station 13 peaked in April 1986 and then dropped slightly in July 1986. Previously recorded data for total body length of *P. yongei* shows an average length of 1.5 to 2.1 mm (Scheltema 1985). The specimens measured in this study are comparable, with average lengths ranging from 1.4 to 2.0 mm. However, the largest individual, with a length of 3.7 mm, was significantly longer than Scheltema's largest individual, which measured 2.8 mm in length (Scheltema 1985).

The reproductive and size frequency data do not suggest any seasonal patterns of reproduction or recruitment. The level of reproductive activity is probably very low in this species at all times, with little change seasonally.

DIASTYLIS STYGIA SARS (FIGURES 12.9–12.12)

The cumacean fauna in the U.S. North Atlantic slope and rise region is rich; a total of 51 species have been identified (Maciolek et al. 1987a). Station 5 featured unusually high cumacean abundances with a total of 409 individuals obtained over the six sampling periods. Of these, 393 individuals belonged to a single species, *Diastylis stygia.* This species ranges as far south as the U.S. Mid-Atlantic study area, but was not found in the U.S. South Atlantic region off the Carolinas (Blake et al. 1987). The species thus appears to be more northern in its distribution.

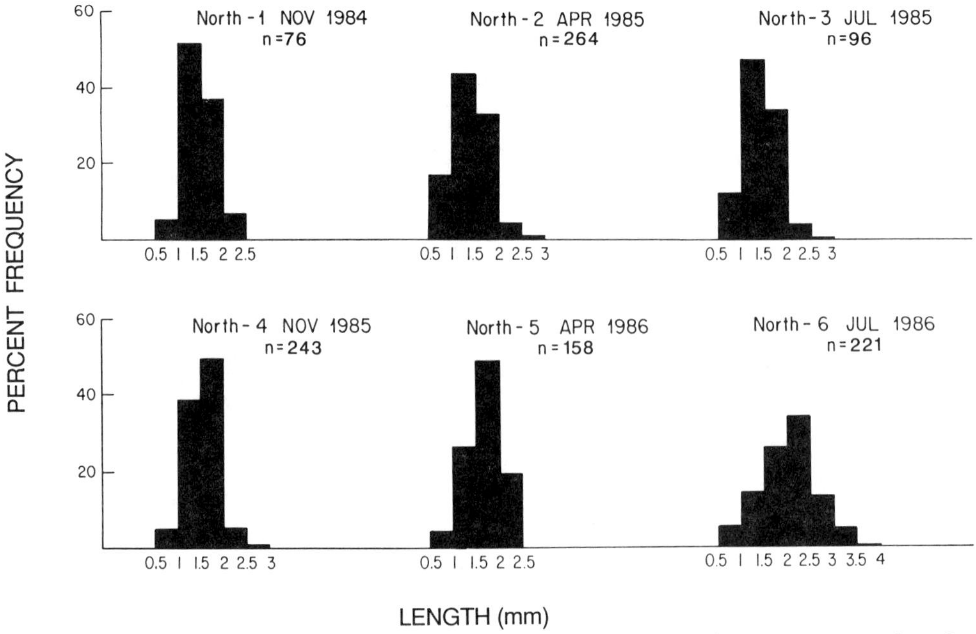

FIGURE 12.7. Frequency distribution of size classes of *Prochaetoderma yongei* combined for four 1220-m to 1350-m U.S. North Atlantic stations (3, 9, 10, 13).

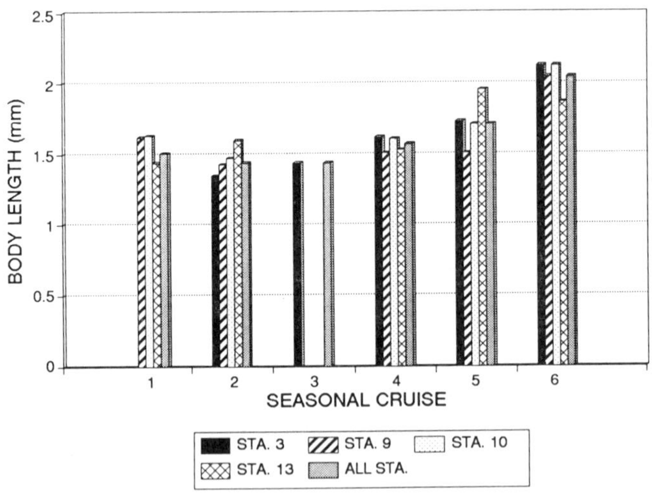

FIGURE 12.8. Histogram showing mean body length of *Prochaetoderma yongei* at U.S. North Atlantic stations 3 (1350 m), 9 (1220 m), 10 (1220 m), and 13 (1250 m) shown separately and combined. Some stations not sampled on Cruise 3.

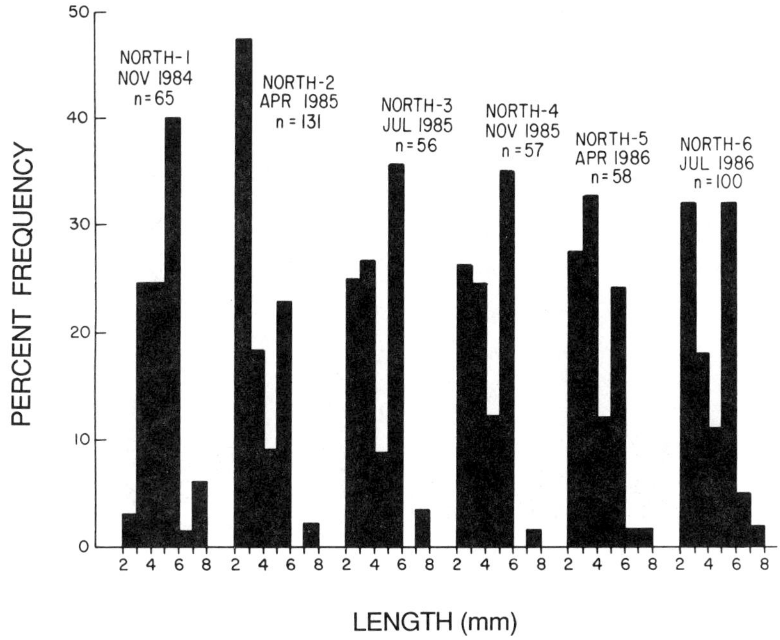

FIGURE 12.9. Frequency distribution of size classes for *Diastylis stygia* from U.S. North Atlantic station 5 (2065 m) and 6 (2115 m).

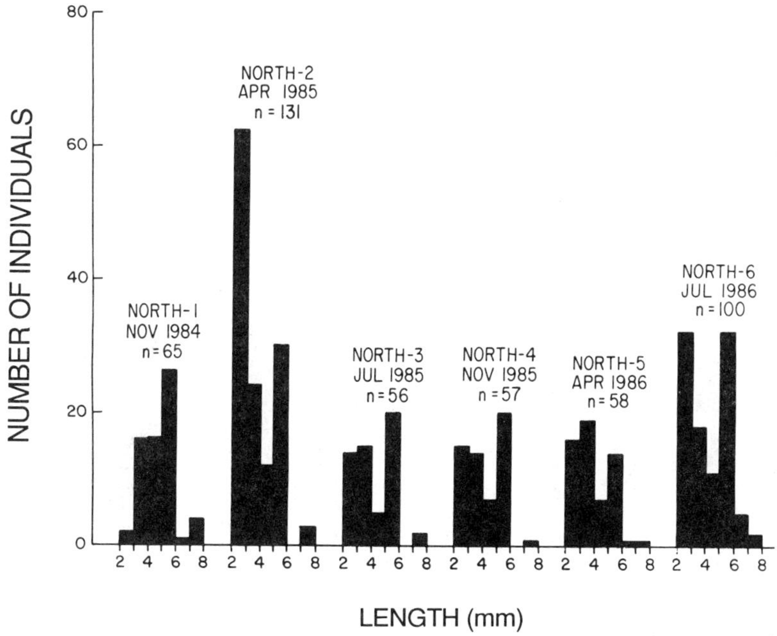

FIGURE 12.10. Distribution of total numbers of individuals by size classes for *Diastylis stygia* from U.S. North Atlantic station 5 (2065 m) and 6 (2115 m).

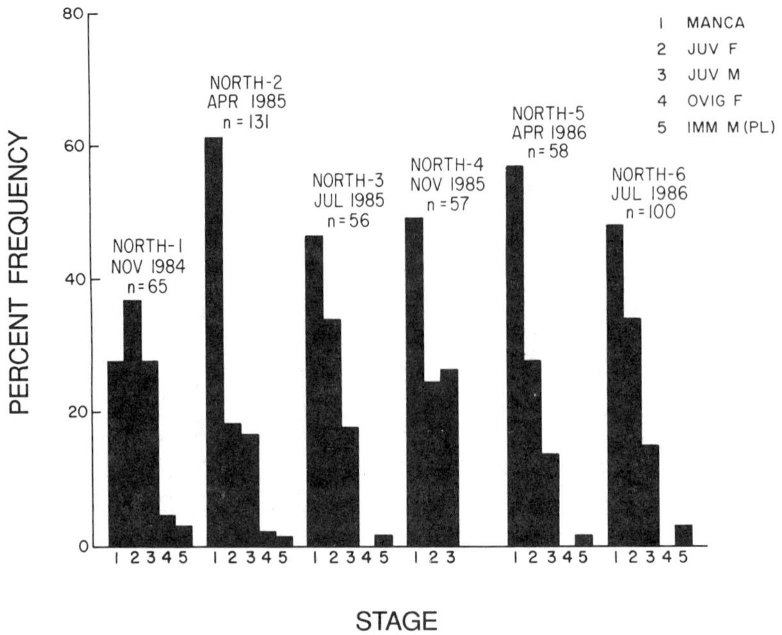

FIGURE 12.11. Frequency distribution of juveniles and mature adults for *Diastylis stygia* from U.S. North Atlantic station 5 (2065 m) and 6 (2115 m).

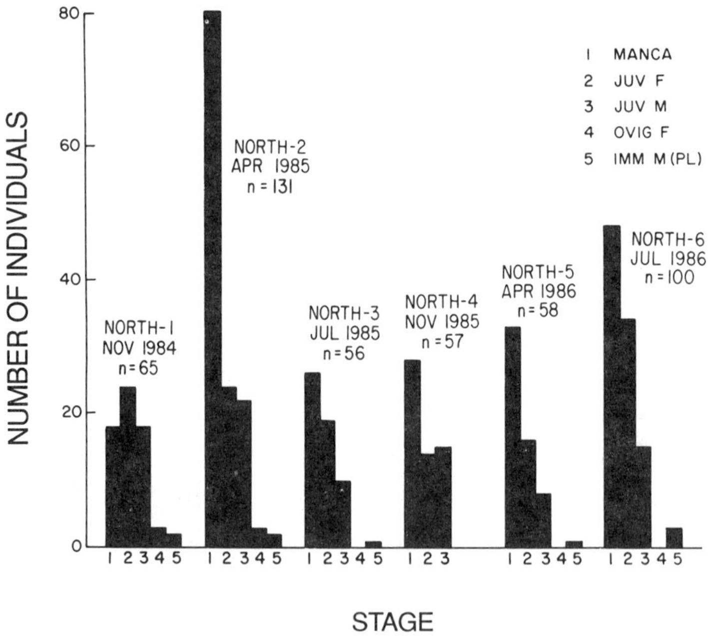

FIGURE 12.12. Distribution of total numbers of juveniles and mature adults for *Diastylis stygia* from U.S. North Atlantic station 5 (2065 m) and 6 (2115 m).

The size frequency data show a relatively constant pattern for all sampling dates (figure 12.9). At least a few individuals of all but the largest size classes were present during every sampling period. Of particular interest is the repeated decline in proportional abundance of the 4- to 5-mm size classes followed by proportionally high abundance of 5- to 6-mm individuals. There was a large pulse of small individuals in April 1985, and a pulse of this magnitude was not seen again (figure 12.10). This pulse then disappeared, suggesting high mortality or emigration of juveniles, because no effect was observed on other size classes over the course of the next twelve months.

The general pattern of population structure of *Diastylis stygia* seems to be one of relatively high initial recruitment followed by moderate mortality or emigration, at each stage, with only a few individuals surviving to reproductive maturity. The higher proportional abundance of 5- to 6-mm individuals relative to the 4- to 5-mm size class (figure 12.10) may be explained by slower growth with increasing size. Because the sampling periods were approximately three to four months apart, it is not possible to infer rate of growth from one molt to the next. Thus, a gradual increase in the total abundance of mancae from July 1985 to July 1986 (figures 12.11, 12.12) could be explained either by a gradual addition of new individuals to a resident population of mancae, with slow growth to the juvenile stage, or by successive increases in recruitment of new mancae to the population, with no holdover of manca-stage individuals from one sampling period to the next. In either case, there is no correlation between the number of mancae at one sampling period and the number of juveniles at some subsequent time.

Diastylis stygia is characterized by variable, but constant, recruitment. The probability of survival from the manca to the adult appears to be small, but variable at each stage.

Discussion

Little is known about the reproductive biology and life history of deep-sea benthic infaunal invertebrates. However, considerable discussion has focused on whether deep-sea invertebrates in general exhibit any form of seasonality or periodicity in their reproduction (Tyler 1988). Given the relatively constant environmental conditions present in the deep sea, continuous reproduction would be predicted. The earliest attempts to address the question of seasonality were based on samples that had not specifically been intended for studies of reproduction, and efforts to synthesize these data led to a conclusion that reproductive periodicity or seasonality was present in some species (George and Menzies 1967; Schoener 1968; Sanders and Hessler 1969; Scheltema 1972). The validity of these conclusions was challenged by Rokop (1974, 1977a, b,

1979), who examined seasonal collections of 11 invertebrate species from several phyla. Nine of these species were found to be in a reproductive condition all year round and two spawned seasonally.

As part of a long-term study of deep-sea benthic invertebrates in the Rockall Trough off Ireland and Scotland conducted by Paul A. Tyler, John D. Gage, and their associates, 29 species of echinoderms and 2 molluscs were assessed for their mode of reproduction, gametogenesis, and life history. Eight species of echinoderms and 2 molluscs were found to reproduce seasonally (Lightfoot et al. 1979; Gage and Tyler 1982; Tyler et al. 1982a; Tyler et al. 1982b; Tyler and Gage 1984a; Tyler 1986), while another 19 species of echinoderms exhibited no evidence of seasonality (Tyler et al. 1982a; Pain et al. 1982a, b; Harvey and Gage 1984; Tyler et al. 1984; Tyler and Gage 1984b; Tyler 1986). These results are discussed by Tyler (1986, 1988) and Gage and Tyler (1991). Blake (1993) found two polychaetes, *Aurospio dibranchiata* and *Cossura longocirrata,* that exhibited seasonal reproduction off North Carolina, while three others (*Pholoe anoculata, Microrbinia linea,* and *Scalibregma inflatum*) exhibited continuous reproduction.

In the present study, some individuals of the polychaete *Aurospio dibranchiata* contained eggs in all seasons, but larger eggs were present only in the fall. This result suggests seasonality in egg maturation, which was also seen for the population at 2000 m off North Carolina (Blake 1993). However, the size class data for *A. dibranchiata* shows little evidence of seasonal recruitment.. A second polychaete species, *Cossura longocirrata,* was studied at a 600-m-deep station off Cape Hatteras (Blake et al. 1987). That species appears to reproduce throughout the year, but exhibits peaks in recruitment during the summer.

Neither the aplacophoran *Prochaetoderma yongei* nor the cumacean *Diastylis stygia* exhibited seasonal or cyclic growth patterns. No gametes were seen in the specimens of *P. yongei* examined in this study, but the species is thought to reproduce year round (Scheltema 1987). A large pulse of small individuals of *D. stygia* seen in April 1985 was not repeated, suggesting that epizodic recruitment occurs at unpredictable intervals.

Isopods, amphipods, tanaidaceans, and cumaceans are extremely abundant and diverse in the deep sea, but there has been little detailed examination of their reproductive biology. For the most part, trends and patterns have been inferred from aggregations of data on a group, for example, asellote isopods, which cover widely spaced localities and years (e.g., George and Menzies 1967). Harrison (1988) was able to refine the problem by examining patterns at the isopod family level for 20 samples taken over a ten-year period. He found that while there was no evidence of seasonal reproductive activity for most isopod families, there was a strong seasonal pattern of breeding intensity.

The problem of obtaining sufficient individuals of any one species has been difficult to overcome, but two isopod and one amphipod species occurred in marginally sufficient numbers at 1240 m in the San Diego Trough to be examined by Rokop (1977b). In fact, at several times of the year, the two isopod species were represented by less than 10 individuals. However, Rokop argued that since at least one of those individuals was in brooding condition, the evidence was compelling enough to suggest that reproduction was continuous throughout the year. A stronger case was made for the amphipod, *Pseudharpinia* (= *Harpiniopsis*) *excavata* (Chevreux) (Rokop 1977b). The present study is unique in that a very large number of individuals of one species of cumacean were obtained from a single station at several times during the year. In this case, both sexually mature adults and newly hatched mancae of *Diastylis stygia* were present in low numbers throughout the year. There was seasonally intense recruitment in April of 1985 and July of 1986.

When the data of this study are combined with those of Blake (1993) from off the Carolinas, at least three patterns of reproduction and recruitment are evident among infaunal invertebrates. First, there is strong evidence of seasonal recruitment in *Cossura longocirrata,* one of two upper slope polychaete species found at 600 m off Cape Hatteras. This is not surprising because *C. longocirrata* is also common on the continental shelf, where it has been shown to exhibit seasonal reproduction and recruitment (Blake and Baptiste 1985). Second, *Pholoe anoculata* is representative of species that range throughout the upper and middle slope. Reproduction occurs in the upper slope depths, and these populations supply the middle slope depths with recruits. Third, the truly deep water middle and lower slope species, with high population abundances at 1200–2000 m or greater, exhibit continuous but asynchronous low intensity reproduction. Examples of this third group include *Aurospio dibranchiata, Microrbinia linea, Prochaetoderma yongei,* and *Diastylis stygia.*

The larger size of individuals of *Aurospio dibranchiata* at station 3 (1350 m) than at station 2 (2100 m) along the U.S./Canadian boundary and the occurrence of a large pulse of juvenile *Diastylis stygia* recruited to the population at station 5 in April 1985 may be caused by the response of these species to inputs of food to the environment. Surface-derived particulate organic matter has been found to sink to the deep seabed and is now known to be correlated with phytoplankton blooms (Deuser and Ross 1980; Deuser et al. 1981; Billett et al. 1983; Lampitt 1985; Gooday and Turley 1990; Hecker 1990; Thiel et al. 1990). Tyler and Gage (1980) and Tyler et al. (1982b) speculated that sinking organic matter might provide a food source for larvae developing in the water column and an energy source for benthic invertebrates to complete gametogenesis. Tyler (1988) developed a hypothesis linking the periodic sinking of organic matter from surface primary production to physiological processes on the seafloor. Direct visual evidence of a plankton bloom settling on the seafloor was observed twice as part of camera sled surveys taken during the present study (Hecker 1990). The first instance was in April 1985, when pictures from camera tows taken four days apart provided evidence that a large amount of greenish-tan flocculent material had settled to the bottom at all depths along the U.S./

Canadian boundary transect. A less substantial flocculent layer was also observed in pictures taken along the 2100-m isobath transect on the slope east of Lydonia Canyon in April 1986.

Acknowledgments

This study was supported by the U.S. Department of the Interior, Minerals Management Service, under contract No. 14–12–0001–30064. This study was part of a larger effort to characterize the benthic infauna for the entire U.S. Atlantic continental slope. As a result, numerous individuals participated in field surveys, sorted the samples, identified the taxa, and organized the data. The life history study was made possible by the dedicated efforts of Ms. Ellen M. Baptiste, who supervised the laboratory work and managed the data.

References

Billett, D. S. M., R. S. Lampitt, A. L. Rice, and R. F. C. Mantoura. 1983. Seasonal sedimentation of phytoplankton to the deep-sea benthos. *Nature* 302:520–522.

Blake, J. A. 1993. Life history analysis of five dominant infaunal polychaete species from the continental slope off North Carolina. *J. Mar. Biol. Ass. U. K.* 73. In press.

Blake, J. A. and E. M. Baptiste. 1985. Chapter 6. Life History Studies on dominant Polychaete species from Georges Bank. In N. Maciolek-Blake, J. F. Grassle, and J. M. Neff, eds., *Georges Bank Benthic Infauna Monitoring Program: Final Report for the Third Year of Sampling*. U.S. Department of the Interior, Minerals Management Service, pp. 140–178.

Blake, J. A., B. Hecker, J. F. Grassle, B. Brown, M. Wade, P. D. Boehm, E. Baptiste, B. Hilbig, N. Maciolek, R. Petrecca, R. E. Ruff, V. Starczak, and L. Watling. 1987. *Study of Biological Processes on the U.S. South Atlantic Slope and Rise: Phase 2*. Final report prepared for the U.S. Department of the Interior, Minerals Management Service. National Technical Information Service (NTIS) No. PB 87- 214359.

Deuser, W. G. and E. H. Ross. 1980. Seasonal change in the flux of organic carbon to the deep Sargasso Sea. *Nature* 283:364–365.

Deuser, W. G., E. H. Ross, and R. F. Anderson. 1981. Seasonality in the supply of sediment to the deep Sargasso Sea and its implication for the rapid transfer of matter to the deep ocean. *Deep-Sea Res.* 28:495–505.

Gage, J. D. and P. A. Tyler. 1982. Growth strategies in deep-sea ophiuroids. In J. M. Lawrence, ed., *Echinoderms: Proceedings of the International Conference, Tampa Bay*, pp. 305–311. Rotterdam: A. A. Balkema.

Gage, J. D. and P. A. Tyler. 1991. *Deep-Sea Biology: A Natural History of Organisms at the Deep-Sea Floor*. Cambridge: Cambridge University Press.

George, R. Y. and R. J. Menzies. 1967. Indication of cyclic reproductive activity in abyssal organisms. *Nature* 215:878.

Gooday, A. J. and C. M. Turley. 1990. Responses by benthic organisms to inputs of organic material to the ocean floor: A review. *Phil. Trans. Roy. Soc. London* Section A 331:119–138.

Harrison, K. 1988. Seasonal reproduction in deep-sea Crustacea (Isopoda: Asellota). *J. Nat. Hist.* 22:175–197.

Harvey, R. and J. D. Gage. 1984. Observations on the reproduction and postlarval morphology of pourtalesiid sea urchins in the Rockall Trough area (NE Atlantic Ocean). *Mar. Biol.* 82:181–190.

Hecker, B. 1990. Photographic evidence for the rapid flux of particles to the sea floor and their transport down the continental slope. *Deep-Sea Res.* 37:1773–1782.

Lampitt, R. S. 1985. Evidence for the seasonal deposition of detritus to the deep-sea floor and its subsequent resuspension. *Deep-Sea Res.* 32:885–897.

Lightfoot, R. H., P. A. Tyler, and J. D. Gage. 1979. Seasonal reproduction in deep-sea bivalves and brittlestars. *Deep-Sea Res.* 26:967–973.

Maciolek, N. J. 1981a. A new genus and species of Spionidae (Annelida: Polychaeta) from the north and south Atlantic. *Proc. Biol. Soc. Wash.* 94:228–239.

Maciolek, N. J. 1981b. Spionidae (Annelida: Polychaeta) from the Galapagos Rift geothermal vents. *Proc. Biol. Soc. Wash.* 94:826–837.

Maciolek, N., J. F. Grassle, B. Hecker, B. Brown, J. A. Blake, P. D. Boehm, R. Petrecca, S. Duffy, E. Baptiste, and R. E. Ruff. 1987a. *Study of Biological Processes on the U.S. North Atlantic Slope and Rise.* Final report prepared for the U.S. Department of the Interior, Minerals Management Service. 362 pp. + 12 appendices. NTIS No. PB 88- 196514/AS.

Maciolek, N., J. F. Grassle, B. Hecker, P. D. Boehm, B. Brown, B. Dade, W. G. Steinhauer, E. Baptiste, R. E. Ruff, and R. Petrecca. 1987b. *Study of Biological Processes on the U.S. Mid-Atlantic Slope and Rise.* Final report prepared for the U.S. Department of the Interior, Minerals Management Service. 310 pp + 13 appendices. NTIS No. PB 88- 183090.

Pain, S. L., P. A. Tyler, and J. D. Gage. 1982a. The reproductive biology of the deep-sea asteroids *Benthopecten simplex* (Perrier), *Pectinaster filholi* Perrier, and *Pontaster tenuispinus* Düben and Koren (Phanerozonia: Benthopectiniidae) from the Rockall Trough. *J. Exp. Mar. Biol. Ecol.* 65:195–211.

Pain, S. L., P. A. Tyler, and J. D. Gage. 1982b. The reproductive biology of *Hymenaster membranaceus* from the Rockall Trough, North-East Atlantic Ocean, with notes on *H. gennaeus. Mar. Biol.* 70:41–50.

Rokop, F. J. 1974. Reproductive patterns in the deep-sea benthos. *Science* 186:743–745.

Rokop, F. J. 1977a. Seasonal reproduction of the brachiopod *Frieleia halli* and the scaphopod *Cadulus californicus* at bathyal depths in the deep sea. *Mar. Biol.* 43:237–246.

Rokop, F. J. 1977b. Patterns of reproduction in the deep-sea benthic crustaceans: a reevaluation. *Deep-Sea Res.* 24:683–691.

Rokop, F. J. 1979. Year-round reproduction in deep-sea bivalve molluscs. In S. E. Stancyk, ed., *Reproductive Ecology of Marine Invertebrates,* pp. 189–198. University of South Carolina Press.

Rowe, G. T. 1983. Biomass and production of the deep-sea macrobenthos. In G. T. Rowe, ed., *Deep-Sea Biology,* pp. 97–121. New York: Wiley.

Sanders, H. L. and R. R. Hessler. 1969. Ecology of deep-sea benthos. *Science* 163:1419–1424.

Scheltema, A. H. 1985. The aplacophoran family Prochaetodermatidae in the North American Basin, including *Chevroderma* n.g. and *Spathoderma* n.g. (Mollusca: Chaetodermomorpha). *Biol. Bull. Mar. Biol. Lab., Woods Hole* 169:484–529.

Scheltema, A. H. 1987. Reproduction and rapid growth in a deep-sea aplacophoran mollusc, *Prochaetoderma yongei. Mar. Ecol. Prog. Ser.* 37:171–180.

Scheltema, R. S. 1972. Reproduction and dispersal of bottom dwelling deep-sea invertebrates: A speculative summary. In R. W. Brauer, ed., *Barobiology and the Experimental Biology of the Deep Sea,* pp. 58–66. University of North Carolina, Chapel Hill: North Carolina Sea Grant Program.

Schoener, A. 1968. Evidence for reproductive seasonality in the deep sea. *Ecology* 49:81–87.

Thiel, H., G. Pfannkuche, G. Schriever, K. Lochte, A. J. Gooday, C. Hemleben, R. F. G. Mantoura, C. M. Turley, J. W. Patching, and F. Riemann. 1990. Phytodetritus on the deep-sea floor in a central oceanic region of the northeast Atlantic. *Biol. Oceanogr.* 6:203–239.

Tyler, P. A. 1986. Studies of a benthic time series: Reproductive biology of benthic invertebrates in the Rockall Trough. *Proc. Roy. Soc. Edinburgh* 88B:175–190.

Tyler, P. A. 1988. Seasonality in the deep sea. *Oceanogr. Mar. Biol. Ann. Rev.* 26:227–258.

Tyler, P. A. and J. D. Gage. 1984a. Seasonal reproduction of *Echinus affinis* (Echinodermata: Echinoidea) in the Rockall Trough, northeast Atlantic Ocean. *Deep-Sea Res.* 31:387– 402.

Tyler, P. A. and J. D. Gage. 1984b. The reproductive biology of echinothuriid and cidarid sea urchins from the deep sea (Rockall Trough, northeast Atlantic Ocean). *Marine Biology* 80:63–74.

Tyler, P. A., S. L. Pain, and J. D. Gage. 1982a. The reproductive biology of the deep-sea asteroid *Bathybiaster vexillifer*. *J. Mar. Biol. Ass. U. K.* 62:57–69.

Tyler, P. A., A. L. Grant, S. L. Pain, and J. D. Gage. 1982b. Is annual reproduction in deep-sea echinoderms a response to variability in their environment. *Nature* 300:747–750.

Tyler, P. A., J. D. Gage, and S. L. Pain. 1983. Reproductive variability in deep-sea echinoderms and molluscs from the Rockall Trough. *Oceanologica ACTA 1983. Proceedings 17th European Marine Biology Symposium, Brest,* pp. 191–195.

Tyler, P. A., S. L. Pain, J. D. Gage, and D. S. M. Billett. 1984. The reproductive biology of deep-sea forcipulate seastars (Asteroidea: Echinodermata) from the N.E. Atlantic Ocean. *J. Mar. Biol. Ass. U. K.* 64:587–601.

CHAPTER THIRTEEN

The Influence of Natural Organic Enhancement on Life Histories and Community Structure of Bathyal Polychaetes

Lisa A. Levin, Gayle R. Plaia, and Cynthia L. Huggett

Abstract

Most studies of infaunal response to organic enrichment in the deep sea have focused on effects of temporally or spatially patchy inputs such as phytodetritus, carcass, or macrophyte falls. The present paper evaluates the responses of benthic polychaetes to persistently high input of particulate carbon resulting from upwelling and lateral transport. Two gradients of organic enrichment are examined. One encompasses four zones along the slopes of Volcano 7, a seamount penetrating the eastern Pacific oxygen minimum zone. A second consists of three 850 m sites along the Carolina continental slope. At both sites increasing availability of organic matter was associated with (1) a sixfold elevation in polychaete abundance, and (2) a general reduction in polychaete diversity, including increased dominance by relatively few species. Striking parallels exist between community-level responses of shallow-water macrofauna to organic pollution (Pearson and Rosenberg 1978; Pearson et al. 1983) and responses of bathyal polychaetes to persistent enrichment.

In contrast, life history responses were more varied. Increased intraspecies body size, reproductive activity, and representation of brooding taxa were observed along a gradient of increasing enrichment on Volcano 7. None of these patterns was observed along a similar gradient on the Carolina slope. Instead, slope populations experiencing highest organic inputs were numerically dominated by juveniles, and brooding behavior was rarely observed. Despite high levels of organic matter availability at the most enriched sites, the classic opportunistic taxa (e.g., *Capitella, Ophryotrocha* and spionids) were not present. The dominant taxa observed in the Atlantic (*Scalibregma inflatum, Aricidea quadrilobata,* and *Cossura* sp.) and in the Pacific (*Cirrophorus lyra, Euchone incolor, Protodorvillea* sp., and *Tharyx* sp.) are cosmopolitan, and several have been previously associated with organically enriched settings.

The powerful effects of organic enrichment on the community structure and population biology of organisms inhabiting shallow-water sediments have been recognized for some time. Organic pollution is known to cause diminished diversity, either drastically reduced or highly elevated densities, increased body sizes, and dominance by opportunistic species, many of which exhibit brooding behavior (Grassle and Grassle 1974; Pearson and Rosenberg 1978; Levin 1986; Ritz et al. 1989; Weston 1990). Detailed investigations of the response of individual species to increased food quantity or quality have revealed increases in survivorship, growth and fecundity, earlier age at first reproduction, and altered egg composition (Levin 1986; Levin and Creed 1986; Gremare et al. 1988; Marsh et al. 1989; Bridges and Levin 1991; Qian and Chia 1991).

Much less is known about the influence of organic matter on the biology of benthic organisms in the deep sea (reviewed by Gooday and Turley 1990) than in shallow-water systems (Jumars and Wheatcroft 1989). Food is generally recognized as the primary limiting resource to deep-sea benthos (Rowe et al. 1974; Rowe 1983; Gage and Tyler 1991). Early investigations focused on organic inputs reaching the seabed as a steady, albeit scarce, rain of particles (Riley et al. 1964). Depth-related changes in infaunal density, biomass, body size, and diversity have been attributed to bathymetric variation in nutrient input and food availability (Rex 1981, 1983). In the past decade attention has turned to episodic inputs of large organic food falls (Smith 1986; Grassle and Morse-Porteous 1987) or plant material (Desbruyères et al. 1982; Smith 1983; Levin and Smith 1984; Grassle and Morse-Porteous 1987; Snelgrove et al. 1990) and their effects on deep-sea community structure. The relatively recent discovery of seasonal deposition of fresh phytodetritus at several places on the deep-sea floor (Rice et al. 1986; Graf 1989; Thiel et al. 1989; Hecker 1990) has stimulated research on the response of deep-sea benthos to undegraded (nonparcel) organic inputs (reviewed in Gooday and Turley 1990). Most of the work on systems receiving phytodetritus has examined colonization by bacteria and small protozoans (Lochte and Turley 1988; Turley et al. 1988), or meiofauna (Gooday 1988; Gooday and Lambshead 1989). The only reported macrofaunal response to phytodetritus is of rapid mixing of chlorophyll-containing fecal pellets of pelagic copepods by sipunculans on the Voring Plateau (Graf 1989), and the apparent absence of short-term response by macrofauna in the NE Atlantic (Gooday and Lambshead 1989).

Studies of benthic responses to temporal or basin-wide spatial gradients in organic matter input are largely limited to investigations of community density, biomass, or respiration (Rowe 1983; Smith et al. 1983), and rarely address phenomena at the population level. In the present paper, we compare the community and population ecology of polychaetes inhabiting two very different deep-sea settings, both of which exhibit distinct gradients in organic matter input. Our objective is to identify the general community, population level, and reproductive consequences of persistent, natural organic matter enhancement of the deep seabed. We ask (1) whether organically-enriched bathyal sites have a distinctive, specialized enrichment assemblage possessing particular life-history traits and (2) whether models of changing community structure along shallow-water enrichment gradients are applicable to bathyal environments.

Study Area and Methods

We present data for polychaete community and population structure along gradients of organic enrichment at two sites: (1) on the slopes of Volcano 7, a 2.7 km-high seamount penetrating the oxygen minimum zone in the eastern Pacific Ocean off Mexico (Wishner et al. 1990), and (2) along the 850 m depth contour of the North and South Carolina continental slope in the northwest Atlantic Ocean (Schaff et al. 1992).

Volcano 7

Volcano 7 is an inactive, conical seamount located at 13°25'N 102°35'W, approximately 360 km west of Acapulco, Mexico (figure 13.1A). The surrounding seafloor is 3400 m deep, and the summit rises to 730 m. A strong oxygen minimum zone (OMZ) persists in the water column surrounding Volcano 7 (Wishner et al. 1990; Levin et al. 1991a). Penetration of the OMZ by Volcano 7 produces dramatic downslope gradients in near-bottom oxygen concentration (< 0.1 ml/l at the summit to > 2.5 ml/l at the base) (figure 13.2A), sediment percent organic carbon (TOC) and sediment chlorophyll a content (figures 13.2B, 13.2C). Downslope variations in these parameters are discussed in detail in Levin et al. (1991a). Physical data indicate the existence of a gradient of decreasing organic input and preservation downslope, with the summit sediments exhibiting unusually high levels of TOC (up to 3.9 percent) and chlorophyll a (up to 25.5 μg/g), and the flank and base exhibiting lower carbon and pigment levels (figure 13.2B, 13.2C) typical of other deep-sea

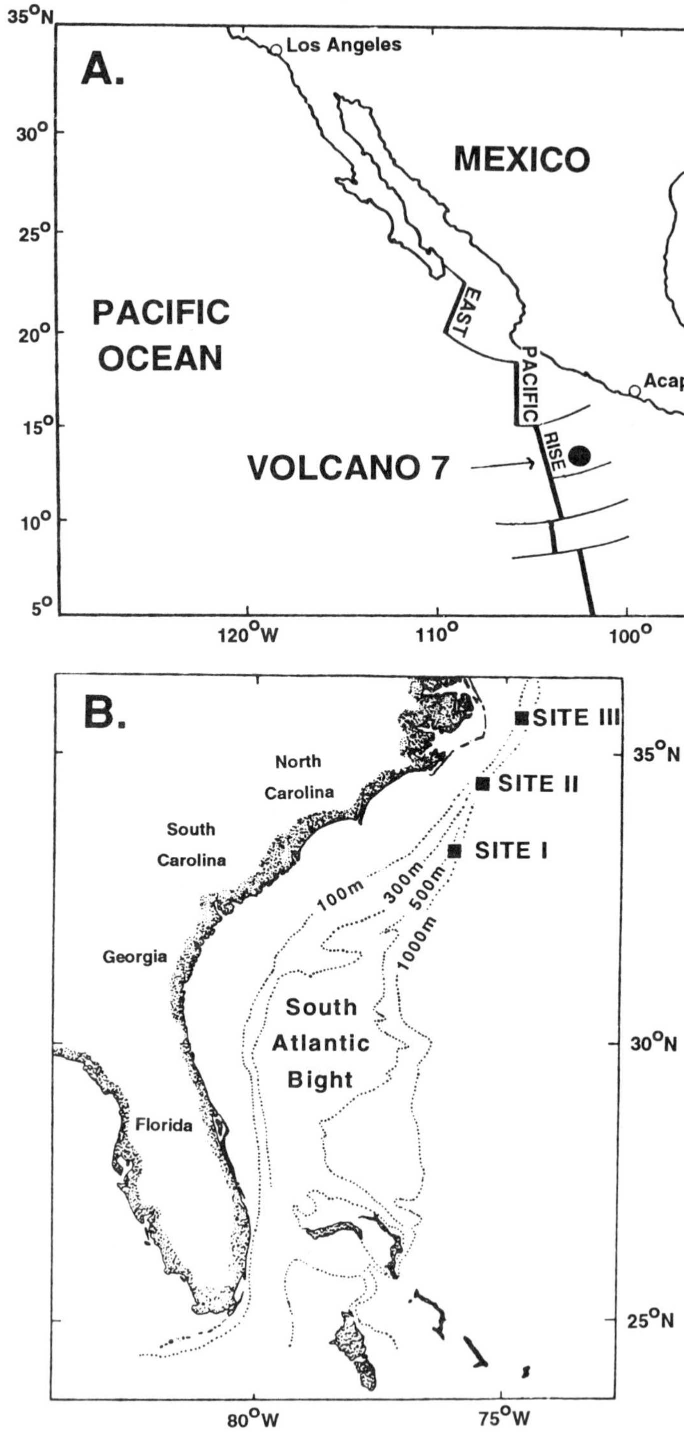

FIGURE 13.1. A. Location of Volcano 7 near the East Pacific Rise. The 25 km-diameter seamount has a summit depth of 733 m and a basal depth of 3400 m. B. Location of the Carolina Slope Sites I, II, and III, all at approximately 850 m water depth.

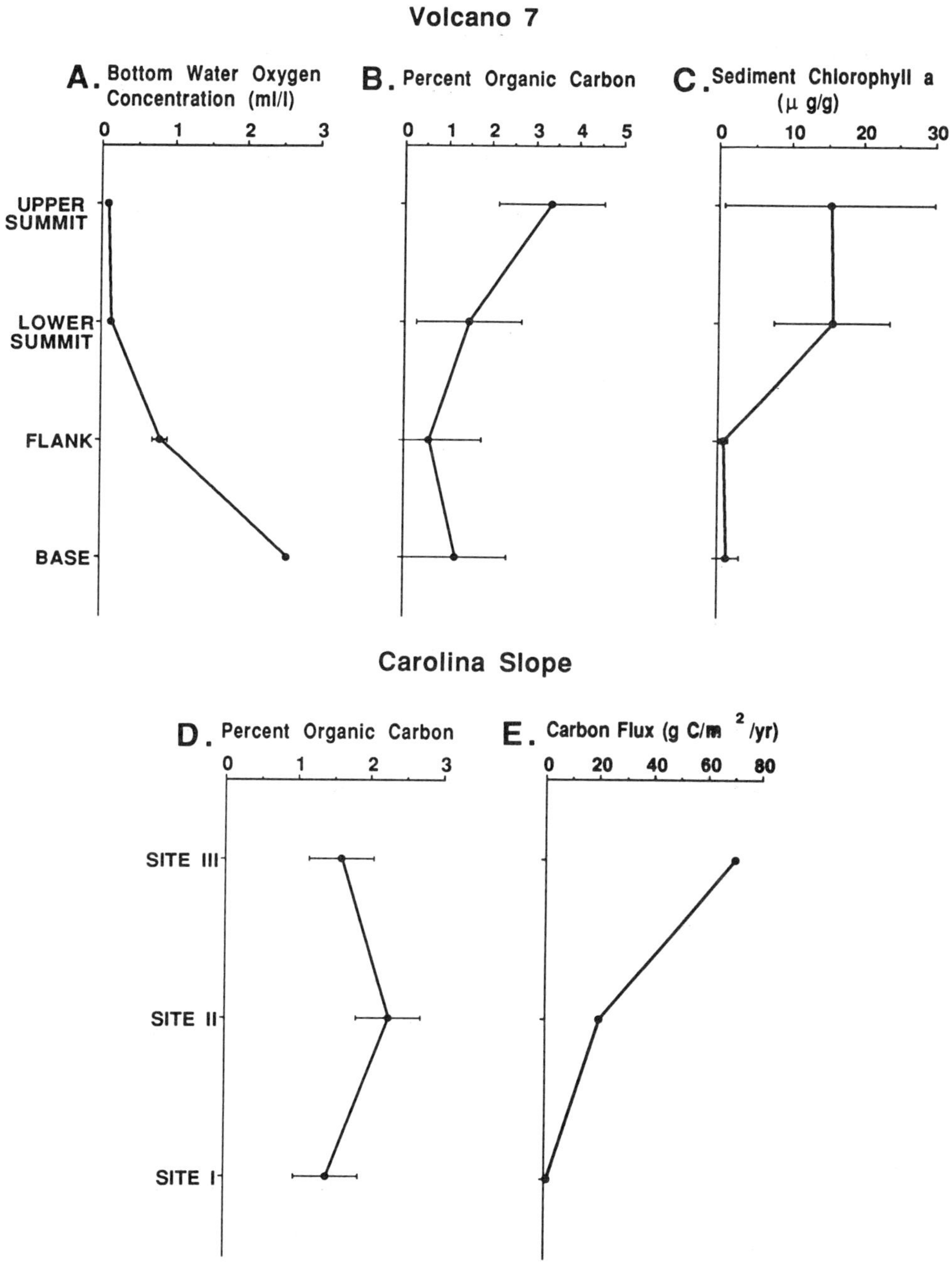

FIGURE 13.2. Chemical concentrations for Volcano 7 depth zones (average ± 1 SD). A. Bottom water oxygen concentration; B. Total sediment organic carbon C. Sediment chlorophyll a. Zone depths are given in table 13.1. Analytical methods are given in Levin et al. (1991a).

Chemical concentrations for the Carolina Slope sites. D. Total sediment organic carbon (average ± 1SD); E. Estimated carbon flux to the seabed based on ^{14}C analyses of sedimentation rate and sediment TOC values. Data are from Schaff et al. (1992).

environments. Wishner et al. (1990) hypothesized that the organic enrichment of summit sediments on Volcano 7 results from reduced consumption and degradation of sinking material in the OMZ.

For purposes of analysis (and based on the depths actually sampled), we consider 4 depth zones representing a gradient of decreasing enrichment: the upper summit (730–770 m), the lower summit (770–900 m), the flank (1300–1800 m), and the base (2800–3350 m). Sediments above 1800 m are composed largely of foraminiferal sands (74–89 percent). On the upper and lower summit sediments appear olive-green in color, and the presence of undegraded flocculent material (probably phytodetritus) has been reported (Wishner et al. 1990; Levin et al. 1991a). Sediments at the base of the volcano (2800–3350 m) are over 50 percent clay and average only 13 percent sand. High concentrations of Mn impart a deep brown color to the mud.

Infauna were collected in 196 cm^2 Ekman cores by the DSRV Alvin during two cruises aboard the RV *Atlantis II*, one in June 1984 and one in November 1988. Cores were taken from the summit region, the western flank, and the northern, western, and southern base regions. Details regarding sample locations, sampling methods, processing, and general patterns of meiofaunal, macrofaunal, and megafaunal community structure are described in Levin et al. (1991a). Polychaetes were measured for the width of the first setiger using a Wild M-5 dissecting microscope fitted with an ocular micrometer. Specimens measured had been initially fixed in formalin and then were preserved in 70 percent ethanol for over one year.

CAROLINA CONTINENTAL SLOPE

Three separate sites were studied on the northwest Atlantic slope, each at approximately 850 m and located 130–150 km apart along a north-south contour (figure 13.1B). Information about the local current regimes, sediment textures, chemistries, and macrofaunal communities inhabiting these sites is given in Schaff (1991), Schaff et al. (1992), and Gooday et al. (1992). The southernmost site (Slope Site I) is located at 32°52′N, 76°27′W, 177 km northeast of Charleston, South Carolina. The seafloor has a gentle slope (2–3°) and appears to be swept periodically by the Gulf Stream. The middle site (Slope Site II) is located at 34°15′N 75°44′W, 82 km east of Cape Lookout, North Carolina. The seafloor has a slope of approximately 15–20° and the sediments are pocked by large numbers of biogenically produced pits and mounds, probably made by holothurians and galatheid crabs. The northernmost site (Slope Site III) is located at 35°24′N 74°48′W, 61 km northeast of Cape Hatteras, North Carolina. In this region the sea floor is cut by canyons, ridges, and gullies oriented across the mean depth contour. Sampling was carried out on a broad but steeply sloped (30°) expanse between canyons. Epifauna are abundant at this site and large numbers of white tubes, belonging to the agglutinated foraminiferan *Bathysiphon filiformis,* are a conspicuous feature protruding above the sediment surface (Gooday et al. 1992). Sediments at slope Sites I, II and III contained 34, 20, and 31 percent sand, respectively (Schaff et al. 1992).

Measurements of pore-water concentrations of SO_4 = , CH_4, ΣCO_2 (Schaff et al. 1992) and depth of O_2 penetration (B. Bebout personal communication), combined with variations in $\delta^{13}C$ ΣCO_2 profiles (N. Blair personal communication) provide evidence of differences in organic matter availability at the three slope sites (Schaff et al. 1992; N. Blair unpublished data). Estimates of organic carbon flux to the seabed, based on sediment TOC (figure 13.2D) and ^{14}C-based sedimentation rates, are 0.6, 20, and > 70 gC/m^2/yr at Sites I, II and III respectively (figure 13.2E) (Schaff et al. 1992). Support for unusually high organic inputs at the northernmost site is provided by the presence of biogenic methane at and below 5 cm in the sediment (Schaff et al. 1992), and isotopically light $\delta^{13}C$ ΣCO_2 values indicative of high respiration rates (N. Blair personal communication). The sources of organic inputs in this region are believed to include Gulf Stream-induced upwelling, lateral transport resulting from mass wasting and from a confluence of deep-water, slope, and western boundary currents that may be focused by bottom topography (Johnson 1989; Schaff et al. 1992). Observations of vascular plant detritus in Site III sediments suggest input of material from terrestrial habitats or coastal wetlands as well.

Infauna were collected in Ekman box cores (196 cm^2) by submersible from Slope Sites I and II during September 1988 (using Pisces II), and from Slope Sites II and III during October 1989 (using Johnson-Sea-Link II), and in a 0.25 m^2 Mark III box corer from the R/V Cape Hatteras in June 1989. Samples were fixed first in formalin, then preserved in 70 percent ethanol. Details of sample processing are given in Schaff et al. (1992). Polychaete community structure (dominance and diversity) was analyzed for the 1988 and 1989 samples combined. Size measurements were made using a Zeiss compound microscope fitted with an ocular micrometer. Setiger numbers were counted directly or for incomplete specimens were obtained through regressions of 1st, 2d, or 3d setiger width on total setiger number.

All polychaetes reported here were retained on a 300 μm screen. Similar core types, sample processing, sorting and identification methods were used for material from Volcano 7 and the Carolina slope. Polychaete abundance patterns were compared across depth zones or sites using ANOVA, where submersible dives were nested within zones or sites. Data were not transformed, as inspection of residuals suggested it to be unnecessary. Rarefaction curves and Shannon-Weiner indices were calculated using the PRARE1 program (Grassle and Sanders 1980). Size-frequency histograms were compared using Kolmogorov-Smirnov tests (Sokal and Rohlf 1969).

Results

Abundance

Polychaete abundance increased along gradients of increasing organic input at both sites (figure 13.3A, 3B), but with one exception. The uppermost summit of Volcano 7, where TOC and chlorophyll a contents were high but measured bottom-water oxygen concentration was < 0.1 ml/l, had markedly reduced polychaete densities compared to sampling sites on the lower summit, only tens of meters deeper. This same phenomenon was observed for total macrofaunal densities as well and was attributed to oxygen limitation (Levin et al. 1991a).

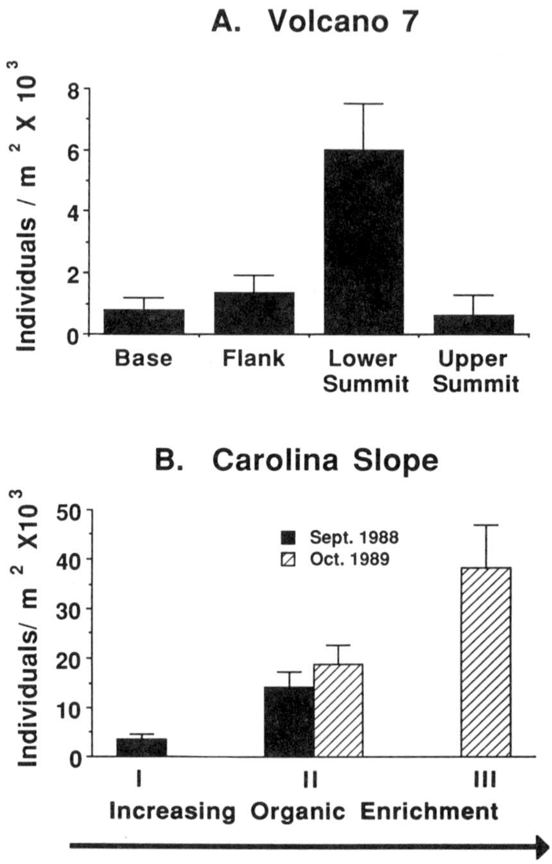

FIGURE 13.3. Polychaete densities (average ± 1SD) for animals retained on a 300 μm screen. A. Volcano 7: Densities on the lower summit were significantly higher than at the other 3 zones, which did not differ from one another (Nested ANOVA + LSD: $F_{3,10} = 26.83$, $p < 0.0001$). B. Carolina Slope: Sites I and II were significantly different (1988 data: Nested ANOVA $F_{1,19} = 93.42$, $p < 0.0001$). Sites II and III were significantly different (1989 data, Nested ANOVA, $F_{1,28} = 51.14$, $p < 0.0001$). Direct statistical comparisons between Sites I and III were not possible since they were sampled in different years.

The similar densities observed on the flank and base lend further support to the idea that organic matter availability dictates abundance patterns below the upper summit. The flank and base sediments have similarly low organic matter availability but very different substrate textures and compositions, and are separated by more than 1500 m.

The peak polychaete densities observed on Volcano 7 (7,926/m^2 on the lower summit) and on the Carolina slope (Site III—38,140/m^2) represent extraordinarily high values for these specific settings. Polychaete densities on other eastern Pacific seamounts, in comparable coarse-grained, carbonate sediments are typically 1,000/m^2 or less (Levin et al. 1991b). The Carolina slope Site III densities are similar to those reported by Blake et al. (1987) for a site 10 km away, and are higher than those reported for any other continental slope region at comparable depths. They are, in fact, more typical of estuarine habitats (Dauer et al. 1984; Woodin 1974).

Composition and Diversity

The Volcano 7 study yielded 880 individual polychaetes belonging to 31 families and 84 species. The Carolina slope study yielded 6,202 polychaete individuals belonging to 38 families and 79 species. Table 13.1 summarizes the dominant polychaete species present in each region.

Both regions exhibited marked decreases in diversity along gradients of increasing organic input. These changes were manifested in patterns of dominance, Shannon-Wiener information diversity indices, and rarefaction estimates of species number. Dominance was high on the upper and lower summit of Volcano 7 (figure 13.4A) and at Slope Site III (figure 13.4B). The top 4 species accounted for 97 percent and 74 percent of the total polychaetes on the Volcano 7 upper and lower summit, respectively, but only 22.3 and 22.6 percent on the flank and base (table 13.1; figure 13.4A). Similarly, the top 4 species accounted for 76 percent of the total polychaetes at Slope-Site III, but only 36.8 and 37.2 percent at Sites I and II, respectively (table 13.1, figure 13.4B).

Pooled rarefaction curves show patterns similar to those of species richness. Diversity (expected number of species for a given sample size) increased downslope on Volcano 7 to the mid slope, but flank and base diversities were indistinguishable (figure 13.4C). Slope Site III diversity was much lower than that of Sites I and II, which were indistinguishable (figure 13.4D). Shannon-Wiener information indices were similarly low for the high organic input sites in each region, the upper (1.79 ± 0.62) and lower (2.83 ± 0.65) summit on Volcano 7, and Slope Site III (2.76 ± 0.54). Information indices were somewhat higher for the Volcano 7 flank and base (both 5.20 ± 0.83) than the Slope Sites I (4.66 ± 0.81) and II (4.57 ± 0.75).

Table 13.1 *Dominant Polychaetes at the Volcano 7 and Carolina Slope Study Sites*

A. Volcano 7

Base (2850–3353 m)	% of Total	Flank (1320–1790 m)	% of Total	Lower Summit (788–857 m)	% of Total	Upper Summit (745–767 m)	% of Total
Chaetozone sp.	5.8	*F. similis*	7.6	*Tharyx* sp. A	24.5	*Cirrophorus lyra*	54.2
Euchone incolor	5.8	*Cirrophorus* sp. 2	5.4	*E. incolor*	22.6	*Protodorvillea* sp. A	23.6
Atelesyllis sp.	5.8	*Ampharete* sp.	4.9	*Cirrophorus lyra*	15.1	*Bonuania* sp.	11.1
Aglaophamus sp.	5.2	*Tharyx* sp. B	4.3	*Protodorvillea* sp. A	12.2	*Tharyx* sp. A	8.3
Tharyx sp. A	4.5	*Syllidia* sp.	4.3	*Caulleriella* sp. 1	11.1	*Caulleriella* sp. 1	1.4
Fabrisabella similis	4.5	*Prionospio* sp.	4.3	*Ampharetidae* sp.	7.3	*Flabelliderma* sp.	1.4
Nereididae sp. A	3.9	Acrocirridae sp. A	3.8	*Bonuania* sp.	2.1		
Aricidea lopezi	3.9	*Protodorvillea* sp. B	3.8	*Flabelligera* sp.	0.8		
		Maldanidae sp. A	3.8				

B. Carolina Slope, 850 m.

Site I (off Cape Fear)	% of Total	Site II (off Cape Lookout)	% of Total	Site III (off Cape Hatteras)	% of Total
Dysponetus sp. 1	15.8	*Meiodorvillea minuta*	14.3	*Scalibregma inflatum*	26.3
Cossura longocirrata	9.6	*Dysponetus* sp. 1	9.7	*Aricidea quadrilobata*	23.6
Paraonis cerruti	6.1	*Cossura longocirrata*	7.6	*Cossura* sp.	12.4
Microrbinis linea	5.3	*Paraonis uncinatus*	5.6	*Chaetozone* sp.	3.8
Hesionura sp. 1	4.4	*Ophelia profunda*	3.9	*Ceratocephale loveni*	3.3
Questa sp. 1	4.4	*Flabelligella* sp. 1	3.8	*Schistomeringos caecus*	2.3
Barantolla sp. 1	3.5	*Barantolla* sp. 1	3.6	Phyllodocidae sp. 2	2.2
Dysponetus sp. 2	3.5	*Tachytrypane* sp. 1	3.1	*Lumbrineris fragilis*	1.6
Meiodorvillea minuta	3.5	*Tharyx* sp. 1	3.1	*Artacama* sp. 1	0.7
Ampharetidae sp. 1	2.6	*Levinsenia gracilis*	2.9	Opheliidae sp. 1	0.5
Dorvillea sp. 1	2.6	Oweniidae sp. 1	2.5	*Levinsenia* sp. 1	0.5
Spionidae sp. 1	2.6	*Paraonis* sp. A	2.2	Dorvilleidae sp. 3	0.5

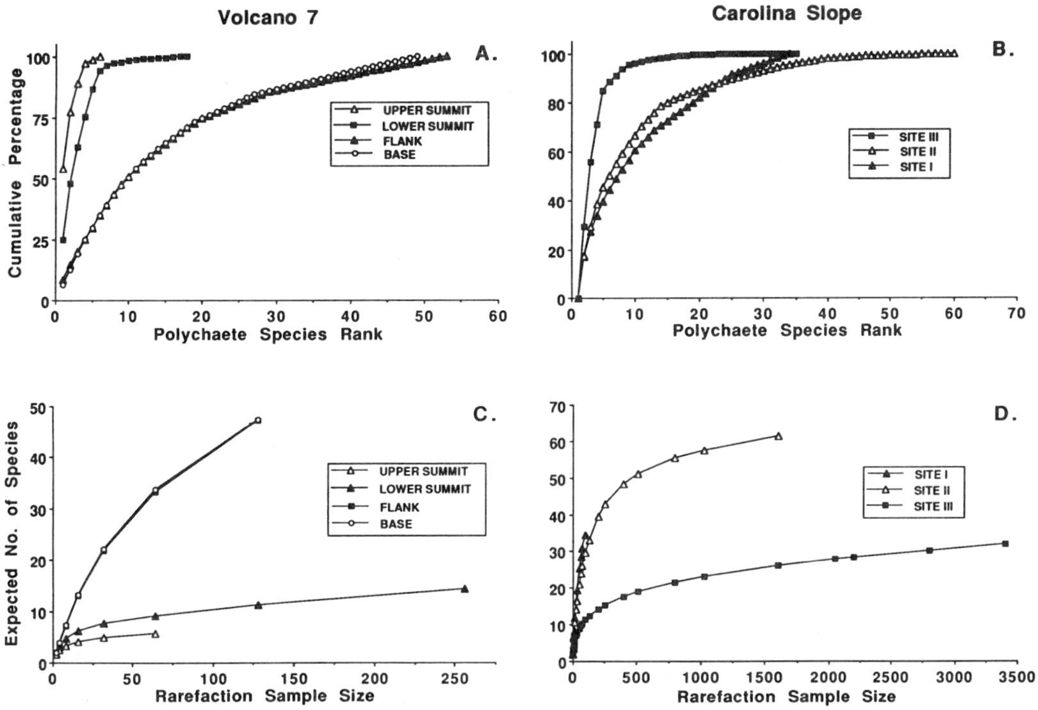

FIGURE 13.4. Species dominance curves for polychaetes (> 300 μm) from: A. 4 zones on Volcano 7; B. 3 sites on the Carolina slope.

Rarefaction curves of expected number of polychaete species for pooled cores from C. Volcano 7; D. Carolina slope. Study site depths are given in table 13.1.

Population Size Structure

Size-frequency distributions were examined for some of the most abundant polychaete species in each habitat. Effects of organic enrichment on population size structure appeared to differ along the Volcano 7 and Carolina slope organic gradients. On Volcano 7 in November 1988, 3 of the 4 dominant polychaetes, *Protodorvillea* sp. A, *Tharyx* sp. A, and *Cirrophorus lyra* exhibited an increase in body size upslope (figure 13.5: $p < 0.01$). *Euchone incolor* size frequencies did not differ significantly between the flank and lower summit (figure 13.5). No other polychaete species were collected in quantities on Volcano 7 sufficient to assess between-zone changes in size structure.

In contrast to the increased body sizes observed in enriched settings on Volcano 7, the organic-rich Slope Site III exhibited polychaete size distributions dominated by juveniles. The histogram for *Ceratocephale loveni* collected from Slope Site III in 1989 was dominated by individuals less than 20 setigers, whereas 1989 collections of this species from Site II contained a fairly uniform distribution of small, mid-sized and large individuals (figure 13.6, K-S tests between October 1989 data for Sites II and III; $p < 0.01$). Size-frequency

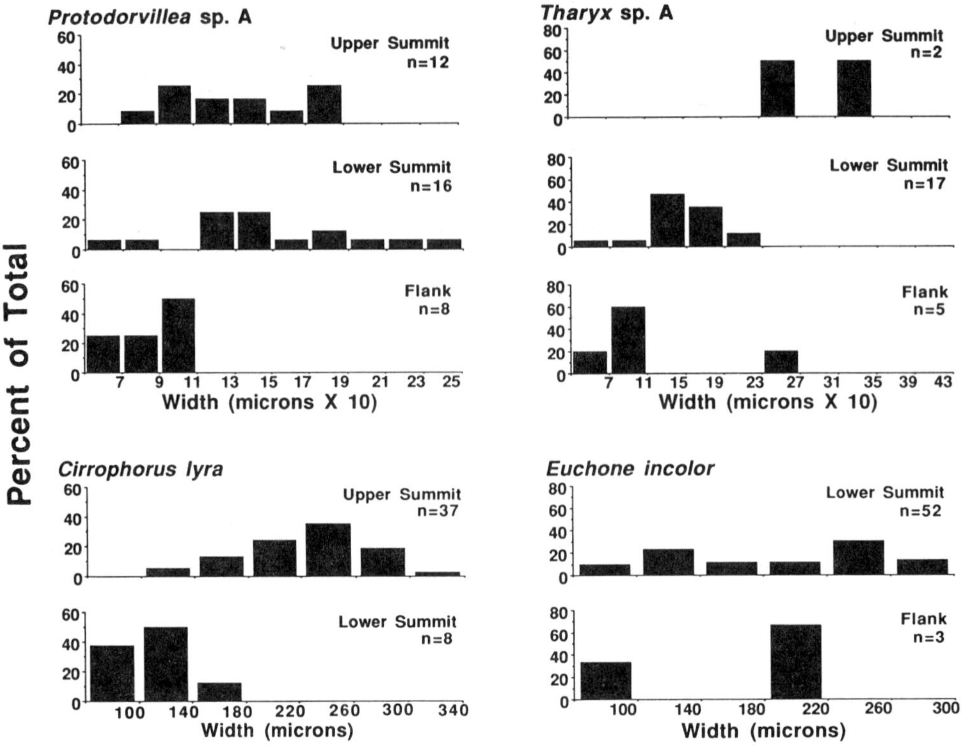

FIGURE 13.5. Size-frequency distributions of 4 dominant polychaete species on Volcano 7, Nov.–Dec. 1988. Widths of the first setiger are given for each species. These 4 species account for 86, 74, and 14% of the polychaetes on the upper summit, lower summit, and flank, respectively. See text for statistical differences among sites within each species.

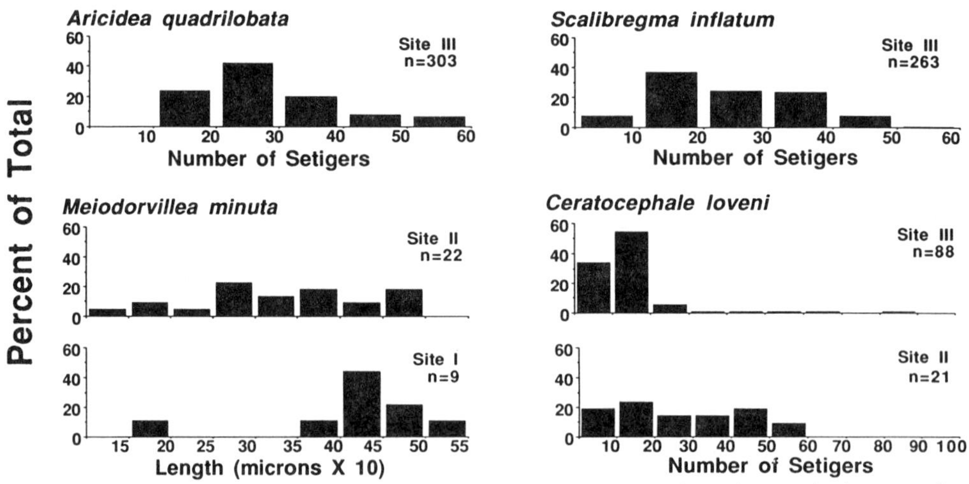

FIGURE 13.6. Size-frequency distributions for 4 common Carolina slope polychaetes collected in October 1989: *Aricidea quadrilobata* and *Scalibregma inflatum*, dominant species at Slope Site III, show distributions skewed toward smaller sizes. Size distribution of *Ceratocephale loveni* from Slope Site III is also left skewed and is significantly different from the Slope Site II distribution (Kolmogorov-Smirnov test, $p < 0.01$). *Meiodorvillea minuta* distribution at Slope Site I is significantly different from that of Slope Site II ($p = 0.012$).

distributions of *C. loveni* collected in September 1988 and October 1989 at Site II did not differ significantly from one another ($p = 0.18$); a single large (62 setiger) individual was collected at Site I. The dominant species at Slope Site II, *Meiodorvillea minuta,* exhibited a distribution skewed toward larger individuals at Slope Site I and many more small individuals at Slope-Site II (figure 13.6) ($p = 0.0125$).

The two most common polychaetes at Site III, the scalibregmatid *Scalibregma inflatum* and the paraonid *Aricidea quadrilobata,* both had skewed size frequencies, suggesting juvenile-dominated populations (figure 13.6). In *S. inflatum* over 50 percent of the individuals were ≤ 25 setigers while the maximum measured size was 52 setigers. In *A. quadrilobata* 62 percent of the individuals were < 30 setigers, with the maximum size measured at 63 setigers. Average size for ovigerous females was 56.3 setigers (SD = 5.4, n = 4) and for ripe males was 59 setigers (SD = 3.1, n = 5). In combination, *S. inflatum* and *A. quadrilobata* composed 50 percent of the polychaetes at Site III (table 13.1). Thus, Slope Site III, experiencing the highest levels of carbon input, had polychaete populations composed largely of young individuals.

Reproductive Patterns

The polychaetes present on Volcano 7 exhibited two notable trends with respect to reproduction. One was an upslope increase in reproductive activity (defined as the presence of maturing ova within the body or brooded embryos and larvae attached to the body or tube) (Nested ANOVA: $F_{3,7} = 6.39$, $p = 0.021$). The average percent of polychaetes reproductive on Volcano 7 in November-December 1988, was 63.4 percent on the upper summit, 25.3 percent on the lower summit, 5.1 percent on the flank, and 2.2 percent on the base. A second observation was an upslope increase in the proportion of polychaete species that brood larvae (table 13.2). The brooding species *Protodorvillea* sp. A, *Cirrophorus lyra,* and *Tharyx* sp. A, accounted for 86 percent of the polychaetes on the upper summit. These same three species, along with the brooders *Euchone incolor* and *Ampharete* sp. B, composed 81 percent of the individuals on the lower summit. Only 26.6 percent of the polychaetes collected on the flank and 15.5 percent of those on the base belong to species observed or known to brood. It is reasonable to expect that these percentages for the flank and base are underestimates of the importance of brooding, as not all species were reproductive at the time of sample collection.

The two patterns described above were not observed on the Carolina slope. Fewer than 2 percent of slope polychaetes were observed to be carrying ova, and only 1 spionid at Site II has been observed to brood. Very small juveniles of *Scalibregma inflatum, Aricidea quadrilobata, Cossura* sp., *Terebellides* sp., and *Lumbrineris* sp. 1 are commonly found at Site III. This pattern may suggest high levels of recruitment at the more organic-rich Slope Site III, but brooding

Table 13.2 *Distribution and Method of Brooding of Infaunal Polychaetes on Volcano 7 that are Known to Brood Larvae*

Species	Zone*	Method of Brooding
Acrocirridae sp. A	US, LS, F, B	Larvae attached dorsally
Ampharete sp. A	LS, F	Larvae brooded in tubes
Ampharetidae sp. B	LS, F	Larvae brooded in tubes, weakly or not attached
Tharyx sp. A	US, LS, F, B	Paired dorsal brood pouches, 1–2 embryos/pouch
Protodorvillea sp. A	US, LS	Dorsal brood pouches, 1–2 embryos/pouch
Euphrosine sp.	F	Larvae attached dorsally
Nephtyidae sp. C	B	Brood pouches
Aricidea (Allia) quadrilobata	F	Embryos brooded dorsally
Cirrophorus lyra	US, LS	Larvae attached laterally (2/segment) for 15–20 setigers
Euchone incolor	LS, F, B	Larvae attached dorsally, and in tubes
Terebellidae	F, B	Larvae brooded in tube

* US = Upper Summit (730 m–770 m)
LS = Lower Summit (770 m–860 m)
F = Flank (1300 m–1800 m)
B = Base (2970 m–3350 m)

is not implicated as a prevalent reproductive strategy in any of the Carolina slope populations.

Discussion

Life History Responses to Enrichment

Assemblages inhabiting the most enriched sites on Volcano 7 exhibited a higher proportion of reproductive females, a prevalence of brooding behavior, and an increase in body size (figure 13.5). Data from Slope Site III show a low proportion of reproductive females, but juvenile size frequencies suggestive of high recruitment (figure 13.6). Whereas abundance and diversity patterns illustrate similar effects of enrichment gradients at Volcano 7 and the Carolina slope, life history responses are more diverse and may be complicated by geographic, genetic, and temporal factors. Though our view of these assemblages is a static one, constrained by limited sampling, the few available life-history data are suggestive of opportunism in enriched settings. High levels of reproductive activity and a preponderance of brooding species on the summit of Volcano 7 (table 13.2) may indicate potential for rapid, localized population growth. Wood borers in the deep sea have developed opportunistic life histories that involve brooding and high reproductive rates, thus permitting exploitation of patchy, highly localized food resources (Turner 1973). Respondents to shallow-water enrichments, such as *Capitella* spp., *Polydora ligni,* and *Streblospio benedicti,* also brood young, at least briefly, and have high reproductive rates (Grassle and Grassle 1974; Pearson and Rosenberg 1978; Levin and Huggett 1990). A review of invertebrate reproductive patterns in the deep sea suggests that, despite a preponderance of lecithotrophic development, brooding is uncommon in deep-sea taxa (with the important exception of peracarid crustaceans) (Gage and Tyler 1991). Little is known, however, about reproductive patterns of deep-sea polychaetes.

An alternative explanation exists for the prevalence of brooding on the summit of Volcano 7. Isolated settings, such as islands, often exhibit high proportions of brooding species (Johannesson 1988). Levin (1984) noted the prevalence of brooding among polychaetes and other taxa inhabiting a variety of shallow, soft-sediment habitats on the U.S. Pacific coast that were isolated by many miles of inhospitable shoreline. The summit of Volcano 7 is isolated from other sediments at comparable depths by topography. It is also potentially isolated from many species occupying the lower slopes of the volcano by the severe oxygen minimum zone that surrounds the upper slopes. Thus, the importance of brooding on the summit may be a manifestation of isolation and the need to reduce wasteful loss of propagules from the habitat.

Our information on reproductive modes of the Carolina slope polychaetes is less complete and may have been constrained by sampling only in June, September, and October. An unusual feature of the enriched Slope Site III, not observed on the Volcano 7 summit, was a tendency for the dominant polychaete populations to have size frequencies skewed toward smaller size classes (figures 13.5 and 13.6). The presence of numerous, small individuals belonging to several species presumably indicates active recruitment near the time of sample collection or year round. Blake et al. (1987) reported size-frequency shifts toward smaller specimens in *Scalibregma inflatum* between May and September at a 600 m station approximately 6 km from Slope Site III, suggestive of summer recruitment. However, size-frequencies reported by Blake et al. (1987) for *Cossura* sp. from this station were clearly dominated by large individuals.

Deep-sea populations were traditionally believed to be dominated by older individuals and characterized by slow reproductive rates (Grassle and Sanders 1973). However, at the 4800 m deep HEBBLE site, disturbance related to episodic sediment transport events leads to juvenile domination in an ampharetid population (Thistle et al. 1985). Radiochemical data from Slope Site III (DeMaster unpublished data) suggest that the slope off Cape Hatteras is subject to episodic sediment transport and mass wasting. Sediment disturbance is generally more typical of shallow-water populations, which also are composed largely of juveniles.

None of the dominant, potentially opportunistic species at Site III have been observed to brood larvae. Though *Aricidea quadrilobata* (= *A. antennata*?)

collected on the Volcano 7 flank was found to brood embryos dorsally, we have seen no evidencc of similar behavior in over 300 otherwise morphologically identical specimens collected from Site III on the Carolina slope. Two distinct species may be involved. We suggest that response to organic enrichment and possibly to disturbance on the slope may consist of increased reproduction and/or recruitment rates by polychaetes with nonbrooding reproductive strategies.

On Volcano 7, intraspecies body sizes of dominant species increased along the enrichment gradient (figure 13.5). Similar responses have been observed for polychaetes inhabiting (1) enriched sites beneath or near salmon pens in shallow water (Weston 1990); (2) mesocosms subject to a eutrophication gradient (Levin 1986); and (3) laboratory treatments enriched with sewage and bluegreen algae (T. Bridges et al., in press). We are unable to distinguish between differential growth rates and variable age structure as causes for the size-frequency shifts observed along the upper slopes of Volcano 7 (figure 13.5). The flank may be occupied primarily by juveniles of the species studied, or may simply have stunted adults. The Carolina slope sites did not exhibit the same trends. Instead, body size diminished along the enrichment gradient due to an increase in numbers of juveniles (figure 13.6).

What Taxa Respond to Organic Enrichment in Deep Water?

Early experimental studies of organic enrichment in the deep sea were often carried out in colonization trays containing defaunated sediments as well as added organic matter. The experiments thus embodied an isolated physical disturbance as well as an enrichment. Grassle and Morse-Porteous (1987) reported on the appearance of *Capitella* spp. in enriched and unenriched colonization trays deployed in the North Atlantic and in the Bahamas. They also observed the dorvilleid polychaete *Ophryotrocha* sp. in control cores that were naturally enhanced with decomposing *Sargassum*. Desbruyères et al. (1980) found high numbers of a spionid, *Prionospio* sp., and dorvilleids, *Ophryotrocha* spp., colonizing enriched trays in the Bay of Biscay. Levin and Smith (1984) reported avoidance of kelp-enriched colonization trays by Santa Catalina Basin infauna. However, a dorvilleid polychaete, resembling an undescribed genus characteristic of pulp mills and sewage outfalls (D. Montagne and I. Oug personal communication), colonized one tray that went anoxic. The primary respondents in these studies were capitellid, spionid, and dorvilleid polychaetes not normally found in samples of the deep-water background communities. They belonged to the same families that typically respond to enrichment and disturbance in shallow water (Pearson and Rosenberg 1978). In contrast, Smith (1986) demonstrated, with experimental studies of organic food falls placed directly on the Santa Catalina Basin floor, that many of the dominant taxa in the background community were capable of rapid exploitation of pulsed food inputs.

The polychaetes inhabiting our enriched Pacific and Atlantic study sites were taxonomically distinct and exhibited different reproductive strategies from polychaetes at the nonenriched settings. Neither set included the traditional opportunistic taxa described above. Rather, the enrichment species on both Volcano 7 and the Carolina slope were cosmopolitan species. Many of them (*Cirrophorus lyra, Euchone incolor, Scalibregma inflatum, Aricidea quadrilobata,* and *Cossura* sp.) are frequently found on the continental shelf or even shallower, but are also reported from bathyal or abyssal depths. *Scalibregma inflatum* has been reported from organically-enriched settings in shallow water (Rosenberg 1976; Rosenberg et al. 1987; Aschan and Skullerud 1990). *Aricidea quadrilobata* is found from 3 to 2269 m (Streslov 1973) but has not been identified with disturbance or enrichment. Natural, persistent enrichment does not attract the disturbance opportunists characteristic of other deep- and shallow-water studies. In fact, there does not appear to be a generalized "enrichment fauna" associated with bathyal settings subject to high organic inputs.

Abundance and Diversity Along Enrichment Gradients

The studies presented here suggest that natural organic enhancement at upper bathyal depths in both the Pacific and Atlantic Oceans leads to (1) elevated polychaete abundance (when oxygen in overlying water is sufficient) (figure 13.3); (2) high dominance by relatively few species (figures 13.4A, 4B); and (3) reduced diversity and species richness (figures 13.4C, 4D). These effects on the two bathyal communities studied are also observed in shallow-water communities.

The abundance and diversity patterns of polychaetes observed along the gradients described for Volcano 7 and the Carolina slope resemble those described by Pearson and Rosenberg (1978) for shallow-water gradients of organic pollution (figure 13.7). Levin et al. (1991a) noted the same pattern for total macrofauna on Volcano 7 and similar changes in community structure were reported by Sanders (1969) along an oxygen gradient in Walvis Bay off NW Africa. Communities subject to no excess enrichment typically exhibit moderate densities and relatively high diversity, e.g., the flank and base of Volcano 7 and Slope Site I. Along a gradient of increasing organic enrichment communities shift toward high-density, low-diversity assemblages where a few opportunistic species flourish (as on the lower summit of Volcano 7 and Slope Site III), and eventually to low-density, low-diversity assemblages, where only a few stress-tolerant individuals survive (as on the upper summit of Volcano 7)

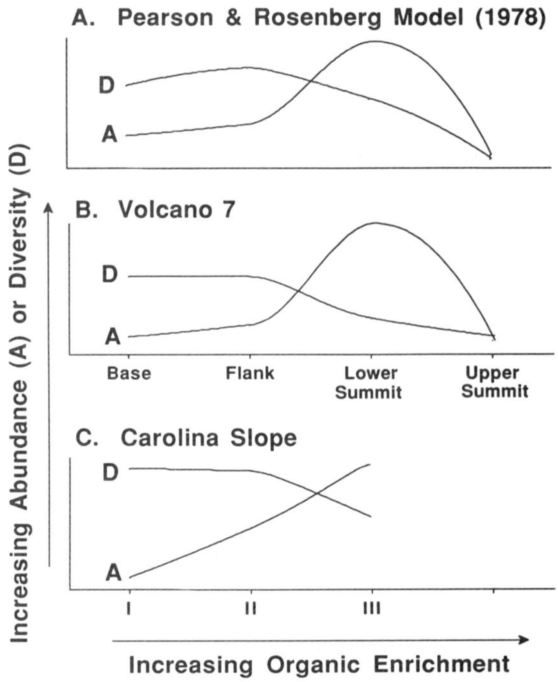

FIGURE 13.7. Changes in abundance (A—average number of individuals) and species diversity (D—Shannon-Wiener diversity index) along three organic enrichment gradients: A. Generalized macrofaunal abundance and diversity patterns for shallow-water pollution gradients, redrawn from Pearson and Rosenberg (1978, fig. 2); B. Polychaete abundance and diversity at 4 depth zones on Volcano 7. Organic contents of flank and base sediments are not significantly different (depths are given in table 13.1); C. Polychaete abundance and diversity at the 3 Carolina slope sites (850 m).

(figure 13.7). Oxygen stress associated with high production in surface waters and significant deposition of undegraded organic matter produces a low density, end-member assemblage on Volcano 7, but we found no comparably stressed setting on the Carolina slope.

The polychaete species abundance distributions also respond as Gray and Pearson (1982) and Pearson et al. (1983) demonstrated for macrofauna along organic pollution gradients at shallow-water sites in the northeast Atlantic. In an undisturbed community the species abundance distribution resembles the right-hand portion of a log normal distribution, as observed for the Volcano 7 flank and base and Slope Site I (figure 13.8). Sewage, china clay effluent, oil, and paper mill pollution were all shown to cause a decline in the number of rare species and an increase in the incidence of high abundance species thriving in disturbed conditions (Pearson et al. 1983). This progression is observed along the enrichment gradients on the Carolina slope and Volcano 7.

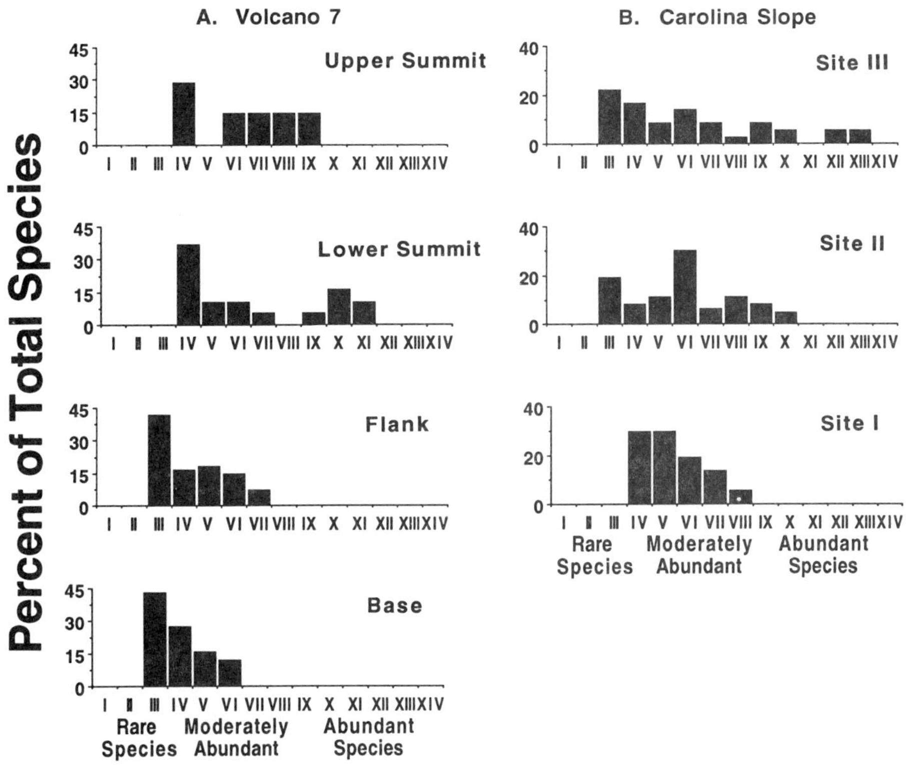

FIGURE 13.8. Percent of total number of polychaete species in geometric abundance classes along organic enrichment gradients represented by: A. Four depth zones on Volcano 7 (depths given in table 13.1); B. Three Carolina slope sites at 850 m. Abundance classes are: I = 1 individual/m^2, II = 2-3, III = 4-7, IV = 8-15, V = 16-31....... XIII = 4,096-8,192, XIV = 8,192-16,384.

TEMPORAL AND SPATIAL PATTERNS OF ORGANIC INPUT

We hypothesize that highly pulsed inputs of organic material, as observed in the NE Atlantic (Lampitt 1985; Thiel et al. 1989) and on parts of the NW Atlantic slope (Hecker 1990), might produce more ephemeral benthic responses than more steady inputs, such as observed beneath upwelling regions. Seasonally varying organic fluxes are likely to regulate the timing of gamete production and larval release or recruitment (Tyler 1988), but may not be sufficient to support persistent, dense assemblages of opportunistic species. The most dramatic responses to pulsed inputs are exhibited by microbial and protozoan populations (Gooday and Turley 1990). Unpredictable inputs, resulting from turbidity current deposition or carcass and plant detritus falls, probably represent an important source of disturbance, and contribute to overall habitat heterogeneity and diversity maintenance (Grassle and Morse-Porteous 1987; Grassle 1989). However, we hypothesize that these spatially and temporally patchy inputs will produce different effects on infaunal community structure than those observed in the Volcano 7 and Carolina slope systems, where organic inputs appear to be more continuous.

The generality of the abundance, diversity, and reproductive patterns observed on Volcano 7 and the Carolina slope could be tested by examining macrofauna in other regions of the ocean subject to continuous enrichment resulting from high primary production. These include other areas in the eastern Pacific, the slope off NW Africa, the Arabian Sea (all regions of OMZ development), the Chukchi and Bering Seas, and the Norwegian and Greenland seas (Berger 1990). We would predict patterns of elevated abundance, reduced diversity and opportunistic life histories among polychaetes at these sites.

Acknowledgments

We thank the ship and submersible crews and many scientists that participated in the field programs that produced the samples upon which the data are based. Appreciation is extended to T. Schaff, N. Blair, and D. DeMaster for sharing data and knowledge of the Carolina slope, and to K. Wishner, for providing oxygen data from Volcano 7. Helpful comments on the manuscript were offered by J. Gage, P. Jumars, M. Rex, C. Young, and an anonymous reviewer. Support for the Volcano 7 research was provided by the U.S. Office of Naval Research, grant N00014–89-J-3072. Support for the Carolina slope work was provided by the NOAA National Undersea Research Program at UNC Wilmington, grant no. NA88AA-D-UR004 and by the North Carolina Duke-UNC Oceanographic Consortium.

References

Aschan, M. M. and A. M. Skullerud. 1990. Effects of changes in sewage pollution on soft-bottom macrofaunal communities in the inner Oslofjord, Norway. *Sarsia* 75:169–190.

Berger, W. H. 1989. Global maps of ocean productivity. In W. H. Berger, V. S. Smetacek, and G. Wefer, eds., *Productivity of the Ocean: Present and Past,* pp. 429–455. New York: Wiley.

Blake, J. A., B. Hecker, J. Grassle, B. Brown, M. Wade, P. Boehm, E. Baptiste, B. Hilbig, N. Maciolek, R. Petrecca, R. Ruff, V. Starczak, and L. Watling. 1987. Study of biological processes on the U.S. South Atlantic slope and rise, Phase 2. Final Report to the U.S. Department of Interior, Minerals Management Service, Contract no. 14–12–0001-30064.

Bridges, T. S. and L. A. Levin. 1991. Individual- and population-level response of *Streblospio benedicti* to different forms of organic enrichment. *Amer. Zool.* 31:144A.

Bridges, T. S., L. A. Levin, D. Cabrera, and G. Plaia. In press. Effects of sediment amended with sewage, algae, or hydrocarbons on growth and reproduction in two opportunistic polychates. *J. Exp. Mar Biol. Ecol.*

Dauer, D. M., T. L. Stokes, Jr., H. R. Barker, Jr., R. M. Ewing, and J. W. Sourbeer. 1984. Macrobenthic communities of the lower Chesapeake Bay. IV. Bay-wide transects and the intercontinental shelf. *Int. Rev. Ges. Hydrobiol.* 69:1–22.

Desbruyères, D., J. Y. Bevas, and A. Khripounoff. 1980. Un cas de colonisation rapide d'un sédiment profond. *Oceanol. Acta* 3:285–291.

Desbruyères, D., P. Crassous, J. Grassle, A. Khripounoff, D. Reyss, M. Rio, and M. Van

Praet. 1982. Données écologiques sur un nouveau site d'hydrothermalisme actif de la ride du Pacifique oriental. *Compte rendus de l'Académie des sciences, Paris* Ser. III 295:489–94.

Gage, J. D. and P. A. Tyler. 1991. *Deep-Sea Biology: A Natural History of Organisms at the Deep-Sea Floor*. Cambridge: Cambridge University Press.

Gooday A. J. 1988. A response by benthic foraminifera to the deposition of phytodetritus in the deep sea. *Nature* 332:70–73.

Gooday, A. J. and P. J. D. Lambshead. 1989. The influence of seasonally deposited phytodetritus on benthic foraminiferal populations in the bathyal northeast Atlantic: The species response. *Mar. Ecol. Prog. Ser.* 58:53–67.

Gooday, A. J. and C. M. Turley. 1990. Responses by benthic organisms to inputs of organic material to the ocean floor: a review. *Phil. Trans. R. Soc. Lond.* A 331:119–138.

Gooday A. J., L. A. Levin, C. L. Thomas, and B. Hecker. 1992. The taxonomy, distribution, and ecology of *Bathysiphon filiformis* and *B. major* (Protista: Foraminiferida) on the continental slope off North Carolina. *J. Foram. Res.* 22:129–146.

Graf, G. 1989. Benthic-pelagic coupling in a deep-sea benthic community. *Nature* 341:437–439.

Grassle, J. F. 1989. Species diversity in deep-sea communities. *Trends in Ecology and Evolution* 4:12–15.

Grassle, J. F. and J. P. Grassle. 1974. Opportunistic life histories and genetic systems in marine benthic polychaetes. *J. Mar. Res.* 32:253–284.

Grassle, J. F. and L. S. Morse-Porteous. 1987. Macrofaunal colonization of disturbed deep-sea environments and the structure of deep-sea benthic communities. *Deep-Sea Res.* 34:1911–1950.

Grassle, J. F. and H. L. Sanders. 1973. Life histories and the role of disturbance. *Deep-Sea Res.* 20:643–659.

Grassle, J. F. and H. L. Sanders. 1980. PRARE 1 program.

Gray, J. S. and T. H. Pearson. 1982. Objective selection of sensitive species indicative of pollution-induced change in benthic communities. 1. Comparative methodology. *Mar. Ecol. Progr. Ser.* 9:111–119.

Grémare, A., A. G. Marsh, and K. R. Tenore. 1988. Short-term reproductive responses of *Capitella* sp. I (Annelida: Polychaeta) fed on different diets. *J. Exp. Mar. Biol. Ecol.* 123:147–162.

Hecker, B. 1990. Photographic evidence for the rapid flux of particulates to the sea floor and their transport down the continental slope. *Deep-Sea Res.* 37:1773–1782.

Johannesson, K. 1988. The paradox of Rockall: Why is a brooding gastropod (*Littorina saxatilis*) more widespread than one having a planktonic larval dispersal stage (*L. littorea*)? *Mar. Biol.* 99:507–513.

Johnson, T. C. 1989. Hydrography and sedimentation on the continental slope and rise off North and South Carolina. North Carolina Coastal Oceanography Symposium. National Undersea Research Program Report 89–2, pp. 151–172.

Jumars, P. A. and R. A. Wheatcroft. 1989. Responses of benthos to changing food quality and quantity, with a focus on deposit feeding and bioturbation. In W. H. Berger, V. S. Smetacek, and G. Wefer, eds., *Productivity of the Ocean: Present and Past,* pp. 235–254. New York: Wiley.

Lampitt, R. S. 1985. Evidence for the seasonal deposition of detritus to the deep-sea floor and its subsequent resuspension. *Deep-Sea Res.* 32:885–897.

Levin, L. A. 1984. Life history and dispersal patterns in a dense infaunal polychaete assemblage: community structure and response to disturbance. *Ecology* 65:1185–1200.

Levin, L. A. 1986. Effects of enrichment on reproduction in the opportunistic polychaete *Streblospio benedicti* (Webster): A mesocosm study. *Biol. Bull.* 171:143–160.

Levin, L. A. and E. L. Creed. 1986. Effect of temperature and food availability on reproductive responses of *Streblospio benedicti* (Polychaeta: Spionidae) with planktotrophic or lecithotrophic development. *Mar. Biol.* 92:103–113.

Levin, L. A. and D. V. Huggett. 1990. Implications of alternative reproductive modes for seasonality and demography in an estuarine polychaete. *Ecology* 71(6):2191–2208.

Levin, L. A. and C. R. Smith. 1984. Response of background fauna to disturbance and enrichment in the deep sea: A sediment tray experiment. *Deep-Sea Res.* 31:1277–1285.

Levin, L. A., C. L. Huggett, and K. F. Wishner. 1991a. Control of deep-sea benthic community structure by oxygen and organic matter gradients in the eastern Pacific Ocean. *J. Mar. Res.* 49:763–800.

Levin, L. A., L. D. McCann, and C. L. Thomas. 1991b. The ecology of polychaetes on deep seamounts in the eastern Pacific Ocean. *Ophelia Suppl.* 5:467–476.

Lochte, K. and C. M. Turley. 1988. Bacteria and cyanobacteria associated with phytodetritus in the deep-sea. *Nature* 333:67–69.

Marsh, A. G., A. Grémare, and K. R. Tenore. 1989. Effect of food type and ration on growth of juvenile *Capitella* sp. I (Annelida: Polychaeta): Macro- and micronutrients. *Mar. Biol.* 102:519–527.

Pearson, T. H. and R. Rosenberg. 1978. Macrobenthic succession in relation to organic enrichment and pollution in the marine environment. *Oceanogr. Mar. Biol. Ann. Rev.* 16:229–311.

Pearson, T. H., J. S. Gray, and P. J. Johannessen. 1983. Objective selection of sensitive species indicative of pollution-induced change in benthic communities. 2. Data analyses. *Mar. Ecol. Progr. Ser.* 12:237–255.

Qian, P. Y. and F. S. Chia. 1991. Fecundity and egg size are mediated by food quality in the polychaete worm *Capitella* sp. *J. Exp. Mar. Biol. Ecol.* 148:11–25.

Rex, M. A. 1981. Community structure in the deep-sea benthos. *Ann. Rev. Ecol. Syst.* 12:331–53.

Rex, M. A. 1983. Geographic patterns of species diversity in the deep-sea benthos. In G. T. Rowe, ed., *Deep-Sea Biology,* pp. 453–472. New York: Wiley.

Rice, A. L., D. S. M. Billet, J. Fry, A. W. G. John, R. S. Lampitt, R. F. C. Mantoura, and R. J. Morris. 1986. Seasonal deposition of phytodetritus to the deep-sea floor. *Proc. R. Soc. Edinb.* 88B:265–279.

Riley, G. A., P. J. Wangersky, and D. van Hemert. 1964. Organic aggregates in tropical and subtropical surface waters of the North Atlantic Ocean. *Limnol. and Oceanogr.* 9:546–550.

Ritz, D. A., M. E. Lewis, and M. Shen. 1989. Response to organic enrichment of infaunal macrobenthic communities under salmonid seacages. *Mar. Biol.* 103:211–214.

Rosenberg, R. 1976. Benthic faunal dynamics during succession following pollution abatement in a Swedish estuary. *Oikos* 27:414–427.

Rosenberg, R., J. S. Gray, A. B. Josefson, and T. H. Pearson, 1987. Peterson's benthic stations revisited. II. Is the Oslofjord and eastern Skagerrak enriched? *J. Exp. Mar. Biol. Ecol.* 20:1–41.

Rowe, G. T. 1983. Biomass and production of the deep-sea macrobenthos. In G. T. Rowe, ed., *Deep-Sea Biology,* pp. 97–121. New York: Wiley.

Rowe, G. T., P. T. Polloni, and S. G. Horner. 1974. Benthic biomass estimates from the northwestern Atlantic Ocean and the northern Gulf of Mexico. *Deep-Sea Res.* 21:641–650.

Sanders, H. L. 1969. Benthic marine diversity and the stability-time hypothesis. *Brookhaven Symposium in Biology* 12:71–81.

Schaff, T. 1991. Spatial heterogeneity of continental slope benthos off the Carolinas. M.S. thesis, North Carolina State University, Raleigh.

Schaff, T., L. Levin, N. Blair, D. DeMaster, R. Pope, and S. Boehme. 1992. Spatial heterogeneity of benthos on the Carolina continental slope: Large (100 km)-scale variation. *Mar. Ecol. Prog. Ser.* 88:143–160.

Smith, C. R. 1983. Enrichment, disturbance, and deep-sea community structure: The significance of large organic falls to bathyal benthos in Santa Catalina Basin. Ph.D. dissertation, University of California, San Diego.

Smith, C. R. 1986. Nekton falls, low-intensity disturbance, and community structure of infaunal benthos in the deep sea. *J. Mar. Res.* 44:567–600.

Smith, K. L., Jr., M. B. Laver, and N. O. Brown. 1983. Sediment community oxygen consumption and nutrient exchange in the central and eastern North Pacific. *Limnol. and Oceanogr.* 28:882–898.

Snelgrove, P. V. R., J. F. Grassle, and R. F. Petrecca. 1992. The role of food patches in maintaining deep-sea diversity: Field experiments with hydrodynamically unbiased colonization trays. *Limnol. Oceanogr.* 37:1543–1550.

Sokal, R. R. and F. J. Rohlf. 1969. *Biometry: The Principles and Practice of Statistics in Biological Research.* San Francisco: Freeman.

Streslov, V. E. 1973. In V. S. Kothekar, ed., *Polychaete Worms of the Family Paraonidae Cerruti, 1909 (Polychaeta, Sedentaria).* Leningrad: Nauka.

Thiel, H., O. Pfannkuche, G. Schriever, K. Lochte, A. J. Gooday, C. Hemleben, R. F. G. Mantoura, C. M. Turley, J. W. Patching and F. Riemann. 1989. Phytodetritus on the deep-sea floor in a central oceanic region of the northeast Atlantic. *Biol. Oceanogr.* 6:203–239.

Thistle, D., J. Y. Yingst, and K. Fauchald. 1985. A deep-sea benthic community exposed to strong near-bottom currents on the Scotian Rise (Western Atlantic). *Mar. Geol.* 66:91–112.

Turley, C. M., K. Lochte, and D. J. Patterson. 1988. A barophilic flagellate isolated from 4500 m in the mid-North Atlantic. *Deep-Sea Res.* 35:1079–1092.

Turner, R. D. 1973. Wood boring bivalves, opportunistic species in the deep sea. *Science* 180:1377–1379.

Tyler, P. A. 1988. Seasonality in the deep sea. *Oceanography and Marine Biology: An Annual Review* 26:227–258.

Weston, D. P. 1990. Quantitative examination of macrobenthic community changes along an organic enrichment gradient. *Mar. Ecol. Progr. Ser.* 61:233–244.

Wishner, K., L. Levin, M. Gowing, and L. Mullineaux. 1990. Multiple roles of the oxygen minimum in benthic zonation on a deep seamount. *Nature* 346:57–59.

Woodin, S. A. 1974. Polychaete abundance patterns in a marine soft-sediment environment: The importance of biological interactions. *Ecol. Monogr.* 44:171–187.

CHAPTER FOURTEEN

The Advantages of Dispersal in a Patchy Environment: Effects of Disturbance in a Cellular Automaton Model

Ron J. Etter and Hal Caswell

ABSTRACT

We use a cellular automaton model to theoretically explore whether long-distance dispersal provides any advantages over local dispersal in a competitive system when the frequency and scale of disturbance varies. We consider a model for a fugitive species capable of persisting regionally if disturbance creates new patches, but excluded by a superior competitor within any patch where they both occur. Only a certain range of disturbance frequencies creates an advantage for global dispersal and this range varies with reproductive output, rate of competitive exclusion, and the extent of disturbance. Local and global dispersal are equally successful when disturbance is rare. If there is no cost to dispersal (global and local dispersers have similar numbers of colonizing propagules), global dispersal is advantageous at intermediate disturbance frequencies. This advantage is considerably higher and extends over a broader range of disturbance frequencies when disturbance varies in size. If there is a cost to dispersal, global dispersal is disadvantageous at high frequencies of disturbance. Bathymetric variation in the proportion of deep-sea prosobranch gastropods with either planktotrophic or nonplanktotrophic larval development are consistent with the predictions of the model.

Most benthic marine organisms have a dispersal phase in their life cycle, frequently involving a specialized planktonic larval stage. The distances larvae disperse vary widely among species, some dispersing only a few centimeters and others hundreds of kilometers. The extent to which an organism disperses is thought to have profound ecological and evolutionary consequences. On an ecological time scale, dispersal ability can influence the outcome of interspecific interactions (Levin 1974; Caswell 1978; Kareiva 1984; Shmida and Ellner 1984; Chesson 1985), the relative importance of regional and local processes (Pacala 1987), a species geographic range (Hansen 1980; Scheltema 1988; but see Jackson 1986; Johannesson 1988; Vermeij et al. 1990), and its scaling response to environmental heterogeneity (Bowers and Dooley 1991; Milne 1992). On an evolutionary time scale, dispersal ability can affect gene flow, population differentiation, speciation, and extinction (reviewed in Jablonski and Lutz 1983; but also see Hedgecock 1986; Jackson 1986), patterns of species diversity (Josefson 1985; Stuart and Rex, this volume) and perhaps macroevolutionary shifts (Valentine and Jablonski 1983a).

In marine invertebrates, dispersal potential is thought to be associated with mode of larval development; species with planktotrophic development generally disperse more widely than those with nonplanktotrophic development (reviewed in Jablonski and Lutz 1983; but see Johannesson 1988 for a counter example). Although several models have addressed the importance of planktotrophic and nonplanktotrophic development (Vance 1973; Christiansen and Fenchel 1979; Roughgarden 1989) and the advantages of dispersal per se (Frank 1981; Palmer and Strathmann 1981), none have included the pervasive effects of disturbance despite considerable paleontological evidence that disturbance may favor species with planktotrophic development (Hansen 1978; Hansen 1980; Jablonski and Lutz 1983).

In this paper we use a new spatio-temporal modeling approach to theoretically explore whether long-distance dispersal provides any advantages over local dispersal in a competitive system when the frequency and scale of disturbance varies. This is the first in a series of papers that will deal with the scale of dispersal in spatially and temporally varying environments. Here, we consider only the extremes in dispersal ability, local and global, in a homogeneous patchy environment. The results indicate that the frequency and scale of disturbance, as well as costs associated with a long planktonic life, influence the advantages of global dispersal. We will compare the predictions of the model to bathymetric variation in the proportion of prosobranch gastropods with planktotrophic or nonplanktotrophic larval development.

Modeling Approach

To explore the advantages of dispersal, we need a model that describes the population in space as well as time. At least four major approaches to spatio-temporal population modeling exist (Caswell and Etter 1992). They can be classified by whether space, time, and population density are described by continuous or discrete variables (Caswell and Etter 1992).

In **reaction-diffusion-advection** models, space, time, and population density are all continuous (see Okubo 1980; Britton 1986; Murray 1989 for reviews). When space in a reaction-diffusion system is cut up into discrete patches, the result is termed a **reaction-diffusion network** (Levin 1974; Rosen 1981). The discrete-time analog of a reaction-diffusion network is a **coupled map lattice:** a discrete array of discrete-time maps, each interacting with its neighbors (see review by Crutchfield and Kaneko 1987; for an ecological example, see Sole and Valls 1991).

Cellular automata (CA) result from simplifying a coupled map lattice by reducing the continuous patch state variable (population density) to a discrete index (e.g., presence/absence). The dynamics, expressed as a partial differential equation in the reaction-diffusion models, as a system of ordinary differential equations in a reaction-diffusion network, and as a system of difference equations in a coupled map lattice, now appear as a transition rule for the local cell state.

A CA model (in its simplest form) is a regular array (in 1, 2, or more dimensions) of cells. Each cell is characterized by a discrete state variable and at each discrete time step the state of each cell is updated as a function of its own state and the states of the neighboring cells using a set of transition rules. We will use a CA model for two competing species to study the effects of dispersal. We consider each cell to represent a patch of habitat in which an assemblage of organisms could live and will use the terms "cell" and "patch" interchangeably.

To construct an ecological CA model we must translate a set of ecological hypotheses into a set of CA transition rules. We use the approach of Caswell and Cohen (1991a, b), in which the transition rules are derived from the time scales of ecological interactions and disturbance, and the spatial scales of organism movement. This approach specifies the model directly in terms of the temporal and spatial scales that are of interest to ecologists, is applicable to any interaction, and permits us to examine the resulting spatial pattern.

Model

Consider a simple two-species (S_1 and S_2) competition system. A patch may be occupied by S_1, S_2, or both. S_1 always excludes S_2 when they co-occur, with a characteristic time scale $\mathbf{t_c}$. Disturbance follows a Poisson process with characteristic time scale $\mathbf{t_d}$; when it occurs it produces an empty patch available for colonization. Colonization is a Poisson process with a mean number of propagules arriving at a patch per unit time proportional to the abundance of patches in the dispersal neighborhood containing the species in question. Let $\mathbf{x_1}$, $\mathbf{x_2}$, $\mathbf{x_3}$, and $\mathbf{x_4}$ denote the proportion of patches containing, respectively, neither species, only $\mathbf{S_1}$, only $\mathbf{S_2}$, and both species. The model for this interaction can be written as a nonlinear Markov chain

$$\mathbf{x}_{t+1} = \mathbf{A}_x\mathbf{x}_t$$

with the transition matrix $\mathbf{A_x}$ given by

$$\begin{pmatrix} (1-C_1)\,(1-C_2) & pd & pd & pd \\ C_1\,(1-C) & (1-pd) & 0 & pc\,(1-pd) \\ C_2\,(1-C_1) & 0 & (1-C_1)\,(1-pd) & 0 \\ C_1C_2 & 0 & C_1\,(1-pd) & (1-pc)\,(1-pd) \end{pmatrix}$$

where $\mathbf{pd} = 1/\mathbf{t_d}$ is the disturbance rate, $\mathbf{pc} = 1/\mathbf{t_c}$ is the competitive exclusion rate, and the $\mathbf{C_i}$ are colonization rates, given by

$$C_i = 1 - e^{-r_i f_i}$$

with $\mathbf{r_i}$ (hereafter referred to as reproductive output) the mean number of propagules produced by each patch available to colonize an open patch (i.e., have survived the dispersal period) and $\mathbf{f_i}$ the proportion of patches containing species **i** (Caswell and Cohen 1991a, b). Dispersal ability of each species is determined by the number of cells over which $\mathbf{f_i}$ is calculated. Local dispersal is modeled by computing $\mathbf{f_i}$ from the eight adjacent cells (Moore neighborhood) of any open patch; global dispersers have $\mathbf{f_i}$ calculated based on their global (entire array) frequency. This strategy allows global dispersers to colonize any open patch within the CA; local dispersers can only colonize immediately adjacent patches.

Species 2 in this model is a fugitive species. It is unable to persist locally, but given the right disturbance, dispersal, and competitive exclusion rates it may persist (and even reach high frequencies) globally (Caswell and Cohen 1991a, b; Caswell and Etter 1992).

Methods

The CA consisted of 65,536 cells (256 x 256) with periodic boundary conditions (i.e., the upper and lower, and left and right edges were connected to form a torus). Simulations were initialized with equal frequencies of the four states in a random spatial distribution and were run for 1000 iterations. Simulations were stopped at 1000 iterations because of time constraints and because nearly all CA reached equilibrium (i.e., show no consistent directional change in the frequencies of any of the four states) within the first 200–500 iterations. To determine if stopping the simulations at 1000 biased the results, we ran 10 replicate runs of several simulations that took the longest to equilibrate. Replicate runs terminated at 1000 iterations showed little variability in the global frequency of species 2 (range in $\mathbf{f}_2 < 0.01$).

Parameters were constant for each simulation, but varied between simulations. We varied the probability of competitive exclusion (pc = 0.01, 0.1, 1), the reproductive outputs $\mathbf{r}_1$ and $\mathbf{r}_2$ (all four combinations of 1 and 10) and the probability of disturbance (from 10^{-3} to almost 1).

The output of these models includes state frequencies, species frequencies, the mean and variance of alpha diversity, beta diversity, interspecific association patterns, a variety of measures of turnover rate and recurrence times as well as spatial patterns. However, we confine our analysis to the average global frequency of $\mathbf{S}_2$ because the intention of this analysis is to explore the relative advantage of global dispersal and only the dispersal ability of $\mathbf{S}_2$ was varied. The average global frequency of $\mathbf{S}_2$ was calculated by averaging over the last 100 iterations (900–1000) the proportion of patches containing $\mathbf{S}_2$.

Disturbance in nature varies in size, as well as frequency, with distributions typically skewed to the right (i.e., many small and few large disturbances). For example, Paine and Levin (1981) report lognormal distributions in the size of patches in the rocky intertidal. Because global dispersers can colonize the center of large patches while local dispersers must colonize in from the edge, variability in patch size may have important consequences for dispersal. To explore this possibility, variable disturbance sizes were incorporated into the model using a simple skewed size distribution. If a cell is to be disturbed, the size of the disturbance is **j** x **j** cells with probability $\mathbf{g}_j$; in our simulation we set $\mathbf{g}_1 = 0.75$, $\mathbf{g}_2 = 0.24$, $\mathbf{g}_{10} = 0.01$. Because the large disturbances effect more than one cell, we replaced the parameter **pd** with a value **pd'** adjusted to give approximately the same overall probability of disturbance as in the single cell disturbance simulations. The adjusted frequency **pd'** is given by **pd** divided by the average disturbance size:

$$\mathbf{pd'} = \mathbf{pd} / \Sigma\, \mathbf{j}^2 * \mathbf{g}_j.$$

The probability of disturbance given by **pd'** is slightly less than **pd**, especially at higher disturbance levels. An exact formula relating **pd'** and **pd** for arbitrary disturbance size distributions is given in Caswell and Etter (1992).

Results

Figure 14.1 shows 3-dimensional time series of the complete landscape for CA with small (figure 14.1A) and variable size disturbance (figure 14.1B). The end face of each figure depicts the spatial patterns in cell states after 85 iterations; the first slice (bottom) shows the random distribution of cell states used to initialize the simulation. The landscape after 85 iterations appears rather homogeneous when disturbance is small, but when disturbance varies in size there is considerable patchiness; large empty patches and patches in various stages of colonization. The sides of the figure show the temporal development (time moving up) of transects through the landscape; the process of disturbance, colonization, and competitive exclusion can be seen by following individual cells upward in the figure. When disturbance size varies, the temporal dynamics of the large patches clearly show the creation and gradual elimination of disturbed patches as S_1 colonizes from the edges; in this example S_2 has global dispersal and can colonize any cell in the disturbance even when it is not adjacent to it. When S_2 is also restricted to local dispersal, the successional sequence involves both species colonizing from the edges. A similar sequence of events can occur when disturbance affects only single cells but is less obvious because the patches are small.

The effects of disturbance and competition on the frequency of S_2 when it has global and S_1 has local dispersal (figure 14.2) are qualitatively similar to patterns when both species have global dispersal (Caswell and Cohen 1991a), or both are restricted to local dispersal (Caswell and Etter 1992). S_2 can persist regionally even though it is excluded locally if disturbance is sufficiently frequent. The frequency f_2 of the fugitive species is maximized at intermediate disturbance frequencies. The maximum f_2 varies inversely with, and the disturbance frequency that maximizes f_2 varies directly with, the competitive exclusion rate. If S_1, the superior competitor, has a reproductive advantage, S_2 is excluded independent of dispersal ability, competitive exclusion rate, or disturbance frequency.

Relative Advantage to Global Dispersal

More disturbance is necessary to maintain S_2 (critical disturbance level) when it is restricted to local dispersal (figure 14.3). This difference is a consequence of the inter-neighborhood variance in the frequency of S_2 when dispersal is local (Caswell and Etter 1992). The colonization rate C_2 is a concave downward function of f_2, thus the inter-neighborhood variance reduces the effective colonization rate necessitating a higher disturbance level to maintain S_2. That is, the average C_i of two or more f_i will be less than the C_i of the average f_i when the function is concave downward. The magnitude of the reduction will be a function of the concavity of the relationship between C_i and f_i.

A reduction in f_2 and an increase in the critical disturbance rate necessary to maintain S_2 are important consequences of the reduction in effective colonization rate. Because the equilibrium f_2 is itself a concave downward function of r_2, it is more sensitive to reductions in r_2 when r_2 is small. This is precisely where local and global dispersal results differ the most (figure 14.3).

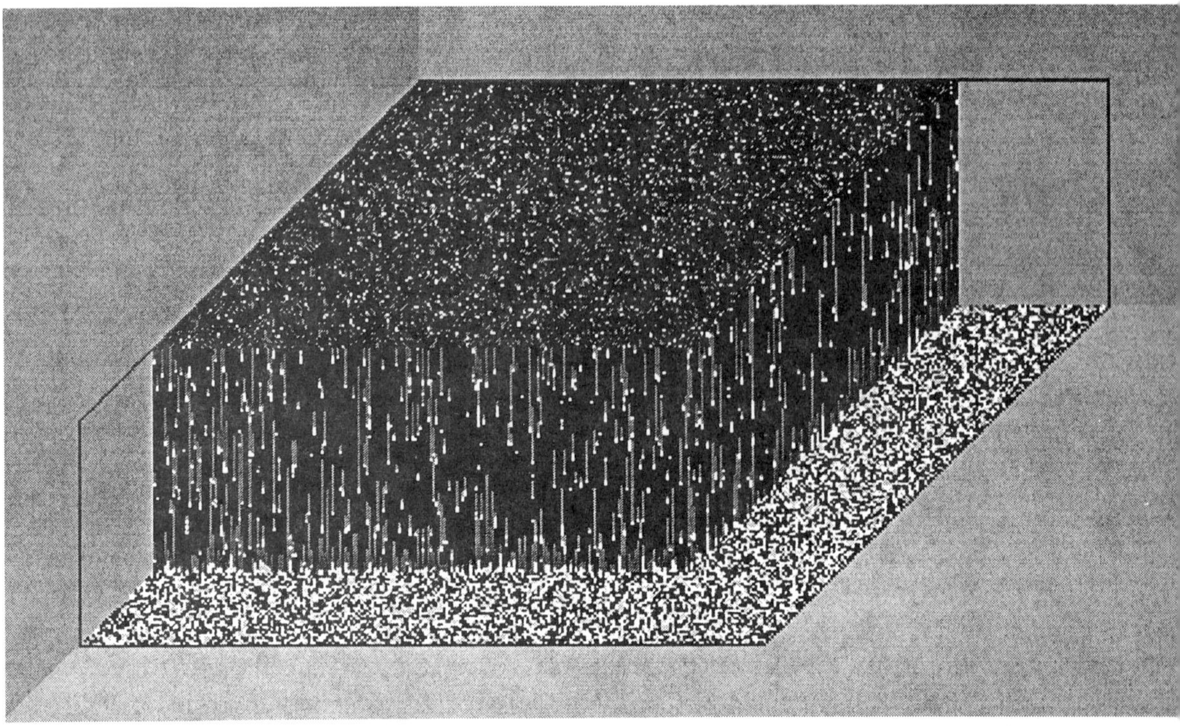

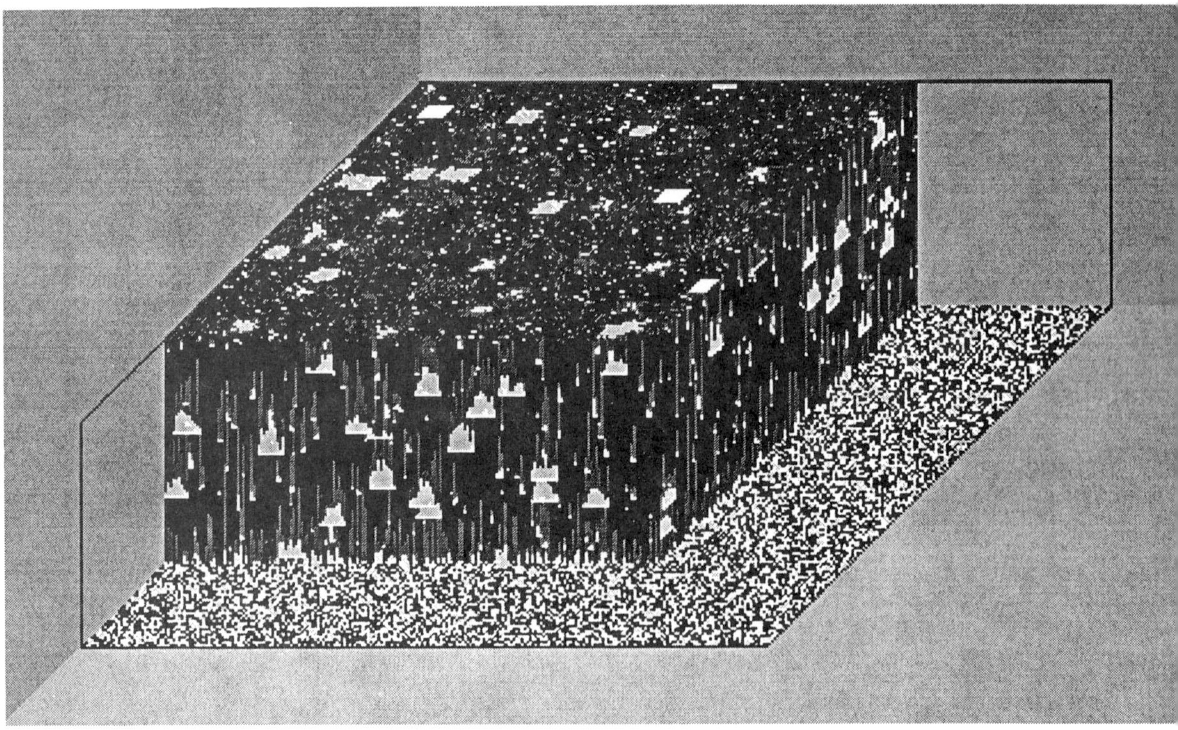

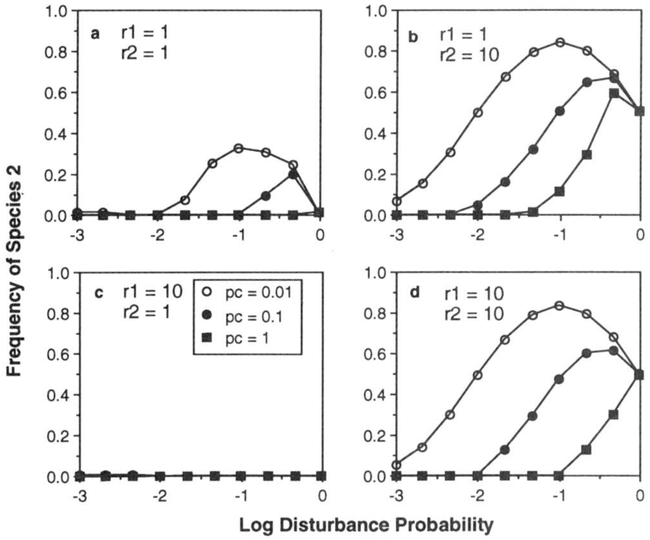

FIGURE 14.2. The frequency of species 2 after 1000 iterations as a function of the disturbance probability for three competitive exclusion rates, **pc**, and four combinations of reproductive output.

To compare the relative advantage of local and global dispersal, we ran two simulations for each parameter combination; one in which S_2 could only disperse locally (into the adjacent 8 cells) and another in which it could disperse to any cell in the CA (global). The relative advantage of global dispersal was measured by the ratio of f_2 in the global dispersal simulation to its value in the local dispersal simulation with the same parameter values. If the relative advantage is > 1 then global dispersal is advantageous, if < 1 local dispersal is better, and if $= 1$ then success is independent of dispersal ability.

If both S_1 and S_2 have low reproductive output (figure 14.4A), global dispersal is most successful at high frequencies of disturbance. When disturbance is low, S_2 is excluded independent of dispersal ability. As the frequency of disturbance increases, global dispersal becomes advantageous because local dispersers are unable to colonize much of the newly opened space; they are not adjacent to it which reduces their effective colonization rate. The disturbance

FIGURE 14.1. A three-dimensional series of a CA simulation with (A) small size disturbances and (B) variable size disturbances. The cube represents successive iterations of the CA stacked on top of one another (time moving up). The end face of each block is the spatial pattern in cell states after 85 iterations; the first slice (bottom) shows the random starting conditions. The sides of the cube show the temporal dynamics of transects through the landscape. The parameters for both simulations were $\mathbf{r}_1 = 1$, $\mathbf{r}_2 = 10$, $\mathbf{pc} = 0.1$, $\mathbf{pd} = 0.09$ and dispersal abilities for S_1 and S_2 were local and global respectively. Color codes for cell states are white = empty patch, black = S_1 only, light gray = S_2 only, dark gray = both species (S_1 and S_2).

level at which global dispersal becomes advantageous increases (shifts right) as the rate of competitive exclusion increases.

When the reproductive output of S_2 is high (figure 14.4B, D), there is a peak in the advantage to global dispersal at intermediate disturbance frequencies. At high **pd** there is no advantage to global dispersal because there is sufficient open space for even the local disperser to avoid competitive exclusion. When disturbance is infrequent, S_2 is competitively excluded even when it is a much better disperser. The intermediate peak in relative advantage and the disturbance level over which global dispersal is advantageous increases (shifts right) with increasing **pc.**

If S_1 has a reproductive advantage, S_2 is excluded even if it has much better dispersal abilities (figure 14.4C). There is no advantage to global dispersal, although time to regional exclusion of S_2 may take longer.

When disturbance varies in size, the relative advantage to global dispersal exhibits similar patterns to models with small-sized disturbances (figure 14.5),

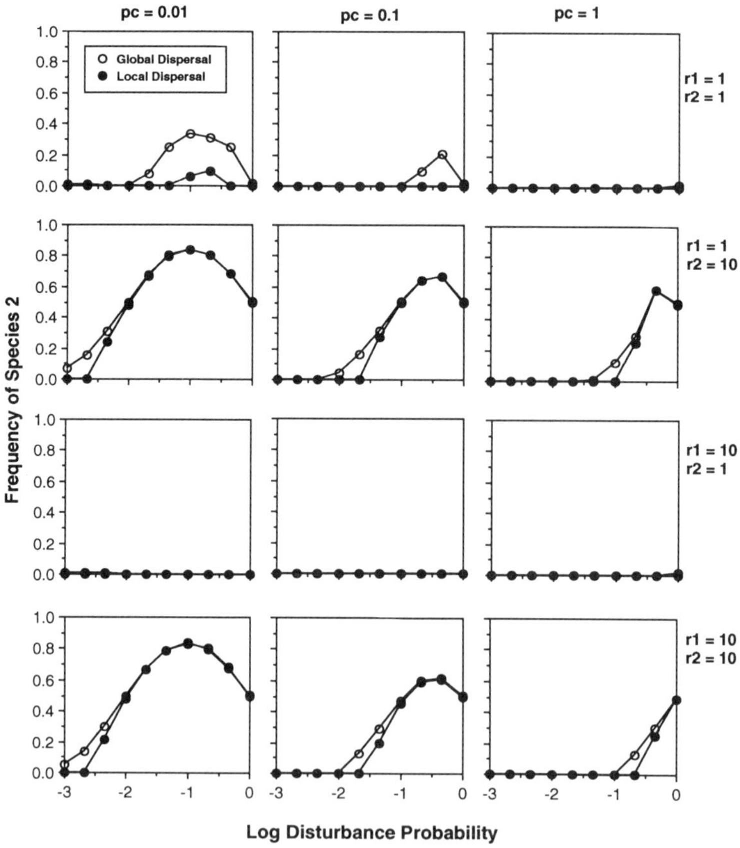

FIGURE 14.3. The frequency of species 2 after 1000 iterations when it disperses either locally or globally as a function of the disturbance probability. Results are shown for three competitive exclusion rates, **pc**, and four combinations of reproductive output.

although the magnitude of the advantage is generally much greater. The difference in relative advantage is especially apparent at high frequencies of disturbance and low reproductive output. This is because when S_2 disperses locally not only is the effective colonization rate reduced by inter-neighborhood variance, but some of the disturbances come in 10 x 10 patches and the cells in the interior of the patch are not immediately available for colonization.

Effects of Reproductive Output

Marine organisms vary enormously in their fecundity, which is usually directly related to the length of their dispersal phase and thus the potential for dispersal. For example, species with planktotrophic development produce many (between 10^3 to 10^9) small larvae per season, whereas those with nonplanktotrophic development produce few (between 10 to 10^3) large offspring (Thorson 1946). It is generally assumed that planktotrophs suffer higher mortality during their planktonic stage and thus their greater fecundity is balanced by the higher mortality resulting in similar numbers of colonists (Strathmann 1985), however there is no data on larval or postlarval mortality to support this notion. Our models thus far also equate net potential colonist between global and local dispersers.

To examine the robustness of our results to this assumption, we apply a cost to global dispersal; despite producing significantly more larvae, the number of propagules available for colonization may be much lower on average for global dispersers if mortality is high during the larval phase or larvae are transported

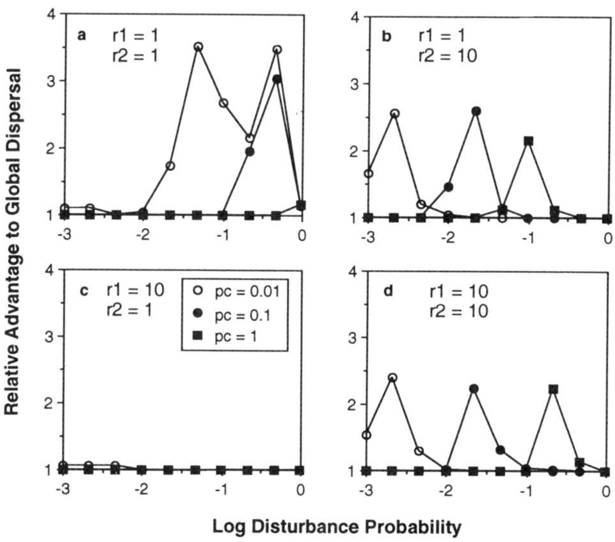

FIGURE 14.4. The relative advantage to global dispersal (calculated as described in text) as a function of the disturbance probability for three competitive exclusion rates, **pc**, and four combinations of reproductive output.

away from suitable habitats. To incorporate this "cost" of global dispersal, the results of CA simulations where S_2 dispersed globally and $r_2 = 1$ were compared to the results of simulations where S_2 dispersed locally and $r_2 = 10$. We also examine a reproductive advantage to global dispersal; low planktonic mortality may result in more propagules available for colonization relative to local dispersers. In this case, the results of CA simulations where S_2 dispersed globally and $r_2 = 10$ were compared to the results of simulations where S_2 dispersed locally and $r_2 = 1$.

When global dispersal is coupled with a reproductive advantage, there is little change in how relative advantage varies with disturbance probability, but the magnitude of the advantage is greater (figure 14.6). The increase in relative advantage is much less for simulations with variable size disturbance (figure 14.6B) because most of the open space is immediately colonized by S_2 even when it has a reproductive output similar to the local disperser. A reproductive advantage therefore provides only a marginal benefit when disturbance varies in size.

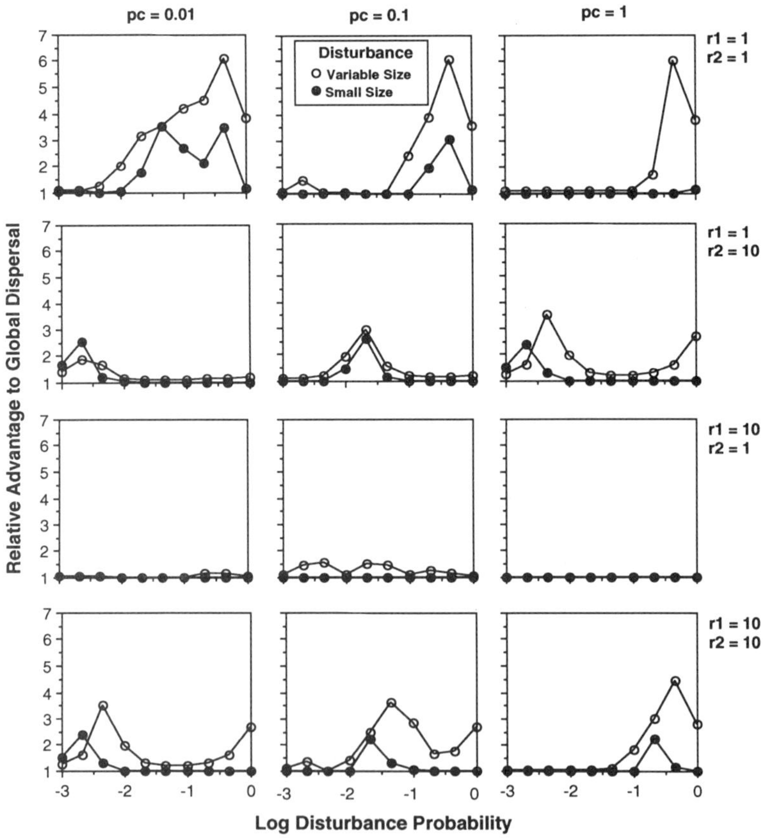

FIGURE 14.5. The relative advantage to global dispersal (calculated as described in text) when disturbance is either small (single cell) or variable in size as a function of the disturbance probability. Results are shown for three competitive exclusion rates, **pc**, and four combinations of reproductive output.

A reproductive cost to global dispersal changes the pattern of relative advantage considerably (figure 14.7). There is little or no advantage to global dispersal at low frequencies of disturbance, and as disturbance increases, global dispersal becomes disadvantageous. Local dispersers are more successful at high frequencies of disturbance in this model because their greater reproductive output allows them to quickly monopolize any newly opened space. Their success is damped to some extent when disturbance size varies because global dispersers partially offset their reproductive disadvantage by colonizing the center of large patches.

Summary

To summarize, global dispersal has an advantage over a range of disturbance frequencies, and this range varies with reproductive output, rate of competitive exclusion, and the scale of disturbance. Global dispersal is most advantageous

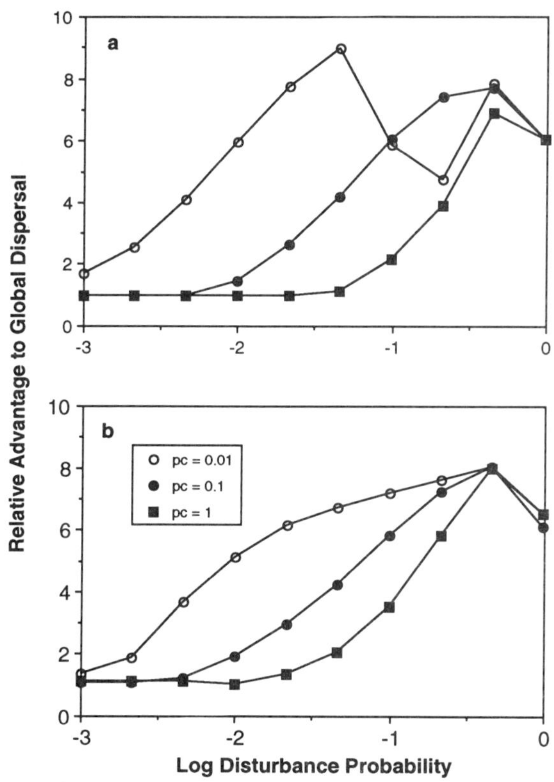

FIGURE 14.6. The relative advantage to global dispersal (calculated as described in text) when the global disperser has an order of magnitude reproductive advantage over the local disperser as a function of the disturbance probability. Results are shown for three competitive exclusion rates, **pc**, and (A) small (single cell) or (B) variable size disturbance simulations.

when reproductive output (the number of larvae that survived the dispersal phase) is low and disturbance varies in size. The advantage to global dispersal peaks at intermediate disturbance frequencies and the disturbance frequency at which it peaks increases with the rate of competitive exclusion. These results were sensitive to the reproductive success of the species. An increase in the net reproductive output of the global disperser had little affect on the outcome, but when the local disperser had a reproductive advantage, global dispersal actually became disadvantageous at high frequencies of disturbance. Interestingly, regardless of costs, local and global dispersal are equally successful when disturbance is infrequent.

Palmer and Strathmann (1981) use a similar approach to modeling the advantages of different scales of dispersal, although they do not consider the effects of disturbance. Dispersal is generally thought to be advantageous because it provides an arithmetic average of the spatial and temporal environmental heterogeneity; nondispersal strategies provide a geometric average that is

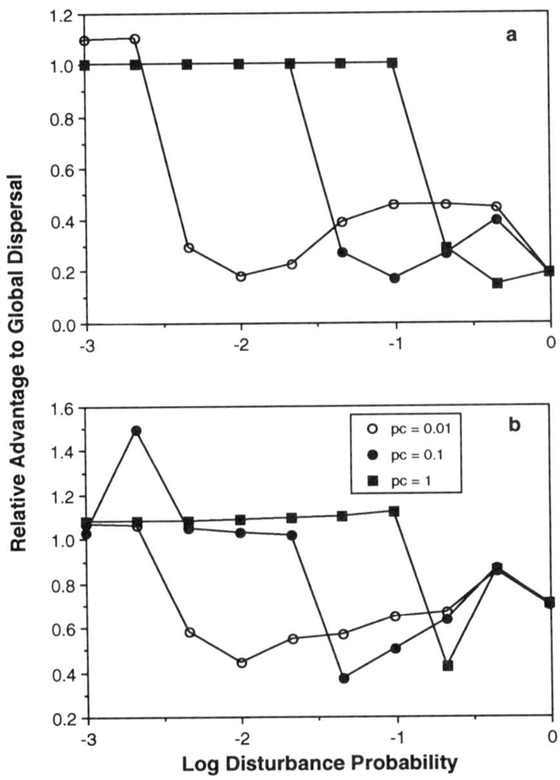

FIGURE 14.7. The relative advantage to global dispersal (calculated as described in text) when the local disperser has an order of magnitude reproductive advantage over the global disperser (cost to global dispersal) as a function of the disturbance probability. Results are shown for three competitive exclusion rates, **pc**, and (A) small (single cell) or (B) variable size disturbance simulations.

less than the arithmetic average (reviewed in Roughgarden 1989). Palmer and Strathmann (1981) point out that the optimal scale of spread to attain an arithmetic average will vary with the spatial and temporal correlations of environmental conditions. They show that increasing the scale of dispersal provides diminishing returns and argue that the distances most planktotrophic larvae are likely to disperse far exceed that necessary to overcome any spatial autocorrelation in environmental favorability. We only consider two scales of dispersal (local is comparable to their 8 site and global [65,556 sites] would be comparable to 256 when the relative increase reaches an asymptote), but we observe much greater advantages to increased spread of larvae between these two scales than their model predicts.

For the most comparable model (no spatial or temporal variability in environmental favorability), the relative increase asymptotes at a scale of spread of approximately 64 sites, changing very little between 8 and 64 sites. Under certain disturbance regimes (e.g., rare disturbance events which would be similar to their model), our results also indicate that there is little advantage dispersing to more than 8 sites, but there are disturbance regimes that favor much greater spread and the relative advantage to increased spread is considerably higher. Thus the frequency and intensity of disturbance may also influence the scale larvae should disperse.

Deep-Sea Example

How do these theoretical predictions compare to observed patterns in the deep sea? Virtually nothing is known about actual dispersal distances in deep-sea species. However, Rex and Warén (1982) related dispersal potential to mode of larval development in deep-sea prosobranch gastropods from the western North Atlantic. Prosobranchs are unusual among deep-sea taxa in having some species with planktotrophic (swimming-feeding) and others with nonplanktotrophic larval development represented at all depths. Mode of development is species specific and can be inferred from larval shell morphology. For most deep-sea prosobranch families, planktotrophic larvae hatch early and have a long planktonic dispersal phase. Nonplanktotrophic larvae develop in an egg capsule and hatch as a benthic snail or have only a brief dispersal phase. Some possible exceptions to this dichotomy are discussed by Rex and Warén (1982) and Stuart and Rex (this volume). With respect to the model, planktotrophs and nonplanktotrophs might be considered global and local dispersers respectively.

Rex and Warén (1982) showed that the proportion of prosobranchs with planktotrophic development increased with depth in the western North Atlantic. Below 2000 m, planktotrophs and nonplanktotrophs were equally abundant (figure 14.8A). This suggests that below 2000 m both modes of development are equally successful. Above 2000 m, the proportion of planktotrophs appears to decline, as Rex and Warén (1982) point out, and nonplanktotrophic species increase. Because an increase in the proportion of planktotrophs from 200–2000 m could result from a decrease in nonplanktotrophs, an increase in planktotrophs or both, we analyzed the change in the number of species possessing each developmental mode (figure 14.8B). below 2000 m, the rates of decline in the number of planktotrophs and nonplanktotrophs are similar and appear to reflect the general decline in diversity with depth. Above 2000 m, where the total number of species decline, there is an increase in nonplanktotrophs and a disproportionate decrease in planktotrophs. These patterns suggest that both modes of development are equally successful below 2000 m. Above 2000 m, nonplanktotrophs are favored and planktotrophs disfavored.

If an increase in depth is associated with a decrease in the rates of competitive exclusion and disturbance, as has been inferred for explaining faunal diversity in the deep sea (Huston 1979; Rex 1983), the CA models indicate that the

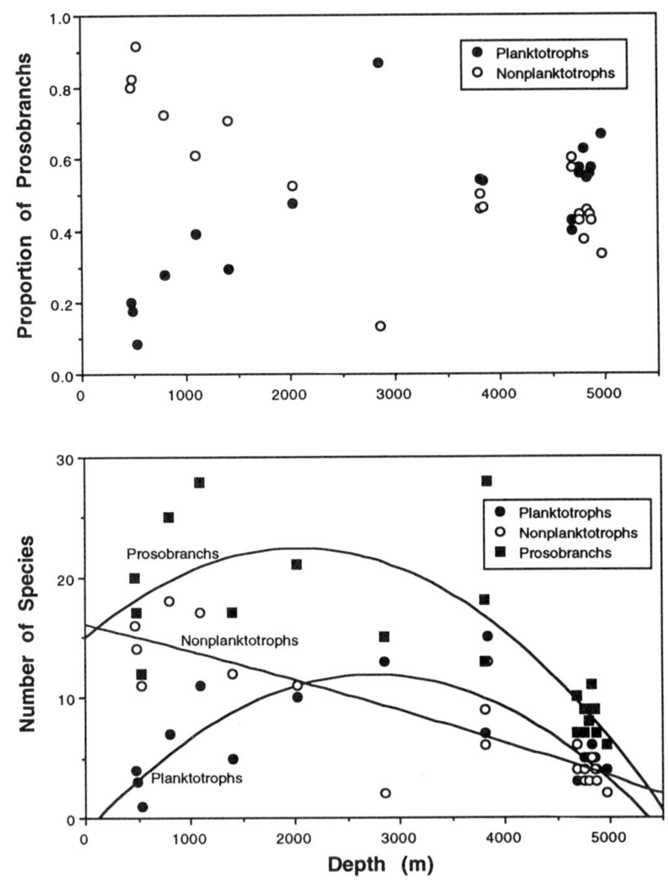

FIGURE 14.8. The (A) proportion and (B) numbers of prosobranch marine gastropods with either nonplanktotrophic or planktotrophic larval development as a function of depth (data from Rex and Warén 1982). Regression lines were fitted using standard least-squares techniques.

advantage to global dispersal should peak at intermediate disturbance frequencies (figure 14.9) for both small and variable size disturbance simulations. Exactly where a depth transect would fall within this plot is unclear, but the model suggests that global dispersal should be most successful at intermediate depths. The peak in the number of prosobranchs with planktotrophic development at intermediate depths is consistent with this result.

The results from our simulations may also help explain the distinct increase in the number of prosobranchs exhibiting nonplanktotrophic development above 2000 m, in spite of the decrease in the number of prosobranchs as a whole. If there is a substantial loss of larvae during the planktonic stage, the model predicts that local dispersers (nonplanktotrophs) should be favored when disturbance is relatively frequent (shallow) (figure 14.10).

Is it plausible that planktotrophic larvae suffer higher mortality at upper bathyal depths? The upper bathyal region (200–2000 m) is extremely heterogeneous; it is topographically complex (MacIlvaine and Ross 1979), has the greatest variability in sediment characteristics (Etter and Grassle in prep), and is transected by submarine canyons in some regions (Rowe 1971). The decrease in population differentiation (Etter and Rex 1990), species diversity (Rex 1983), and zonation (Grassle et al. 1975; Rex 1977; Grassle et al. 1979) of the deep-sea biota with depth as well as an increase in bathymetric and geographic ranges (Etter and Rex 1990) also suggest that the selective regime is most diverse at upper bathyal depths and becomes progressively more uniform with depth. The biotic and abiotic environmental heterogeneity at upper bathyal depths may impose a cost to planktotrophy; larvae that remain in the water column for extended periods are likely to be transported away from suitable habitats (Rex and Warén 1982). This may be especially true for those deep-sea larvae that develop in the surface waters (Killingley and Rex 1985).

At lower bathyal and abyssal depths, the loss of larvae from suitable habitats would be less likely because the environment is, over broad expanses, more homogeneous (Rex and Warén 1982). Indeed, if the cost to dispersal decreases with depth because the environment is becoming more uniform, there may be a shift from the model where there is a cost to dispersal to the one in which both development modes had equal numbers of potential colonist. Depending on where this shift took place, the results could account for the increase in non-planktotrophy above 2000 m, the increase in planktotrophy between 200 and 2000 m, and the equal success of both strategies below 2000 m.

The fundamental assumptions in the CA models involve the processes of dispersal competition and disturbance in a patchy environment. These processes are poorly understood in the deep sea, but evidence exists that dispersal ability varies among snails with different development modes (reviewed in Jablonski and Lutz 1983), that competition is important in structuring deep-sea gastropod assemblages (Rex et al. 1988), that the deep sea is patchy (Jumars 1976; Thistle 1983; Grassle and Morse-Porteous 1987; Levin 1991), and it is often inferred that disturbance frequency decreases with depth (Huston 1979; Rex 1983).

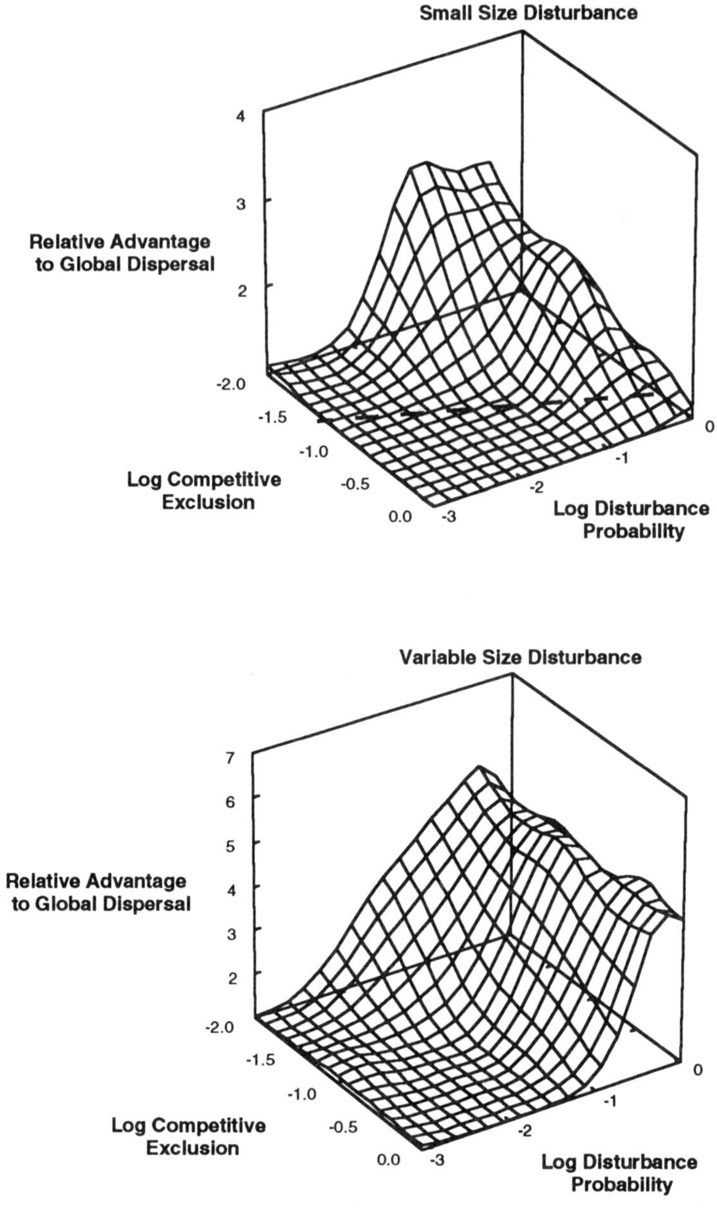

FIGURE 14.9. The relative advantage to global dispersal (calculated as described in text) when there is no cost to dispersal as a function of the disturbance probability and the rate of competitive exclusion. Results are shown for small (single cell) or variable size disturbance simulations. Surfaces were fit to the data from figure 14.6 using a distance weighted least squares algorithm. The broken line through the bottom of the small size disturbance plot represents a hypothetical bathymetric transect where the frequency of disturbance and the rate of competitive exclusion are decreasing with depth.

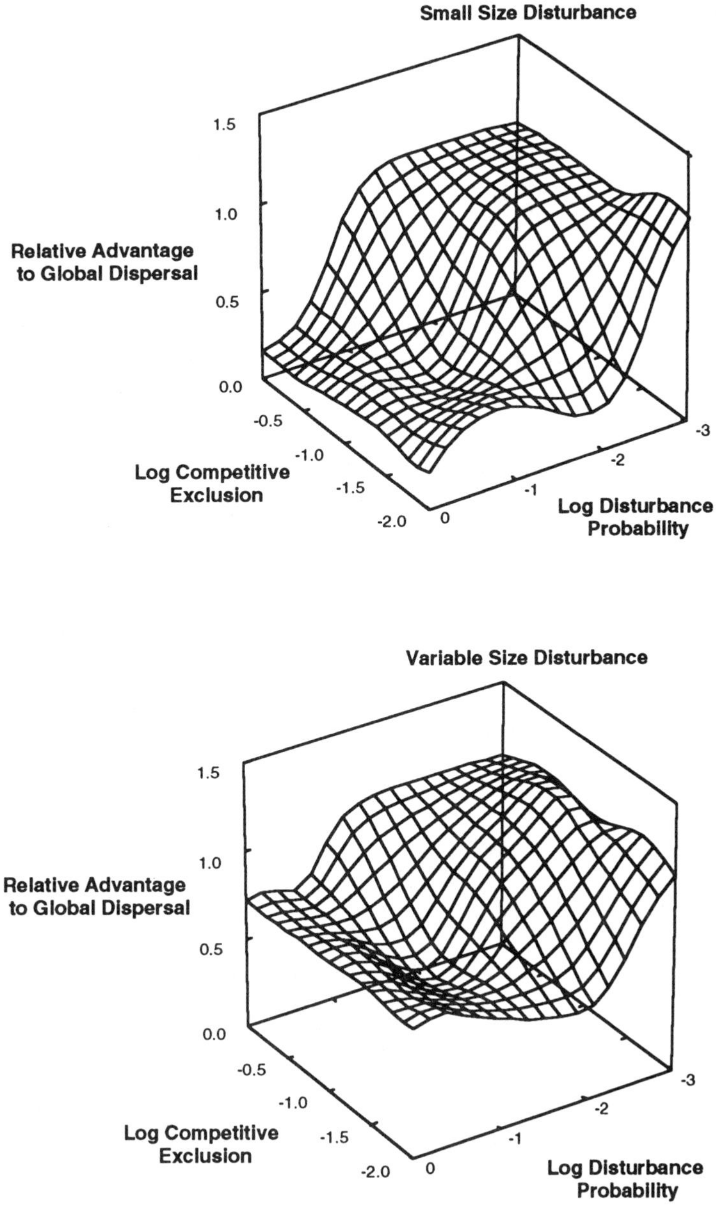

FIGURE 14.10. The relative advantage to global dispersal (calculated as described in text) when there is a cost to dispersal as a function of the disturbance probability and the rate of competitive exclusion. Results are shown for small (single cell) or variable size disturbance simulations. Surfaces were fit to the data from figure 14.7 using a distance weighted least squares algorithm.

Because these are consistent with the assumptions of the model, we felt that the CA was a reasonable theoretical approach to explore the relative advantages of different dispersal strategies. We do not claim that this simplified model is an accurate representation of the ecological and evolutionary forces that have shaped the depth-related patterns in development types, rather the model simply shows the potential integrative effects of changes in rates of disturbance, competitive exclusion, and reproductive success on the success of particular dispersal strategies.

Other plausible explanations for the spatial patterns in development types exist and are not mutually exclusive. For instance, nonplanktotrophs may be more prone to speciation because gene flow among populations would tend to be restricted (Scheltema 1977; Scheltema 1978; Valentine and Jablonski 1983b). The increase in the number of nonplanktotrophs at upper bathyal depths may simply reflect their greater propensity to speciate in a heterogeneous environment. Phylogenetic constraints undoubtedly play a role in the evolution of development strategies (Strathmann 1978), but are unlikely to provide a complete explanation of the depth-related variation in development modes in deep-sea prosobranchs (Rex and Warén 1982).

Our modeling approach can easily incorporate various scales of dispersal, spatial and temporal environmental variability, and other biotic interactions (e.g., mutualism, predation, parasitism, etc.). Future work will incorporate these aspects into our models to explore how they influence the optimal scale of dispersal.

ACKNOWLEDGMENTS

This work benefited from discussions with Mike Rex and his comments on a previous version of this paper. We thank Mike and two anonymous referees for helpful comments on improving this paper. We also thank Craig Young and Kevin Eckelbarger for organizing this symposium and the invitation to participate. Support was provided by NSF Grants DEB-9119420, OCE-8900231, ONR Grant N00014–92-J-1527, and DOE Grant DE-FG0289ER6088W. Woods Hole Oceanographic Institution Contribution 7990.

REFERENCES

Bowers, M. A. and J. L. Dooley. 1991. Landscape composition and the intensity and outcome of two-species competition. *Oikos* 60:180–186.

Britton, N. 1986. *Reaction-Diffusion Equations and Their Applications to Biology*. New York: Academic Press.

Caswell, H. 1978. Predator-mediated coexistence: A nonequilibrium model. *American Naturalist* 112:127–153.

Caswell, H. and J. E. Cohen. 1991a. Communities in patchy environments: A model of disturbance, competition, and heterogeneity. In J. Kolasa and S. Pickett, eds., *Ecological Heterogeneity,* pp. 97–122. New York: Springer-Verlag.

Caswell, H. and J. E. Cohen. 1991b. Disturbance, interspecific interaction, and diversity in metapopulations. *Biological Journal of the Linnean Society,* pp. 193–218.

Caswell, H. and R. J. Etter. 1992. Ecological interactions in patchy environments: From patch-occupancy models to cellular automata. In J. H. Steele and S. Levin, eds., *Patch Dynamics.* New York: Springer-Verlag. In press.

Chesson, P. 1985. Coexistence of competitors in spatially and temporally varying environments: A look at the combined effects of different sorts of variability. *Theoretical Population Biology* 28:263–287.

Christiansen, F. B. and T. M. Fenchel. 1979. Evolution of marine invertebrate reproductive patterns. *Theoretical Population Biology* 16:267–282.

Crutchfield, J. and K. Kaneko. 1987. Phenomenology of spatio-temporal chaos. In B. Hao, ed., *Direction in Chaos.* Singapore: World Scientific.

Etter, R. J. and M. A. Rex. 1990. Population differentiation decreases with depth in deep-sea gastropods. *Deep-Sea Research* 37:1251–1261.

Frank, P. W. 1981. A condition for a sessile strategy. *American Naturalist* 118:288–290.

Grassle, J. F. and L. S. Morse-Porteous. 1987. Macrofaunal colonization of disturbed deep-sea environments and the structure of deep-sea benthic communities. *Deep-Sea Research* 34:1911–1950.

Grassle, J., H. Sanders, R. Hessler, G. Rowe, and T. McLellan. 1975. Pattern and zonation: A study of the bathyal megafauna using the research submersible Alvin. *Deep-Sea Research* 13:5–262.

Grassle, J. F., H. L. Sanders, and W. K. Smith. 1979. Faunal changes with depth in the deep sea benthos. *Ambio, Special Report* 6:47–50.

Hansen, T. A. 1978. Larval dispersal and species longevity in lower tertiary gastropods. *Science* 199:885–887.

Hansen, T. A. 1980. Influence of larval dispersal and geographic distribution on species longevity in neogastropods. *Paleobiology* 6:193–207.

Hedegcock, D. 1986. Is gene flow from pelagic larval dispersal important in the adaptation and evolution of marine invertebrates? *Bulletin of Marine Science* 39:550–564.

Huston, M. 1979. A general hypothesis of species diversity. *American Naturalist* 113:81–101.

Jablonski, D. and R. A. Lutz. 1983. Larval ecology of marine benthic invertebrates: paleobiological implications. *Biol. Rev.* 58:21–89.

Jackson, J. B. C. 1986. Modes of dispersal of clonal benthic invertebrates: consequences for species' distributions and genetic structure of local populations. *Bulletin of Marine Science* 39:588–606.

Johannesson, K. 1988. The paradox of Rockall: Why is a brooding gastropod (*Littorina saxatilis*) more widespread than one having planktonic larval dispersal stage (*L. littorea*)? *Marine Biology* 99:507–513.

Josefson, A. 1985. Distribution of diversity and functional groups of marine benthic infauna in the Skagerrak (eastern North Sea). Can larval availability affect diversity? *Sarsia* 70:229–249.

Jumars, P. A. 1976. Deep-sea species diversity: Does it have a characteristic scale? *J. Mar. Res.* 34:217–246.

Kareiva, P. 1984. Predator-prey dynamics in spatially structured populations: Manipulating dispersal in a coccinellid-aphid interaction. In S. Levin, ed., *Lecture Notes in Bio-Mathematics,* pp. 368–389. New York: Springer-Verlag.

Killingley, J. S. and M. A. Rex. 1985. Mode of larval development in some deep-sea gastropods indicated by oxygen-18 values of their carbonate shells. *Deep-Sea Research* 32:809–818.

Levin, L. A. 1991. Interactions between metazoans and large, agglutinating protozoans: Implications for the community structure of deep-sea benthos. *Amer. Zool.* 31:886–900.

Levin, S. 1974. Dispersion and population interactions. *American Naturalist* 108:207–228.

MacIlvaine, J. C. and D. A. Ross. 1979. Sedimentary processes on the continental slope of New England. *Journal of Sedimentary Petrolology* 49:563–574.

Milne, B. T. 1992. Spatial aggregation and neutral models in fractal landscapes. *American Naturalist* 139:32–57.

Murray, J. 1989. *Mathematical Biology*. New York: Springer-Verlag.

Okubo, K. 1980. *Diffusion and Ecological Problems: Mathematical Models*. Berlin; Heidelberg; New York: Springer-Verlag.

Pacala, S. 1987. Neighborhood models of plant population dynamics. III: Models with spatial heterogeneity in the physical environment. *Theor. Pop. Biol.* 31:359–392.

Paine, R. T. and S. A. Levin. 1981. Intertidal landscapes, disturbance, and the dynamics of pattern. *Ecol. Monogr.* 51:145–178.

Palmer, A. R. and R. R. Strathmann. 1981. Scale of dispersal in varying environments and its implications for life-histories of marine invertebrates. *Oecologia* 48:308–318.

Rex, M. A. 1977. Zonation in deep-sea gastropods the importance of biological interactions to rates of zonation. In B. F. Keegan, ed., *Biology of Benthic Organisms*. New York: Pergamon Press.

Rex, M. A. 1983. Geographic patterns of species diversity in the deep-sea benthos. *The Sea*. New York: Wiley.

Rex, M. A. and A. Warén. 1982. Planktotrophic development in deep-sea prosobranch snails from the western North Atlantic. *Deep-Sea Research* 29:171–184.

Rex, M. A., M. Watts, R. Etter, and S. O'Neill. 1988. Character variation in a complex of rissoid gastropods from the upper continental slope of the western North Atlantic. *Malacologia* 29:325–329.

Rosen, R. 1981. Pattern generation in networks. *Progress in Theoretical Biology* 6:161–209.

Roughgarden, J. 1989. The evolution of marine life cycles. In M. W. Feldman, ed., *Mathematical Evolutionary Theory*. Princeton, N.J.: Princeton University Press.

Rowe, G. 1971. Observations on bottom currents and epibenthic populations in Hatteras Submarine Canyon. *Deep-Sea Research* 18:569–581.

Scheltema, R. 1977. Dispersal of marine invertebrate organisms: paleobiogeographic and biostratigraphic implications. In E. G. Kauffman and J. E. Hazel, eds., *Concepts and Methods of Biostratigraphy*, pp. 73–108. Stroudsburg, Pa.: Dowden, Hutchinson, and Ross.

Scheltema, R. 1978. On the relationship between dispersal of pelagic veliger larvae and the evolution of marine prosobranch gastropods. In B. Battaglia and J. A. Beardmore, eds., *Marine Organisms: Genetics, Ecology, and Evolution*, pp. 303–322. New York: Plenum Press.

Scheltema, R. 1988. Planktonic and nonplanktonic development among prosobranch gastropods and its relationship to the geographic range of species. In J. S. Ryland and P. A. Tyler, eds., *Genetics, Reproduction, and Distributions of Marine Organisms*. Fredensborg, Denmark: Olsen and Olsen.

Shmida, A. and S. Ellner. 1984. Coexistence of plant species with similar niches. *Vegetatio* 58:29–55.

Sole, R. and J. Valls. 1991. Order and chaos in a 2D Lotka-Volterra coupled map lattice. *Physics Letters* A 32:330–336.

Strathmann, R. R. 1978. The evolution and loss of feeding larval stages of marine invertebrates. *Evolution* 32:894–906.

Strathmann, R. R. 1985. Feeding and nonfeeding larval development and life history evolution in marine invertebrates. *Ann. Rev. Ecol. Syst.* 16:339–361.

Thistle, D. 1983. The role of biologically produced habitat heterogeneity in deep-sea diversity maintenance. *Deep-Sea Research* 30:1235–1245.

Thorson, G. 1946. Reproduction and larval development of Danish marine bottom invertebrates with special reference to the planktonic larvae in the Sound. *Medd. Komm. Danm. Fiskeriog. Havunders., Ser. Plankton* 4:1–523.

Valentine, J. and D. Jablonski. 1983a. Larval adaptations and patterns of brachiopod diversity in space and time. *Evolution* 37:1052–1061.

Valentine, J. and D. Jablonski. 1983b. Speciation in the shallow sea: General patterns and biogeographic controls. In R. W. Sims, J. H. Price, and P. E. S. Whalley, eds., *Evolution, Time, and Space: The Emergence of the Biosphere,* pp. 201–226. London: Academic Press.

Vance, R. R. 1973. On reproductive strategies in marine benthic invertebrates. *American Naturalist* 107:339–352.

Vermeij, G., A. Palmer, and D. Lindberg. 1990. Range limits and dispersal of mollusks in the Aleutian Islands, Alaska. *The Veliger* 33:346–354.

CHAPTER FIFTEEN

Ecological Patterns in the Deep-Sea Benthos: How Are They Related to Reproduction, Larval Biology, and Recruitment?

J. Frederick Grassle

> *All organisms face the same dilemma: for particular life history stages, the realized environmental variability will be a consequence of the scales of experience. Various life history adaptations, such as dispersal, dormancy, and iteroparous reproduction, have the effect of modifying the scales of observation, and hence the realized variability.* S. A. Levin (1992)

The contributions in this volume exemplify the progress that has been made in the last two decades in demystifying a region that has alternately been portrayed as hostile, inaccessible, impoverished, and boringly homogeneous. Many of the advances have been a direct consequence of the enhanced accessibility provided by submersibles, which have allowed biologists to witness episodic and seasonal events, and the consequences of the intrusion of scarce resources. Submersibles have not only permitted scientists to witness previously unsuspected communities, patterns and behaviors, but have also allowed precise sampling with respect to physical and biological features, and experimentation. Advancing technologies such as remotely operated and autonomous vehicles, new chemical sensors, and improved cameras will permit enhancement of observational frequencies and acuities.

Recent comparatively extensive quantitative studies of the deep-sea benthos have revealed exceedingly diverse communities of small invertebrates. A single small sediment sample may contain hundreds of species from more than a dozen phyla. We have proposed (Grassle and Maciolek 1992) that deep-sea diversity is maintained in part by microhabitat heterogeneity created by a combination of biogenic disturbance and food resources that are divided into patches of varying size. In an environment where food is generally in short supply, the observed persistence of certain food patches and disturbances, sometimes for several years, contrasts sharply with the rapid obliteration of such heterogeneities in shallow water (Grassle and Grassle, in press).

Studies of deep-sea communities of the continental slope and rise off the mid-Atlantic coast of the United States have revealed diverse assemblages that changed little over two years or along a 176-km transect at 2100 m depth (Grassle and Maciolek 1992). Although there were many rare species, about one-fifth of the species occurred at all stations along the transect. Changes in species composition with depth changes of just 500 m were more marked, a common observation in the deep sea. Many abundant species were broadly distributed along and across depth contours, evidence for the comparative "openness" of the system to species that have the means for dispersal (Ricklefs 1987). Other parts of the deep sea show much less homogeneity in community structure along depth contours. Blake and Grassle (in preparation) found that while communities off the coasts of North and South Carolina could be grouped into upper slope, middle slope, lower slope, and continental rise assemblages, the upper slope stations were highly variable in density and in species composition (see also Schaff et al. 1992). These differences can be attributed to differences in the physical regime and the concomitant sediment distributions, and to the fluxes and types of organic matter arriving at the bottom. These studies (Blake and Grassle, in preparation) gave evidence for a partial zoogeographic boundary in the deep sea between Cape Lookout and Cape Fear, as Cutler (1975) had previously suggested. The fact that this boundary exists for only part of the fauna serves to emphasize the fact that a system that is "open" to one species, is not "open" to another.

A number of the contributions in this symposium have focused on the proportion of species in particular latitudes, depths, habitats, or taxa that have a greater or lesser potential for dispersal. Those who have studied bivalves and gastropods have inferred modes of development from larval shells, always being careful to state that in studying the corresponding prodissoconch and protoconch morphologies they are using criteria based on shallow-water relatives of the deep-sea taxa (e.g., Scheltema; Gustafson and Lutz; Bouchet and Warén; Stuart and Rex), and that the applicability of these criteria has not been tested. For those who study taxa that lack the "history" recorded in larval shells, mode of development has been inferred from the size of, and amount of yolk present in, the oocytes (e.g., Pearse; Young; Tyler et al.) with due regard to the range in oocyte diameters that characterize higher level taxa. Most of these authors are explicit in declaring that for the deep sea, pelagic stages, whether planktrotrophic or lecithotrophic, may not experience temperatures above ~2°C, so that inferring how long they may remain pelagic and available for dispersal cannot be assessed without direct experimentation on those stages. Recent work by Hadfield and Strathmann (1990) has revealed very similar protoconch morphologies in nonfeeding larvae of trochid gastropods, although one species has a pelagic dispersal phase and the other lacks such a stage. Moreover, Manahan and colleagues (discussed by Stuart and Rex) have shown that an Antarctic asteroid, which has comparatively small oocytes (200 μm) produces larvae that remain pelagic for at least six months, surviving at low temperatures largely on low concentrations of amino acids in the ambient water.

Direct studies on the larval development of deep-sea species can now be contemplated. The feasiblity of such studies is due in no small measure to the increasing amount of information on life histories of deep-sea species emerging from a commitment to long-term studies at single sites such as Rockall Trough (Gage; Tyler et al.) and the Caribbean (Eckelbarger; Young), and from extensive quantitative studies (Blake and Watling). These studies encourage continued work on biomass-dominant or numerically dominant species that can be classifed as to the periodic or aperiodic nature of their reproduction. Direct studies on the larval development of these species to competency obviously require close attention to the provision of realistic environmental conditions, and thus studies on the buoyancy of oocytes and early larval stages, and their swimming behavior, will also be useful since they will indicate whether these stages are all likely to complete their development at very low temperatures, or whether, as has been suggested for a few species, some are completed in warmer, near-surface waters.

In addition to studies on deep-sea species that occupy habitats that exemplify the major part of the deep ocean floor, studies on unique habitats that occupy isolated "islands" or "archipelagos" (hydrothermal vents, seep areas, seamounts) are of obvious comparative value. In this volume Mullineaux has considered the kinds of mesoscale flows that occur around seamounts, and that result from hydrothermal vent activity, which might produce more or less coherent "patches" of dispersing larvae. By contrasting environmental conditions at different levels on a seamount, Levin et al. have emphasized how these flows might differentially serve to disperse or retain propagules from the different seamount communities. In all of these considerations, it is clearly of interest to study the levels of genetic differentiation among populations since, as in shallow water, the "cosmopolitan morphospecies" may be the cloak for unsuspected levels of genetic subdivision. A small number of such studies are in progress but are not reported in this volume.

If it were possible to identify a single theme that emerges from the contributions in this volume, it might be embodied in the quotation from Bartholomew (1986) that Pearse chose as his keynote, viz., that most of the generalizations that have been applied to deep-sea organisms and communities are either wrong or of limited applicability. Even the most bizarre ultrastructural features of gametogenesis and gametes in deep-sea species cannot be shown to be unique adaptations to deep-sea life, but have their parallels in related shallow-water species, i.e., phylogenetic constraints play a large role (Eckelbarger; Young). Yet, the fact remains that deep-sea, soft-bottom communities are more highly diverse than their shallow-water counterparts. Possible exceptions to even this generalization can be found in those parts of the deep sea that have a high frequency of physical disturbance, or in areas where there is intense organic enrichment, whether it occurs "naturally" (Levin et al.) or is induced experimentally (Grassle and Morse-Porteous 1987; Snelgrove et al. 1992).

In earlier discussions of the sources of high deep-sea diversity, Smith et al. (1986) and Grassle and Morse-Porteous (1987) have regarded the patchy input of organic material to the deep sea as both a disturbance and a source of divided and ephemeral food resources. Although it is often the case that massive organic inputs do cause mortality in resident individuals, and in that case constitute a disturbance in the sense that Etter and Caswell use the term in their cellular automaton model, it now seems useful to distinguish between a disturbance that adds labile organic matter to the sea floor, and other forms of disturbance that effectively reduce competition but do not enhance food availability (Grassle and Grassle, in press). We have proposed that the metapopulations that occupy widely separated patches of ephemeral food resources, linked as they are by pelagic dispersal stages, are not spatially autocorrelated. This makes possible the coexistence of a large number of species (Grassle and Maciolek 1992).

Early experiments on recolonization of small azoic patches in the deep sea were justifiably criticized because the trays of sediment were placed on the bottom so that the trays themselves constituted a major disturbance to the flow (Smith et al. 1986), and more recent studies have used hydrodynamically unbiased trays (Snelgrove et al. 1992). Smith et al. (1986) also criticized the early deep-sea experiments because the tray walls isolated the experimental sediments from surrounding sediments and thus prevented the migration of adult colonists through the sediments into the patch. Smith et al. (1986) and Snelgrove et al. (in press) have now experimented with "patches-without-walls" (mounds and depressions) that allow colonization by both pelagic stages from the water column and immigrants moving through the immediately surrounding sediments. In considering the results of deep-sea colonization experiments, it is worth asking the question: what sorts and sizes of patches are these experiments trying to mimic?

The first deep-sea colonization experiments (Turner 1973, 1977), in which islands of wood blocks were placed on the bottom, were clearly intended to mimic small, widely separated patches of a rare resource. The early colonists, the planktotrophic larvae of wood- boring species, would come from the water column and not from the surrounding community where wood-boring species were rare or absent. These experiments were akin to experiments conducted on intertidal rocky shores, where the consequences of physical disturbance were mimicked by clearing small patches. Recolonization by species with largely planktotrophic larvae (e.g., barnacles and mussels) was then followed. Deep-sea experiments in trays that prevent most lateral immigration by adult infauna are designed to mimic the recolonization sequence in a relatively large patch, i.e., where most of the colonists come as pelagic larvae from outside the patch. Where patches are small and also permit lateral immigration of neighboring adults, then they are mimicking the small patch situation, where colonization may occur from a variety of sources, both near and far. Generalizing the results of these experiments to larger patch sizes is impossible without knowing the limits of adult migrations in the deep sea.

Observations on the deep-sea floor and of large-scale blooms of planktonic organisms suggest that patch size, and the quality of the organic material in those patches, may be correlated. Blooms of diatoms, coccolithophores, and salps, for example, are likely sometimes to appear at the deep-sea floor as very large patches, although topographic irregularities on the sea floor may cause some subsequent subdivision into smaller patches. Such blooms may also have a certain degree of temporal predictability. On the other hand, organic enrichments produced by falls of wood, macroalgae such as *Sargassum* spp., and carcasses of large organisms will constitute relatively small patches that lack any temporal predictability. We are just at the beginning of trying to understand how to count the numbers and kinds of resources in soft-sediment environments, but it may be that deep-sea observations will be especially illuminating. As Scheltema points out in this volume, there exists an interesting contrast between areas of high productivity (e.g., the North Atlantic), where there is a strong seasonal input of organic matter to the deep sea, and oligotrophic areas of low productivity (e.g., the North Pacific gyre), where the seasonal signal is weak or absent. On a slightly smaller scale, Blake and Grassle (in preparation) indicate that both physical disturbance by bottom currents and the kinds and amounts of organic enrichment play a role in determining community structure. In this volume Levin et al. have compared areas with different patterns of organic enrichment, at both seamount sites and at three 850-m depth sites off the Carolina coasts (sites that are included in the region covered by Blake and Grassle, in preparation). All of these "natural" experiments provide communities that are extremely useful for further *in situ* experimentation.

Recent deep-sea experiments (Snelgrove et al. 1992; Snelgrove, et al., in press) suggest that deep-sea species may indeed specialize on particular organic resources. Recruitment of different suites of species to small, experimental patches of *Sargassum* sp. and *Thalassiosira* sp. may indicate a degree of specialization by deep-sea species on different resources. That this specialization may be set up at the time of larval settlement in some species is suggested by recent flume experiments on related shallow-water species (Butman et al. 1988;

Grassle and Butman 1989; Bachelet et al. 1992; Butman and Grassle 1992; Grassle et al. 1992a, b; Snelgrove et al. 1993). These experiments demonstrate that larvae of certain species can choose among substrates in flows that mimic those seen in the natural environment, and that changes in near-bottom flow (e.g., changes caused by small depressions) affect different species to different degrees. Although there has been no chemical characterization of the chemical cues in soft-sediments, there are several precedents in research on hard-surface organisms that indicate that larvae of different species are stimulated to settle and metamorphose by different chemical compounds, and that some compounds may be inhibitory. (For a review of some of this research, the reader is referred to papers in the symposium edited by Pawlik and Hadfield 1990.)

As the title indicates, the papers in this volume do not explicitly consider larval settlement. Recruitment is defined operationally in soft-bottom studies, recruits being considered as individuals that have reached a large enough size to be retained on a sieve of a particular mesh size. Numerous studies in shallow water have shown that newly settled individuals of macrofaunal species may not be retained on the 1.0 mm and 0.5 mm sieves in common use in macrobenthic studies, and indeed may require sieves as fine as 0.1 mm, or even less, for their capture (e.g., Grassle et al. 1985; Buchanan and Moore 1986; Bachelet 1990; Emerson and Grant 1992). Since taxa differ in their size at settlement, and in their subsequent growth rates and the effective size they present to the meshes of a sieve, studies of the population dynamics of single species or communities should use appropriate sieves so as to provide proper estimates of abundances of individuals in all size classes.

Frequency of sampling in population dynamics studies also should be appropriate for the life history of individual species. As Blake and Watling show in this volume (see also Blake 1993), for some continuously reproducing and fast-growing deep-sea species even seasonal sampling is not frequent enough to detect early post-settlement stages. This kind of frequent sampling of postlarvae is also necessary in order to obtain some measure of postlarval mortality, so that we can begin to judge what the tradeoffs are between the high fecundity/planktotrophic larvae/high mortality and low fecundity/lecithotrophic/low mortality life history modes that Etter and Caswell have contrasted in their modeling of the effects of disturbance.

In conclusion, these contributions demonstrate that deep-sea communities provide a fertile place for investigation of community dynamics and, indeed, may be especially amenable in that the relative absence of frequent physical disturbance over large expanses permits the persistence of "natural" or experimentally induced manipulations of structure or food inputs. There is, moreover, a sufficient heterogeneity of environments within the compass of the as yet undefined zoogeographic provinces in the deep sea for mesoscale experiments to be designed. In defining broad-scale patterns of community dynamics it will also be necessary to focus attention on the particular characteristics of individual species and how they respond to environmental variability, as larvae or other dispersing stages in the water column, and as adults.

ACKNOWLEDGMENTS

It was a privilege to hear the presentations of the research embodied in this volume and I thank the organizers of the symposium, Craig Young and Kevin Eckelbarger, for the opportunity to summarize and reflect upon them. My own research in the deep sea has been supported by NSF, MMS, and NOAA. My thanks go to the many colleagues who have contributed to my evolving ideas on deep-sea diversity, beginning with Howard Sanders. My thanks also go to Judith Grassle for her major contribution to the completion of this manuscript (Contribution No. 9308 from the Institute of Marine and Coastal Sciences, Rutgers University).

REFERENCES

Bachelet, G. 1990. The choice of a sieving mesh size in the quantitative assessment of marine macrobenthos: A necessary compromise between aims and constraints. *Mar. Env. Res.* 30:21–35.

Bachelet, G., C. A. Butman, C. M. Webb, V. R. Starczak, and P. V. R. Snelgrove. 1992. Non-selective settlement of *Mercenaria mercenaria* (L.) larvae in short-term, still-water laboratory experiments. *J. Exp. Mar. Biol. Ecol.* 161:241–280.

Bartholomew, G. A. 1986. The role of natural history in contemporary biology. *BioScience* 36: 324–329.

Blake, J. A. 1993. Life history analysis of five dominant infaunal polychaete species from the continental slope off North Carolina, *J. Mar Biol. Ass. U.K.* 73:123–141.

Blake, J. A. and J. F. Grassle. In preparation. Benthic community structure on the U.S. South Atlantic slope off the Carolinas: Spatial heterogeneity in a current-dominated system.

Blake, J. A. and L. Watling. 1993. Life history studies of deep-sea benthic infauna: Polychaeta, Aplacophora, and Cumacea from the continental slope off Massachusetts. This volume, chapter 12.

Bouchet, P. and A. Warén. 1993. Ontogenetic migration and dispersal of deep-sea gastropod larvae. This volume, chapter 5.

Buchanan, J. B. and J. J. Moore. 1986. Long-term studies at a benthic station off the coast of Northumberland. *Hydrobiologia* 142:121–127.

Butman, C. A. and J. P. Grassle. 1992. Active habitat selection by *Capitella* sp. I larvae. I. Two-choice experiments in still water and flume flows. *J. Mar. Res.* 50:669–715.

Butman, C. A., J. P. Grassle, and C. M. Webb. 1988. Substrate choices made by marine larvae settling in still water and in a flume flow. *Nature* 333:771–773.

Cutler, E. B. 1975. Zoogeographical barrier on the continental slope off Cape Lookout, North Carolina. *Deep-Sea Res.* 22:893–901.

Eckelbarger, K. J. 1993. Ultrastructural features of gonads and gametes in deep-sea invertebrates. This volume, chapter 7.

Emerson, C. W. and J. Grant. 1992. The control of soft-shell clam (*Mya arenaria*) recruitment on intertidal sandflats by bedload sediment transport. *Limnol. Oceanogr.* 36:1288–1300.

Etter, R. J. and H. Caswell. 1993. The advantages of dispersal in a patchy environment: Effects of disturbance in a cellular automaton model. This volume, chapter 14.

Gage, J. D. 1993. Recruitment ecology and age structure of deep-sea invertebrate populations. This volume, chapter ll.

Grassle, J. F. and J. P. Grassle. In press. Notes from the abyss: The effects of a patchy supply of organic material and larvae on soft-sediment benthic communities. In P. Giller, D. Rafaelli, and A. Hildrew, eds. London: Blackwell Scientific Publications.

Grassle, J. F. and N. J. Maciolek. 1992. Deep-sea species richness: regional and local diversity estimates from quantitative bottom samples. *Amer. Nat.* 139:313–341.

Grassle, J. F. and L. Morse-Porteous. 1987. Macrofaunal colonization of disturbed deep-sea environments and the structure of deep-sea benthic communities. *Deep-Sea Res.* 34:1911–1950.

Grassle, J. F., J. P. Grassle, L. S. Brown-Leger, R. F. Petrecca, and N. J. Copley. 1985. Subtidal macrobenthos of Narragansett Bay: Field and mesocosm studies of the effects of eutrophication and organic input on benthic populations. In J. S. Gray and M. E. Christiansen, eds., *Marine Biology of Polar Regions and Effects of Stress on Marine Organisms,* pp. 421–434. New York: Wiley.

Grassle, J. P. and C. A. Butman. 1989. Active habitat selection by larvae of the polychaetes, *Capitella* spp. I and II, in a laboratory flume. In J. S Ryland and P. A. Tyler, eds., *Reproduction, Genetics, and Distributions of Marine Organisms,* pp. 107–114. Fredensborg, Denmark: Olsen and Olsen.

Grassle, J. P., C. A. Butman, and S. W. Mills. 1992a. Active habitat selection by *Capitella* sp. I larvae. II. Multiple-choice experiments in still water and flume flows. *J. Mar. Res.* 50:717–743.

Grassle, J. P., P. V. R. Snelgrove, and C. A. Butman. 1992b. Larval habitat choice in still water and flume flows in the opportunistic bivalve *Mulinia lateralis. Neth. J. Sea Res.* 30:33–44.

Gustafson, R. G. and R. A. Lutz. 1993. Molluscan life history traits at deep-sea hydrothermal vents and cold methane/sulfide seeps. This volume, chapter 4.

Hadfield, M. G. and M. F. Strathmann. 1990. Heterostrophic shells and pelagic development in trochoideans for classification, phylogeny, and paleoecology. *J. Molluscan Studies* 56:239–256.

Levin, L. A., G. R. Plaia, and C. L. Huggett. 1993. The influence of natural organic enhancement on life histories and community structure of bathyal polychaetes. This volume, chapter 13.

Levin, S. A. 1992. The problem of pattern and scale in ecology. *Ecology* 73:1943–1967.

Mullineaux, L. S. 1993. Implications of mesoscale flows for dispersal of deep-sea larvae. This volume, chapter 10.

Pawlik, J. R. and M. G. Hadfield. 1990. A symposium on chemical factors that influence the settlement and metamorphosis of marine invertebrate larvae: Introduction and perspective. *Bull. Mar. Sci.* 46:450–454 (and papers therein).

Pearse, J. S. 1993. Cold-water echinoderms break "Thorson's Rule." This volume, chapter 2.

Ricklefs, R. E. 1987. Community diversity: Relative roles of local and regional processes. *Science* 235:167–171.

Schaff, T., L. Levin, N. Blair, D. DeMaster, R. Pope, and S. Boehme. 1992. Spatial heterogeneity of benthos on the Carolina continental slope: Large (100 km)-scale variation. *Mar. Ecol. Prog. Ser.* 88:143–160.

Scheltema, R. S. 1993. Adaptations for reproduction among deep-sea benthic molluscs: An appraisal of the existing evidence. This volume, chapter 3.

Smith, C. R., P. A. Jumars, and D. J. DeMaster. 1986. *In situ* studies of megafaunal mounds indicate rapid turnover and community response at the deep-sea floor. *Nature* 323:251–253.

Snelgrove, P. V. R., J. F. Grassle, and R. F. Petrecca. In press. Macrofaunal response to artificial enrichments and depressions in a deep-sea habitat *J. Mar. Res.*

Snelgrove, P. V. R., J. F. Grassle, and R. F. Petrecca. 1992. The role of food patches in maintaining high deep-sea diversity: field experiments using hydrodynamically unbiased colonization trays. *Limnol. Oceanogr.* 37:1543–1550.

Snelgrove, P. V. R., C. A. Butman, and J. P. Grassle. 1993. Hydrodynamic enhancement of larval settlement in the bivalve *Mulinia lateralis* (Say) and the polychaete *Capitella* sp. I in microdepositional environments. *J. Exp. Mar. Biol. Ecol.* 168:71–109.

Stuart, C. T. and M. A. Rex. 1993. The relationship between developmental pattern and species diversity in deep-sea prosobranch snails. This volume, chapter 6.

Turner, R. D. 1973. Wood-boring bivalves, opportunistic species in the deep sea. *Science* 180:1377–1379.

Turner, R. D. 1977. Wood, mollusks, and deep-sea food chains. *Bull. Am. Malacol. Union for 1977*:13–19.

Tyler, P. A., L. S. Campos-Creasey, and L. A. Giles. 1993. Environmental control of quasi-continuous and seasonal reproduction in deep-sea benthic invertebrates; Preliminary observations. This volume, chapter 8.

Young, C. M. 1993. The biology of external fertilization in deep-sea Echinoderms. This volume, chapter 9.

TAXONOMIC INDEX

Abatus cavernosa, 7
Abatus nimrodi, 41
Abatus shackletoni, 43
Abyssochrysidae, 101, 103
Abyssochrysos cf. eburneum, 152
Aclididae, 50, 101, 124
Acodontaster conspicuus, 43
Acodontaster elongatus, 43
Acodontaster hodgsoni, 31, 43
Acrocirridae, 270, 274
Actiniaria, 17, 150
Aliceia, 104, 106
Alvania subsoluta, 107
Alvania testae, 107
Alvinelidae, 226
Alviniconcha, 107, 109
Alvinocaris, 214
Ampharete sp., 270, 273–74
Ampharetidae, 270, 274–75
Amphigymnas bahamensis, 140
Amphiodia occidentalis, 42
Amphipholis squamata, 42
Amphiplica (*Gordabyssia*) *gordensis,* 88
Amphipoda, 256–57
Amphissa acutecostata, 103
Amphissa haliaeeti, 59
Anatoma crispata, 107
Annelida, *see* Polychaeta
Anomura, 16
Antedon bifida, 139
Anthenoides piercei, 141
Anthozoa, 112, 137–38, 150
Aplacophora, 79, 92, 225, 243–44, 250, 256
Apoda, 189
Aporocidaris milleri, 32
Araeosoma belli, 140, 144, 146
Araeosoma fenestratum, 41, 140, 144, 146, 179, 181, 188, 195
Araeosoma owstoni, 144, 146
Arbacia punculata, 195
Arbaciidae, 140, 181, 193
Archaeogastropoda, 57, 91–93, 98–100, 101–2, 108, 110–12, 114, 125, 131–32
Archaeopneustes hystrix, 140, 181, 193
Architectonicidae, 101
Aricidea lopezi, 270
Aricidea quadrilobata, 262, 270, 272–75, 277
Artacama sp., 270
Arthropoda, 14, 16, 19–20
Ascidiacea, 17–18, 19–20, 180
Asellota, 256
Aspidodiadema, 179, 182, 196
Aspidodiadema arcitum, 153, 183, 195
Aspidodiadema jacobyi, 93, 140, 144, 153, 181–83, 195
Aspidodiadematidae, 140
Asterias lincki, 30, 40

Asteriidae, 189
Asterina gibbosa, 32
Asterina miniata, 42
Asterina minor, 32
Asterinidae, 141, 189
Asterinopsis pilosa, 141
Asteroidea, 10, 29–34, 40–43, 137, 141–42, 144–45, 148, 158, 160, 180–82, 184, 189, 193–94, 308
Asterostomatidae, 140
Asthenosoma ijimai, 144, 146
Astropecten californicus, 41
Astylus subviridis, 17
Atelesyllis sp., 270
Auricularia nudibranchiata, 11, 20
Aurospio dibranchiata, 243–50, 256, 257

Bacillariophyceae, 310
Bacteria, 159, 262
Barantolla sp., 270
Bathyarca sp., 49
Bathybembix bairdii, 116
Bathybiaster loripes, 42
Bathybiaster vexillifer, 40–41, 160–61, 169–70, 172–74, 177
Bathymargarites symplector, 88–89, 91
Bathymodiolus sp., 102, 109, 214, 219
Bathymodiolus thermophilus, 49, 66, 79, 91–92, 102, 109
Bathyoncus mirabilis, 18
Bathypecten eucymatus, 80
Bathypecten vulcani, 80, 81, 92
Bathyplotes natans, 140, 142, 144, 146, 148
Bathysciadiidae, 110
Bathysiphon filiformes, 267
Benthodytes sordida, 41
Benthomangelia antonia, 59, 60, 116
Benthomangelia marca, 58, 103
Benthonella tenella, 49, 58–60, 102–6, 110
Benthopecten simplex, 41
Benthothuria funebris, 140
Benthothuria sp., 146
Bivalvia, 8, 14, 18, 28, 47, 49–54, 56–57, 59, 61, 63–64, 66, 68–69, 76–82, 91–92, 100, 102, 153, 159–66, 172–74, 180, 214, 224, 226–27, 229–30, 232, 236, 307
Bonellia sumbi, 18
Bonuania sp., 270
Boretrophon abyssorum, 59
Brachiopoda, 8, 13–14, 18, 19–20, 138, 227
Brevinucula verrilli, 49
Brisinga endecacnemos, 41
Brisingella coronata, 41
Brisingidae, 141–42, 184
Brissidae, 140
Bryozoa, 11, 13–14, 20
Buccinidae, 101
Bythograea, 214

Cadulus californicus, 49
Caenogastropoda, 98–102, 111–12, 152
Calcarea, 18, 20
Calliactis parasitica, 17
Calliotropis ottoi, 49
Calocidaris micans, 140
Calveriosoma histrix, 41
Calyptogena magnifica, 49, 80, 82, 91
Calyptogena sp., 50, 82, 214
Cancellariidae, 101
Capitella spp., 262, 274, 276, 282, 314
Capitellidae, 153, 276
Capulidae, 101
Cassidae, 101
Cassiduloida, 193
Cataegis, 97
Cataegis toreuta, 89
Caulleriella sp., 270
Caymanostella sp., 141, 154
Caymanostellidae, 141
Cephalopoda, 61, 79, 110, 152
Ceratocephale loveni, 270–73
Cerithiellidae, 101
Cerithiidae, 101
Chaetoceros, 166, 168
Chaetozone sp., 270
Cidaris blakei, 140, 181, 194
Cidaris cidaris, 11, 19–20, 24, 31, 41, 181
Cidaris rugosa, 140
Cidaroida, 8, 11, 31, 140, 193
Cingula leptalea, 59
Cirripedia, 16, 19, 20, 310
Cirrophorus lyra, 262, 270–74, 277
Cirrophorus sp., 270
Cladodactyla crocea, 7
Clypeasteroida, 9, 11–12
Clypeosectide, 87–88
Clypeosectus, 87
Clypeosectus curvis, 87
Clypeosectus delectus, 87
Cnidaria, 10, 17, 19–20, 138, 180, 204, 214
Coccolithophoraceae, 310
Coccolithus pelagicus, 166
Cocculinidae, 110
Cocculiniformia, 91
Coelenterata, *see* Cnidaria
Coelocormus huxleyi, 18
Coelopleurus floridanus, 140, 179, 181, 193
Collosendis, 17
Columbellidae, 103
Colus jeffreyianus, 49
Concentricycloidea, 140–41, 154, 182
Conolampus sigsbei, 140, 181, 188, 193
Copepoda, 263
Corallimorphus rigidus, 17
Coscinodiscus, 166
Cossura longocirrata, 256–57, 270
Cossura sp., 262, 270, 273, 275, 277
Crassostrea virginica, 60
Crinoidea, 6, 13, 30, 40, 43, 139, 227
Crossaster papposus, 30, 40
Crustacea, 8, 10, 14, 16, 19–20, 138, 214, 226
Cryptohelia affinis, 19

Cryptohelia pudica, 10, 17
Ctenodiscus crispatus, 30, 40
Ctenodiscus perrieri, 31, 43
Cucumaria curata, 42
Cucumaria frondosa, 29, 40
Cucumaria lubrica, 139, 154
Cucumaria miniata, 42, 192
Cucumaria piperata, 42
Cucumaria pseudocurata, 42, 139, 154
Cuenotaster involutus, 31, 43
Cumacea, 243–44, 251, 256–57
Cyanophycota, 276,
Cyathermia naticoides, 83
Cyathermidae, 83
Cyclococcolithis leptoporus, 166
Cyclopectin ambiannulatus, 80
Cystodytes draschii, 19

Daphnellidae, 105, 108
Deima sp., 141
Deima validum, 41
Deimatidae, 141
Deminucula atacellana, 49, 225, 231, 234
Dendraster excentricus, 42
Dendrochirota, 189
Depressigyra, 86
Depressigyra globulus, 84
Depressigyra planispira, 84
Dermasterias imbricata, 41
Desbruyeresia, 107, 109
Diastylis stygia, 243–45, 251, 253–57
Diplasterias brucei, 43
Diplocirrus sp., 173
Dorocidaris papillata, see Cidaris cidaris
Dorvilleidae, 270, 276,
Drilliola loprestiana, 103
Dysponetus sp., 270
Dytaster grandis, 31, 40, 169
Dytaster insignis, see Dytaster grandis

Echinasteridae, 189
Echinodermata, 6, 7, 12–13, 19, 26–35, 40–43, 137–49, 179–82, 193–94, 227, 229, 256
Echinoidea, 7–8, 11, 19, 29, 30, 31, 32, 33, 40–43, 93, 137, 139–48, 150–51, 153–54, 158, 179, 180–82, 186–89, 193, 195–96
Echinolampadidae, 140
Echinopelta fistulosa, 84–85
Echinosigra phiale, 41
Echinothuridae, 8, 140, 144–47, 153, 179, 181–82, 188, 194–95
Echinus acutus, 11, 20, 31, 41
Echinus acutus norvegicus, 181, 240
Echinus affinis, 31, 38, 41, 160–61, 166–68, 174–75, 177, 181, 190–92
Echinus alexandri, 31, 41
Echinus elegans, 176, 181, 240
Echinus esculentus, 11
Echiura, 18, 20, 225
Elachisinidae, 103
Elasipoda, 7, 11, 20
Elasmobranchii, 61
Elpidia glacialis, 40
Emiliana huxleyi, 166
Enypniastes eximia, 140, 148
Eostichopus regalis, 140, 166
Epitoniidae, 50, 100–101, 104, 112
Epitoniinae, 106, 125
Errina labiata, 17
Eryoneicus sp., 19
Eryonidae, *see* Polychelidae
Euchone incolor, 262, 270–74, 277
Eulamellibranchia, 49, 53, 56
Eulepetopsis vitrea, 88
Eulimidae, 50, 58, 101, 106–7, 124
Eupentacta quinquesemita, 42
Euphausiacea, 236

Fabrisabella similis, 270
Famelica, 103, 104, 106
Famelica mirmidina, 104
Ficidae, 101
Flabelligera sp., 270
Foraminiferida, 267
Frigidoalvania brychia, 110
Fusinidae, 101

Galatheidae, 16, 214, 266
Gastropoda, 8, 14, 20, 28, 44, 49–51, 53–61, 64, 67, 69, 76–79, 82–94, 98–114, 118–34, 152–54, 214, 226, 299, 307
Goniasteridae, 141, 181
Goniobasis, 154
Goniocidaris canaliculata, 7–8
Gorgoleptis, 87
Gorgoleptis emarginatus, 87
Gorgoleptis patulus, 87
Gorgonocephalidae, 140
Gorgonocephalus eucnemis, 42
Gorgonocephalus sp., 140
Gymnobela subaraneosa, 59

Haloceras tricarinata, 104
Haloceratidae, 100–101, 106
Hansenothuria benti, 140, 146
Hapalosoma gemmiferum, 144, 146
Harpiniopsis, see Pseudharpinia
Hathrometra prolixa, 40
Helicoradomenia juani, 79, 92
Heliometra glacialis, 40
Hemiaster cavernosa, see Abatus cavernosa
Hemiaster expurgitus, 172–74
Henricia leviuscula, 42
Henricia sanguinolenta, 30, 40
Henricia sp., 42
Hesionura sp., 270
Heterobranchia, 99, 102
Heterostropha, 100, 112
Hexactinellida, 6, 18–19
Hipponicidae, 101
Holasteroidea, 31

Holothuria occidentalis, 140, 146, 148
Holothuriidae, 140, 184, 188–89, 194, 266
Holothuroidea, 7, 12, 13, 20, 29, 31, 32, 40–42, 139–40, 142, 144, 146, 148–49, 153–54, 179, 184, 189
Hyalogyrinidae, 101
Hyalopecten undatus, 80
Hydrozoa, 17
Hygrosoma petersi, 41
Hygrosoma sp., 140
Hymenaster membranaceus, 30, 32, 41
Hymenaster pellucidus, 10, 30, 32, 40
Hymenaster praecoquis, 10
Hypobythius calycodes, 17–18
Hypsiechinus coronatus, 32

Insecta, 139
Iphitus tuberatus, 104
Isometra vivipara, 139
Isopoda, 256

Lacunoides exquisitus, 83
Laeviphitus verduini, 103
Laganum diplopora, 9, 11–12, 19–20
Lamellariidae, 101
Laubierina sp., 113
Laubierinidae, 100–101, 106, 112–13
Ledella pustulosa, 49, 50, 71, 160, 162–63, 165–66, 172, 174, 229, 231–35
Lepetella laterocompressa, 111
Lepetodrilidae, 87
Leptasterias groenlandica, 30, 40
Leptasterias hexactis, 42
Leptasterias pusilla, 42
Leptoclinum tenue, 18
Leptodrilus cristatus, 86–87
Leptodrilus elevatus elevatus, 86–87
Leptodrilus elevatus galriftensis, 86–87
Leptodrilus fucencis, 86–87
Leptodrilus guaymasensis, 86–87
Leptodrilus ovalis, 86–87
Leptodrilus pustulosus, 86–87
Leptosynapta albicans, 42
Leptychaster kergulensis, 10
Levinsenia gracilis, 270
Levinsenia sp., 270
Linckia nodosa, 141
Linopneustes longispinus, 140, 181
Lirapex granularis, 86
Lirapex humata, 86
Lissothuria nutriens, 42
Lophaster gaini, 43
Lophiiformes, 11,
Luidia foliolata, 41
Luidia sp., 141
Luidiidae, 141
Lumbrineris fragilis, 270
Lumbrineris sp., 273
Lusitanops, 113
Lyonsiella abyssicola, 53
Lytechinus euerces, 181

Macroptychaster accrescens, 43
Maldanidae, 270
Malletia abyssorum, 63
Malletia cuneata, 49, 50, 160, 163, 164, 165, 166, 174, 231, 234
Malletia johnstoni, 172
Marginellidae, 101
Mathildidae, 101
Mediaster aequalis, 10, 21, 42
Meiodorvillea minuta, 273
Meiodorvillea sp., 270, 272
Melanodrymia aurantiiaca, 86, 94, 96
Mesogastropoda, 57
Mesothuria sp., 140, 146, 148
Metazoa, 142
Microbinis linea, 270
Microgloma turnerae, 52, 53
Microgloma yongei, 52, 53
Micromestistius poutassou, 173
Microrbinia linea, 256, 257
Mohnia abyssorum, 111
Mollusca, 14, 20, 44–45, 48–69, 76–94, 137–38, 152–54, 159–66, 172–74, 227, 256
Molpadia sp., 141
Molpadiidae, 141
Munida, 16
Munidopsis, 214
Muricidae, 101
Mytilidae, 61, 66, 79–81, 92, 109, 180, 214, 219, 310

Nassariidae, 101
Natica sp., 172
Naticidae, 101
Neogastropoda, 57–58
Neolepetopsidae, 88, 92
Neolepetopsis densata, 85, 88
Neolepetopsis gordensis, 88
Neolepetopsis occulta, 88
Neomphalidae, 82–83
Neomphalus fretterae, 82, 83
Neopleurotomoides, 104, 106
Nephtyidae, 274
Nereididae, 270
Neritacea, 89
Neritimorpha, 99–102, 109–10
Neritoidea, 92
Nitzschia, 168
Nodopelta heminoda, 84
Nodopelta subnoda, 84
Notasterias armata, 43
Novodinia antillensis, 141–42, 144
Nucula, 162
Nucula cancellata, 53, 71
Nucula darella, 49
Nucula granulosa, 53
Nucula pontonia, 49
Nucula verrilli, 51
Nuculoidea bushae, 49
Nuculoma granulosa, 49
Nymphaster sp., 141

Nymphon, 17
Nystiellinae, 104, 106

Ocnus sacculus, 189
Octocorallia, 17
Odinella nutrix, 184
Odontaster meridionalis, 30, 42
Odontaster validus, 30, 32, 42
Oenopota graphica, 59
Oenopota levindensis, 60
Oenopota ovalis, 59, 108
Oneirophanta mutabilis, 7, 32, 41, 184, 189
Oocorys, 103
Ophelia profunda, 270
Ophelidae, 270
Ophiacantha bidentata, 30, 38, 40–41
Ophiacanthidae, 140
Ophidiaster alexandri, 141
Ophidiasteridae, 141
Ophiocten gracilis, 13, 19–20, 31, 41, 229
Ophiocten robusta, 29
Ophiocten sericium, 29, 40
Ophiomitra valida, 140
Ophiomusium lymani, 41, 229, 230, 239
Ophiomusium sp., 140
Ophiomyxa tumida, 140
Ophiomyxidae, 140
Ophionotus victoriae, 30, 43
Ophiopholis aculeata, 42
Ophiopleura borealis, 40
Ophioplocus esmarki, 42
Ophiopluteus ramosus, see Ophiocten gracilis
Ophiopteris papillosa, 42
Ophiopus arcticus, 40
Ophioscolex glacialis, 40
Ophiosparta gigas, 31, 43
Ophiothrix, 179
Ophiothrix spiculata, 42
Ophiura ljungmani, 31, 40, 229–30, 232, 237
Ophiura robusta, 40
Ophiura sarsi, 40
Ophiuridae, 140,
Ophiuroidea, 13, 19, 29–31, 140, 142, 179, 188, 194, 224, 229–30, 232
Ophryotrocha sp., 262, 276
Opisthobranchia, 124
Orthasterias koehleri, 42
Osteopeltidae, 110
Ostracoda, 8
Ovulidae, 101
Oweniidae, 270

Pachydermia laevis, 84
Pachythyone rubra, 42
Palaeobrissus hilgardi, 140, 181
Palaeopneustes cristatus, 140, 188, 193
Palaeopneustes tholoformis, 140
Palaeopneustidae, 140
Paracalliactis stephensoni, 150
Paragonaster subtilis, 41
Paralepetopsis floridensis, 88
Paraonidae, 273
Paraonis cerruti, 270
Paraonis sp., 270
Paraonis unicinatus, 270
Paroriza spp., 194
Paroriza pallens, 188
Pectinaster filholi, 41
Pectinidae, 47, 50, 80
Pedicellaster, 6-radiatus, 12,
Pelagodiscus atlanticus, 13, 19–20
Pelagothuriidae, 141
Peltaster nidarosiensis, 141
Peltaster placentra, 141, 179, 181
Peltospira delicata, 84
Peltospira operculata, 84
Peltospiridae, 83–86
Pennatulidae, 17
Peracarida, 227
Perknaster fuscus, 43
Phelliactis hertwigi, 150
Phelliactis robusta, 150
Phenacolepadidae, 101, 108
Philbertia linearis, 105
Pholadidae, 61
Pholoe anoculata, 256
Phormosoma placenta, 25, 41, 144, 146, 160–61, 167–68, 174, 181
Phrissocystis multispina, 140, 142, 144, 154, 183, 196, 197
Phymorhynchus sp., 90, 91, 105, 109
Pisanianurinae, 101
Pisaster brevispinus, 42
Pisaster giganteus, 42
Pisaster ochraceus, 42
Pleurotomaria africana, 152
Pleurotomella bureauri, 105
Pleurotemella packardi, 59
Pleurotemella sandersoni, 59
Plutonaster bifrons, 31, 40, 160, 161, 169, 170, 172, 173, 175
Poecilochaetus cf. fulgens, 173
Pogonophora, 148,
Polychaeta, 18, 138, 153, 214, 226–27, 245, 256–57, 261, 263, 266–69, 271–73, 275–78
Polycheles sp., 14, 19–20
Polychelidae, 14, 19–20
Polyclinum molle, 18
Polydora ligni, 274
Polyplacophora, 79
Pontaster tenuispinus, 40–41, 181
Porania spp., 35
Porania antarctica, 30, 32, 43
Poraniomorpha tumida, 40
Porcellanaster ceruleus, 169
Porcellanasteridae, 8
Porifera, 6, 18–20
Poriocidaris purpurata, 41, 166
Porocidaris elegans, 8
Portalesia jeffreysi, 30, 40–41
Portalesia miranda, 41
Prionospio sp., 270, 276
Pristiglomidae, 52,
Pristogloma alba, 53

Pristogloma nitans, 53
Prochaetoderma yongei, 225, 243–45, 250–52, 256–57
Promachocrinus kerguelensis, 43
Prosobranchia, 56–59, 77, 94, 118–34, 154, 284–85, 297–99, 302
Protista, 159
Protobranchia, 49, 50–53, 56, 63, 69, 158, 160–61, 162–66, 172–74, 225, 230, 234
Protodorvillea sp., 262, 270–74
Protozoa, 262, 279
Provanna, 89
Provanna ios, 89
Provanna muricata, 89
Provanna n. sp., 152
Provanna sculpta, 89
Provanna variabilis, 89–90
Provannidae, 89–90, 101, 103, 108,
Pseudarchaster parelii, 41
Pseudarchaster sp., 141
Pseudharpinia (= *Harpiniopsis*) *excavata,* 257
Pseudococculinidae, 88
Pseudorimula marianae, 87
Pseudostichopus sp., 140
Pseudotindaria erebus, 53
Psilaster andromeda, 41
Psilaster charcoti, 42
Psolus ephippifer, 5, 7
Psychropotes depressa, 41
Psychropotes longicauda, 41
Psychropotes semperiana, 41
Pterasteridae, 10, 184
Pteraster militaris, 30, 40
Pteraster tesselatus, 42
Pycnogonida, 17
Pycnopodia helianthoides, 42
Pyramidellidae, 101
Pyropelta corymba, 88
Pyropelta musaica, 88
Pyropeltidae, 88, 92, 110

Questa sp., 270

Rectilabrum lanceolatum, 107
Rhizosolenia, 168
Rhynchopelta concentrica, 83–84
Riftia pachyptila, 148, 150, 181, 219
Rissoidae, 101, 103, 105,
Rissoinidae, 58

Salenia goesiana, 140, 181, 182, 183
Saleniidae, 140
Salpida, 46, 310
Sargassum, 276, 310
Scalibregma inflatum, 256, 262, 270, 272, 273, 275, 277
Scalibregmatidae, 273
Scalpellum, 16, 20
Scalpellum stroemi, 16, 19
Scaphopoda, 49,
Schistomeringos caecus, 270
Scissurellidae, 82, 94
Scleractinia, 17,
Septibranchia, 50, 53, 56
Shinkailepas, 107
Shinkailepas kaikatensis, 89, 92
Siboglinum ekmani, 148
Silicula filatova, 52, 53
Silicula fragilis, 53
Silicula mcalesteri, 53
Siliculidae, 52
Sipuncula, 263
Skeletonema, 166
Skeneidae, 92
Solaster dawsoni, 42
Solutigyra reticulata, 86, 111
Spatangoida, 31, 193
Sperosoma antillense, 144, 146, 195
Sperosoma grimaldi, 41
Spinula flatovae, 53
Spionidae, 243–44, 262, 270, 273, 276
Sporadophora dichotoma, 17
Stellophaera mirabilis, see Pedicellaster, 6-radiatus
Stephanasterias albula, 29
Sterechinus neumayeri, 30–31, 43
Stichopodidae, 140
Stichopus californicus, 42
Stichopus parvimensis, 42
Streblospio benedicti, 274
Strongylocentrotus droebachiensis, 29, 40
Strongylocentrotus franciscanus, 42
Strongylocentrotus purpuratus, 42
Stylocidaris lineata, 140, 181, 186, 188, 193–94
Sycon arcticum, 18
Syllidia sp., 270
Symmetromphalus regularis, 83
Synallactidae, 140
Synaptidae, 11

Tachytrypane sp., 270
Tanaidacea, 256
Temnocinclis euripes, 82
Temnozaga parilis, 82
Terebellidae, 274
Terebellides sp., 173, 273–74
Thalassionema, 166, 168
Thalassiosira, 166, 303
Tharyx sp., 262, 270–73
Theta chariessa, 59
Theta lyronuclea, 59
Thyasira alleni, 53
Thyasira brevis, 53
Thyasira excavata, 53
Thyasira incrassata, 53
Thyasira ovata, 53
Thyasira subovata subovata, 53, 63
Thyasira succisa succisa, 53
Thyasira transversa, 53
Thyasiridae, 52, 63,
Tindaria callistima, 51
Tindaria hessleri, 53
Tonnoidea, 103, 112
Trochidae, 88, 91, 110, 214, 308
Turbinellidae, 101

Turridae, 50, 55, 58, 60, 67, 91, 100–101, 103, 104, 105, 108–9, 113, 214

Urechinus mortenseni, 197
Uroptychidae, 16

Vampryroteuthis infernalis, 152
Vanikoridae, 101
Vermetidae, 112
Verticordiidae, 53
Vesicomyidae, 80, 82
Vestimentifera, 137–38, 148, 150, 181, 213, 219
Volutidae, 101

Xylodisculidae, 101
Xylophagainae, 238
Xylophaga sp., 53, 63, 69
Xyloplacidae, 140
Xyloplax turnerae, 140, 154

Yoldiella jeffreysi, 49–50, 162, 172, 230–34
Ypsilothuria talismani, 41, 173

Zoroaster fulgens, 41

AUTHOR INDEX

Abbott, D. P., 41
Adiyodi, K. G., 138
Adiyodi, R. G., 138
Afzelius, B. A., 139
Agassiz, A., 6–9
Akabane, S., 32
Albert I of Monaco 14,
Allen, J. A., 52–53, 75, 162–63, 165, 226
Allmann, G. J., 17, 20
Alred, R. G., 193
Amemiya, S., 146, 153
Anderson, D. T., 141, 154
Anderson, R. F., 257
Andrews, J. C., 203
Ansell, A. D., 225
Aschan, M. M., 277
Ashworth, J. H., 13, 15, 19–20
Astheimer, H., 236
Atwood, D. G., 139, 144, 154
Auerbach, M., 119
Baccetti, B., 138
Bachelet, G., 232, 311
Bailey, R. S., 175
Baker, A. N., 141, 154
Baker, E. T., 64, 214, 216–18
Baldwin, B. S., 60
Ballard, R. D., 62, 214
Bandel, K., 110
Baptiste, E. M., 244–46, 250–51, 256–57, 269, 275
Barker, H. R., Jr., 269
Barnes, H., 48
Barnes, M., 48
Baross, J. A., 236
Barry, J. P., 234
Bartholomew, G. A., 27, 35, 308
Basch, L. V., 33–34
Batten, R. L., 83
Battisti, T. M., 217
Beach, D. H., 194
Beauchamp, K. A., 30
Becker, K., 62, 214
Beddington, J. R., 236
Behrens-Yamada, S., 110
Berg, C. J., 49, 65, 77, 79–80, 82–84, 89, 91–93, 111, 214, 216–17, 219, 226
Berger, W. H., 280
Bernard, R. T., F., 153
Bevas, J. Y., 224, 276
Bickell, L. R., 142
Billett, D. S., M., 32, 47, 50, 105, 139, 159–60, 169, 174, 184, 193–94, 231–32, 236–37, 256–57, 262
Bingham, B. L., 185
Birkeland, C. E., 10
Biscaye, P. E., 62
Blair, N., 263, 265–66, 267
Blake, J. A., 244–47, 250–51, 256–57, 269, 275, 307–8, 310–11
Blaume, F., 47

Boehm, P. D., 244, 246, 250–51, 256, 269, 275
Boehme, S., 263, 265–67
Bosch, I. M., 29–34, 42, 100, 110
Boschma, H., 193
Bossellmann, A., 232
Bouchet, P., 14, 56–60, 64–65, 67, 77, 83–84, 86, 88–89, 91–94, 99, 100–103, 106–9, 112, 114, 120, 124–25, 133, 180, 203, 214, 307
Boudrias, M. A., 226
Bouvier, E. L., 16
Bowers, M. A., 285
Boyer, D. L., 210
Braithwaite, L. F., 184
Brazeau, D. A., 180
Brichet, E., 214
Bridges, T. S., 262, 276
Brink, K. H., 210–11
Britton, N., 286
Brown, B., 244, 246, 250–51, 256, 269, 275
Brown, N. O., 263
Brown-Leger, L. S., 311
Brumsickle, S. J., 224–25
Brunn, A. F., 159
Bryan, K., 207
Buchan, A., 27
Buchanan, J. B., 311–12
Buchanan, J. Y., 20, 25
Buckland-Nicks, J., 59, 216
Bucklin, A., 219
Burke, R. D., 102
Bush, K. J., 162
Busk, G., 14
Butman, C. A., 47, 66, 185, 187, 203–4, 310–11
Byrne, R. H., 217

Cabrera, D., 276
Cadet, J.-P., 193–94
Calloway, C. B., 28
Calow, P., 52
Cameron, J. L., 10, 32, 93, 142, 144, 146, 153–54, 166, 181–82, 186, 193–96
Cameron, R. A., 30, 32
Campbell, R. D., 17
Campos, L. S., 168, 307–8
Campos-Creasey, L. S., *see* Campos, L. S.
Carey, A. G., Jr., 169
Carney, R. S., 1
Carpenter, P. H., 13, 15
Carrozza, F., 103
Carson, B., 63
Carson, S. F., 29
Cary, S. C., 148
Caswell, H., 133, 285–89
Cavanaugh, C. M., 50
Cerrato, R. M., 52, 226
Chaffee, C., 189
Chang, K.-H., 193
Chang, P., 185
Chapman, D. C., 207–9
Charnov, E. L., 16
Chave, E. H., 204
Chen, C.-P., 193
Chesson, P. L., 120, 285
Chia, F. Lyph S., 21, 59, 139, 142, 144, 154, 180, 184, 192–93, 197, 216, 262
Childress, J. J., 226
Christiansen, F. B., 28, 34, 285
Chun, C., 9, 11
Clark, A. M., 30, 32
Clark, H. E., S., 141, 154, 182
Clarke, A. H., Jr., 14
Cloern, J. E., 234
Cochran, D. P., 226
Cochran, J. K., 52, 226–27
Cochrane, G. R., 63
Cohen, J. E., 286–89, 303
Cohn, E. J., 195
Cole, L. C., 61
Colman, J. G., 49, 55, 100, 124
Converse, D. R., 216–17
Copley, N. J., 82, 93, 214, 217, 311
Cornell, H. V., 119
Cowen, J. P., 217
Cox, G. W., 119
Craddock, C., 202
Craig, C. H., 205
Crassous, P, 262
Crawford, B., 139, 144
Creed, E. L., 262
Crisp, D. J., 4, 27, 78, 93, 110, 115
Crump, R. G., 32
Crutchfield, J., 286
Curl, H. C., Jr., 217
Cutler, E. B., 307
Czihak, G., 185

Dade, B., 244, 246
Darlington, E., 193
Dauer, D. M., 269
Dautzenberg, P., 14, 20
David, B., 182
Davidson, T., 14
Davies, G., 49–50
Davies, P. A., 64
Davis, G., 49
Dayton, P. K., 119, 134, 185, 204
Delaney, J. R., 216
DeMaster, D. J., 119, 225, 263, 265–66, 267, 275, 309
Deming, J. W., 62, 65, 236
DeMorgan, W., 11, 20
Dengler, A. T., 62
Denny, M. W., 179, 184–86, 189
Desbruyères, D., 65, 77, 224, 226, 262, 276
Deuser, W. G., 46, 159, 174, 257
Dinet, A., 65
Dix, T. G., 193
Doflein, F., 16
Dooley, J. L., 285
Downey, M. E., 30, 32
Drew, G. A., 56
Duco, A., 227
Duffy, S., 244, 246, 251
Dymond, J., 217

Eckelbarger, K. J., 93, 138, 141–42, 144, 146, 153, 181–82, 194–96, 308–9
Eckman, J. E., 185, 187
Edmond, J. M., 216, 217
Edmonds, S. J., 18

Eichler, P., 13, 20
Elderfield, H., 62
Elliott, A. J., 185, 187
Ellner, S., 285
Emerson, C. W., 311
Emlet, R. B., 3, 7, 30–32, 34, 40–41, 181, 195
Emson, R. H., 32, 159, 187, 193, 196
Eriksen, C. C., 206
Etter, R. J., 110, 115, 119, 132–33, 286–89, 299, 309
Ewing, M., 62
Ewing, R. M., 269

Faaborg, J., 119
Factor, J. R., 217
Fairweather, P. G., 119
Falk-Petersen, I. B., 30, 40
Fauchald, K., 63, 275
Fautin, D. G., 150
Feely, R. A., 217–18
Felbech, H., 148, 150
Fenchel, T. M., 28, 34, 285
Field, G. W., 139, 156
Fioroni, P., 99
Fisher, H., 14, 20
Fisher, W. K., 184
Flierl, G. R., 213, 216
Fontes, J.-C., 59, 94, 106, 124
Forbes, E., 2, 13
Fournier, D. A., 234
Frank, P. W., 285
Franzen, A., 138–39, 152–53
Fretter, V., 77, 83–84, 87–88, 92
Fritz, L. W., 91, 226
Fry, J., 159, 174, 232, 262
Fuchs, H. M., 11, 20
Fustec, A., 226
Fujioti, K., 193

Gage, J. D., 4, 8, 13, 27–32, 46–47, 49–50, 52, 55, 91, 93–94, 100, 124, 138–39, 159–60, 162–63, 166–69, 172–74, 184, 187, 193, 226–34, 237–38, 256–57, 262, 275, 308
Gaines, S., 119
Gardiner, S. L., 148, 150, 217
Gardner, W. D., 62
Garstang, W., ix
Geiger, S. R., 229
Gemmill, J. F., 30, 195
Genin, A., 185, 197, 204
George, R. Y., 3, 160, 255–56
Ghiselin, M. T., 34, 92
Giard, A., 27
Giese, A. C., 3, 146, 180
Giles, L. A., 49–50, 166, 194, 307–8
Gilkinson, K. D., 52, 227
Gill, I. P., 159
Girard, D., 193–94
Gooday, A. J., 159–60, 174, 257–58, 262–63, 266, 279
Gordon, J. D. M., 175
Gould, W. J., 210
Gowing, M., 204, 263, 266
Graf, G., 47, 237, 262–63
Graham, A., 77, 83
Grant, A., 28, 31–32, 34, 48, 231, 256–57
Grant, J., 311
Grant, W. D., 47
Grassle, J. F., 52, 62, 64–65, 72, 74, 77, 110, 119, 133, 187, 189, 191–93, 224–26, 244, 246, 250–51, 256, 262, 267, 269, 275–76, 279, 299, 307, 309–11
Grassle, J. P., 66, 72, 110, 138, 153, 219, 262, 275, 307, 309, 310–11
Gray, J. S., 261, 277–78
Grémare, A., 262
Grigg, R. W., 204
Gross, T. F., 62, 159
Gurney, R., 14
Gustafson, R. G., 53, 56, 64, 78, 91–93, 109, 111, 120, 124, 180, 214, 307

Haderlie, E. C., 41
Hadfield, M. G., 93, 131, 308, 311
Haedrich, R. L., 1, 50, 52, 226–27
Haidvogel, D. B., 207–9
Hamilton, S. C., 184
Hampton, J., 234
Hannah, F. J., 162–63, 165
Hanscomb, N. J., 194
Hansen, B., 7, 32, 139, 184, 189, 193
Hansen, T. A., 110, 285
Harbison, G. R., 46
Harbour, D. L., 168
Harrison, K., 3, 256, 258
Harvey, R., 30, 31, 49–50, 166, 256
Hashimoto, J., 77, 89
Haszprunar, G., 77, 82, 86, 88, 92
Haury, L. R., 46
Hayes, S. P., 185, 187
Healy, J. M., 77, 89, 141, 152, 154, 182
Heath, G. R., 46
Hecker, B., 63, 244, 246, 250–51, 256–57, 262, 266–67, 269, 275, 279
Hedgecock, D., 285
Heezen, B. C., 62
Helfrich, K. R., 217
Hemleben, C., 159–60, 174, 257, 262, 279
Henderson, J. R., 16
Hendry, R., 210
Herdman, W. A., 2, 6, 10, 15, 17–20, 37
Hertwig, R., 15, 17
Heslinga, G. A., 93
Hessler, R. R., 62, 65, 119–20, 187, 189, 191, 193, 203, 218, 224, 226, 255, 299
Hickman, C. S., 27
Hilbig, B., 244, 250–51, 256, 269, 275
Hines, A. H., 78
Hinga, K. R., 46
Hoagland, K. E., 92, 99
Hodgson, A. N., 154
Hoeg-Guldberg, O., 13
Hoek, P. P. C., 15–17, 19–20
Hogg, N. G., 210, 217
Höisaeter, T., 92

Holland, H. D., 216–17
Holland, N. D., 30–31, 40, 139, 148, 150
Hollister, C. D., 62, 159
Horner, S. G., 262
Houbrick, R. S., 5,
Hubbard, D. K., 159
Hubrecht, A. A., W., 18
Hudson, A., 217
Huggett, C. L., 204, 263, 265–66, 268, 275, 309–10
Huggett, D. V., 275
Huppert, H. E., 207, 210
Hurlbert, S. H., 120, 123, 126–28
Hurley, P. J., 8
Huston, M., 298, 299,
Hutchings, J. A., 50, 52, 72, 227
Hyman, L. H., 79, 96, 184, 189

Iiyama, T., 193–94
Ishino, M., 204

Jablonski, D., 28, 58–59, 64, 77–80, 83–84, 86–87, 91, 99–100, 102, 109–10, 124, 214, 285, 299, 302
Jackson, J. B. C., 213, 285
Jaeckle, W., 185
Jannasch, H. W., 50, 213
Janson, K., 110, 115
Johannesson, K., 275, 285
Johannesson, P. J., 261, 278
John, A. W., G., 159, 174, 232, 262
Johnson, H. P., 216,
Johnson, T. C., 267
Jones, M. L., 63, 148, 150, 217
Josefson, A. B., 119, 120, 277, 285
Jumars, P. A., 62, 119, 225, 236, 262, 299, 309
Juniper, S. K., 63, 226

Kanatani, H., 3, 180
Kaneko, K., 286
Kano, Y. T., 32
Kareiva, P., 285
Kaufmann, R. S., 204
Keller, C. H., 226
Kenk, V. C., 79
Kharkar, D. P., 52, 226
Khripounoff, A., 47, 169, 224, 262, 276
Killingley, J. S., 59–60, 94, 106, 124, 203, 299
Klein, H., 185, 187
Klinkhammer, G., 62, 217
Knauss, J. A., 187
Knudsen, J., 53, 56, 63, 162
Kobayeshi, K., 193–94
Kohn, A. J., 112
Kolliker, A. von, 17
Komatsu, M., 32, 193
Kowalevsky, 18
Krishnan, S., 189
Kukert, H., 62, 225
Kulm, L. D., 63
Kuzirian, A. M., 77, 79, 92

Lallemand, S., 193–94
Lalli, C. M., 225
Lalou, C., 214
Lambshead, P. J., D., 262–63
Lampitt, R. S., 46–47, 50, 105, 159, 173–74, 184, 231–32, 236–37, 257, 262, 279
Landahl, J., 204
Lasker, H. R., 180
Laubier, L., 77, 169, 226
Laursen, D., 116
Lavelle, J. W., 217
Laver, M. B., 263
Lawrence, J., 29, 189
Leal, J. H., 64–65
Leeder, M. R., 159
Lemche, H., 124
Le Pennec, M., 79
Levin, L. A., 204, 262–63, 265–69, 275–77, 299, 309–10
Levin, S., 285–86, 288, 306
Levitan, D. R., 180, 192
Lewis, J. B., 167, 225
Lewis, M. E., 262
Lewison, M., 217
Lightfoot, R., 28, 49, 162, 227–30, 232, 237
Lima, G. M., 78
Lindberg, D., 285
Linke, P., 47
Lipps, J. H., 27
Little, S. A., 216
Littlewood, D. T. J., 53, 78, 91–93, 102, 109, 120, 124
Llewellyn, C., 237
Lochte, K., 159–60, 174, 257, 262, 279
Long, J. A., 13
Lonsdale, P. F., 62, 185, 204
Lucas, A., 79
Lupton, J. E., 216
Lutz, R. A., 28, 53, 58–59, 64, 77–80, 83–84, 86–87, 91–93, 99–100, 102, 109–11, 120, 124, 180, 202, 214, 226, 285, 299, 307
Lyman, T., 6

MacArthur, R., 202
MacBride, E. W., 11
MacDonald, K. C., 214
Macdonald, P. D., M., 234
MacIlvaine, J. C., 299
Maciolek, N. J., 119, 133, 244, 250–51, 256, 269, 275, 307, 309
Madin, L. P., 46
Madsen, F. J., 4, 8
Majowski, J., 234
Malahoff, A., 204
Manahan, D. T., 93, 132, 308
Manrique, F., 82, 93, 214, 217
Mantoura, R. F. C., 159–60, 173–74, 231–32, 236–37, 257, 262, 279
Mariscal, R. N., 150
Marsh, A. G., 262
Marshall, B., 110
Marshall, B. A., 61
Marshall, N. B., 4
Massoth, G. J., 216–18
Mauchline, J., 175
Mauviel, A., 194
Maxwell, W. L., 152
Mayer, L. M., 185, 236

McCann, L. D., 269
McCave, I. N., 46, 62, 159
McClary, D. J., 30
McDonald, K. C., 62, 73
McEdward, L. R., 30–32, 34, 41, 181, 184
McEuen, F. S., 32, 189
McHugh, D., 226
McLean, J. H., 77, 82–84, 87–89, 92
McLellan, T., 187, 189, 191, 193, 299
McClintock, J. B., 29, 31, 34, 42, 100, 110
Meadows, P. S., 47
Menge, B. A., 119
Menzies, R. J., 2–4, 46, 159–60, 255–56
Mileikovsky, S. A., 10, 27, 229
Milne, B. T., 285
Miller, J. E., 181
Miller, R. L., 194
Mills, E. L., 2, 11, 14
Mills, S. W., 311
Milne-Edwards, H., 14
Mita, M., 194–95
Mittelstaedt, E., 185, 187
Mladenov, P. V., 29–30
Moji, T., 46
Monniot, C., 169
Montagne, D., 276
Mooi, R., 182
Moore, J. C., 63
Moore, J. J., 311
Morris, R. H., 41
Morris, R. J., 159, 174, 232, 262
Morse-Porteous, L. S., 65, 119, 224–25, 262, 276, 279, 299, 309
Mortensen, T., 9, 11–13, 19–20, 27–28, 30, 32, 159, 161, 166–67, 181
Morton, B. R., 216
Moseley, H. N., 3, 5–6, 10, 17, 19–20
Mullineaux, L. S., 64–66, 93, 185–87, 203–4, 206, 214, 226, 263, 266, 308
Munk, W., 186
Murray, J., 4–5, 13, 32
Murray, J., 286
Muus, K., 232

Nakamura, K., 193–94
Nakamura, M., 195
Nance, J. M., 184
Nelsen, T. A., 62
Nesis, K. N., 229
Newell, R. I., E., 60
Nichols, D., 166, 238
Nichols, F. H., 234
Noble, M., 204
Noda, E. K., 62
Normark, W. R., 62
Nowell, A. R. M., 62, 195
Nozaki, Y., 226

Ockelmann, K. W., 54–56, 230
O'Foighil, D., 54
Ogden, J. C., 159
Oguro, C., 32
Okubo, K., 286
Okutani, T., 77, 89
Oliver, J. K., 192
Olson, R. R., 34
O'Neill, S., 110, 119, 299
Ormond, R. F., G., 194
Orton, J. H., 1, 3, 45, 193
Oshel, P. E., 52, 227
Oug, I., 276,
Owens, W. B., 185, 187, 210

Pacala, S., 285
Pain, S. L., 31–32, 169, 231, 256–57
Paine, R. T., 288
Palmer, A. R., 285, 296–97
Parulekar, A. H., 225
Patching, J. W., 159–60, 174, 257, 262, 279
Paterson, G. J., L., 172–73
Patterson, D. J., 262
Pawlik, J. R., 311
Pawson, D. L., 139, 181, 184, 189, 193
Payne, C. M., 53,
Pearse, J. S., 3, 6, 10, 20, 29–34, 42, 100, 110, 133, 146, 159, 180, 189, 307–8
Pearse, V. B., 34, 146
Pearson, M., 227–28
Pearson, T. H., 261–62, 275–78
Pechenik, J. A., 28
Pelseneer, P., 14
Pennington, J. T., 180, 192
Petit, H., 79
Petrecca, R. F., 244, 246, 250–51, 256–57, 262, 269, 275, 309–11
Pfannkucke, O., 159–60, 174, 262, 279
Philbin, L. M., 46
Pitcher, T. J., 234
Plaia, G. R., 276, 309–10
Polejaeff, N., 18, 20
Polloni, P. T., 262
Ponder, W. F., 77, 89
Pope, R., 263, 265–67
Popper, K., 3
Possingham, H., 119
Potter, E., 101, 133
Pratt, R. M., 185, 187
Prouho, H., 9, 11, 19–20

Qian, P. Y., 262
Quinn, J. F., Jr., 89

Rachor, E., 225
Reid, D. G., 112
Reid, R. G. B., 56
Reiswig, H. M., 225
Revelas, E. C., 226
Rex, M. A., 28, 49, 56–57, 59–60, 94, 100–101, 106, 110, 119, 121, 124, 126, 131–33, 203, 262, 285, 297–99, 302, 307–8
Reyss, D., 262
Rhoads, D. C., 79, 91, 102, 109, 226
Rice, A. L., 47, 50, 159, 172–74, 184, 193, 231–32, 236–37, 257, 262
Richards, K. J., 225
Richardson, M. J., 185
Richter, G., 58, 102, 104, 114
Ricklefs, R. E., 119, 121, 307

Riemann, F., 159–60, 174, 257, 262, 279
Riley, G. A., 262
Rio, M., 262
Ritger, S. D., 63
Ritz, D. A., 262
Ritzran, W., 47
Robert-Baldo, G., 217
Robertson, R., 99
Rocchini, R., 103
Roden, G. I., 206
Roe, H. S., J., 105
Rohlf, F. J., 267
Rokop, F. J., 3,49, 160, 162, 255, 257
Rona, P. A., 62, 216
Rosen, R., 286
Rosenberg A. A., 236
Rosenberg R., 261–62, 276–78
Ross, D. A., 299
Ross, E. H., 46, 257
Ross, J. C., 27
Roth, S. E., 217
Roughgarden, J., 28, 34, 119, 285, 297
Roux, M., 227
Rowe, F. W. E., 141, 154, 182
Rowe, G. T., 1, 159, 184, 187, 189, 191, 193, 262–63, 299
Ruff, R. E., 244, 246, 250–51, 256, 269, 275
Rumohr, J., 47
Rumrill, S. S., 166, 186, 193–94
Run, J.-Q., 193

Saito, H., 77, 89
Saleh, S., 225
Sammarco, P. W., 203
Sanders, H. L., 52, 53, 113, 119–20, 162, 165, 187, 189, 191, 193, 224, 226, 255, 267, 275, 277, 299
Sandwell, D. T., 205
Saunders, P. M., 185, 187
Schaff, T., 266, 282, 307
Schein, E., 56, 92
Schein-Fatton, E., 80
Scheltema, A. H., 92, 250–51, 256
Scheltema, R. S., 3, 4, 49, 53, 58, 77, 79, 110, 112, 162, 225, 238, 255, 285, 302, 307, 310
Schenck, H. G., 49
Schelz, A., 47
Schmitz, W. J., Jr., 187
Schnute, J., 234
Schoener, A., 3, 24, 176, 229–30, 232, 237, 255
Schriever, G., 159–60, 174, 257, 262, 279
Schulze, F. E., 19
Schutz, S., 202
Schwinghamer, P., 236
Selbie, C. M., 14, 19, 20
Selenka, E., 18, 20
Semenova, T. N., 13, 239
Sewell, M. A., 180, 192
Shearer, C., 11, 13, 20
Shen, M., 262
Shibata, M. F., 184
Shields, W. M., 213
Shilling, F. M., 132
Shimek, R. L., 60
Shmida, A., 119, 285
Shrader, E., 150
Shuto, T., 28, 55
Sibert, J. R., 234
Sibuet, M., 63, 194
Siebenaller, J. F., 110
Sieburth, J. McN., 46
Sigurdsson, J. B., 64
Simroth, H., 13, 20
Skalkin, V. A., 227
Skullerud, A. M., 277
Sladen, W. P., 6, 10
Smiley, S., 11, 189
Smith, C. R., 61–62, 65, 119, 159, 184, 224–25, 262–63, 276, 309
Smith, D. K., 204, 206
Smith, K. L., Jr., 216
Smith, W. K., 299
Smithey, W. M., 218, 226
Snelgrove, P. V. R., 262, 309, 311
Sokal, R. R., 267
Sokolova, M. N., 60
Sole, R., 286
Sourbeer, J. W., 269
Southward, A. J., 51
Southward, E. C., 51
Speer, K. G., 216, 217
Spiess, F. N., 62, 185, 204, 214
Spight, T. M., 133
Starczak, V., 244, 250–51, 256, 269, 275, 311
Staresinic N., 159
Steinhauer, W. G., 244, 246
Stearns, S. C., 52
Steele, M. E., 30
Stephen, A. C., 18
Stephens, G. C., 51
Stokes, T. L., 269
Stolzenbach, D., 47, 216
Stommel, H. M., 217
Strathmann, M. F., 30, 41, 93, 131, 308
Strathmann, R. R., 30–32, 34, 41, 58, 78, 112, 131, 181, 285, 293, 296–97, 302
Streslov, V. E., 277
Stricker, S. A., 13
Stuart, C., 110, 285, 297, 307, 308
Suchanek, T. H., 159
Suess, E., 63
Sutherland, J. P., 119
Suyemitsu, T., 146, 153
Svane, I., 16, 18

Tabunkov, V. D., 227
Tagami, D. T., 204
Taira, A., 193–94
Tait, J., 47
Tamse, A., 82, 93, 214, 217
Taviani, M., 67, 109
Taylor, G. I., 216
Tegner, M. J., 234
Tenore, K. R., 262
Terborgh, J. W., 119
Theel, H., 6, 7, 9, 13, 20
Thiel, H., 159–60, 174, 257, 262, 279
Thistle, D., 63, 185, 187, 275, 299

Thomas, C. L., 266–67, 269
Thomsen, L., 47
Thomson, C. W., 2, 4–7, 10–11, 13, 27, 180, 182, 189
Thorpe, S. A., 185, 187
Thorson, G., 1, 3–5, 8, 10, 12, 14, 16, 19–20, 26–30, 32, 34, 40, 48, 53–54, 58, 61, 93, 100, 102, 104, 120, 125, 131–33, 179–80, 189, 293
Thurston, M. H., 159, 174, 232
Titman, C. W., 64
Tivey, M. K., 216
Tizard, T. H., 6, 7
Trefrey, J. H., 62
Tucholke, B. E., 62
Tunnicliffe, V., 64, 68, 77, 92, 226
Turekian, K. K., 52, 226
Turley, C. M., 159–60, 174, 257, 262, 279
Turner, J. F., 53
Turner, J. S., 216
Turner, R. D., 28, 49, 53, 58–59, 64–65, 77, 79–80, 83–84, 86–87, 91, 93, 102, 109–10, 159, 214, 219, 225–26, 242, 274, 309
Tyler, A., 139
Tyler, B. S., 139
Tyler, P. A., 3, 4, 8, 13, 27–33, 40, 46, 48–50, 52, 55, 91, 93–94, 100, 124, 138–39, 146, 159–60, 162–63, 166–69, 173–74, 180, 184–87, 193–94, 196, 227–30, 232–33, 236–38, 255–57, 262, 275, 279, 307–8

Uchida, R. N., 204
Uda, M., 204
Uemura, I., 146, 153
Underwood, A. J., 119

Vaisnys, J. R., 226
Valentine, J., 285, 302
Valls, J., 286
Vance, R. R., 16, 28, 34, 285
Van Dover, C. L., 65, 68, 82, 93, 181, 214, 216–17, 219, 226
Vangriesheim, A., 47
van Hemert, D., 262
Van-Praet, M., 150, 262
Van Ummersen, C. A., 28, 49
Vermeij, G. J., 285
Verrill, A. E., 162
Vinogradova, N. G., 120
Visnys, J. R., 52
Vogel, H., 185
Von Bodungen, B., 46
Von Damm, K. L., 217
von Herzen, R. P., 216
Vrijenhoek, R. C., 110, 202

Wade, M., 244, 250–51, 256, 269, 275
Walker, C. W., 29
Waller, T. R., 54,
Wangersky, P. J., 262
Warén, A., 28, 56–58, 60, 77, 83–84, 86, 88–89, 91–94, 100–103, 106–9, 112, 114, 120, 124–25, 131–33, 203, 214, 297–99, 302
Warner, R. R., 120
Warren, B. A., 185, 187
Watling, L., 244, 250–51, 256, 269, 275, 308, 311
Watson, J., 237
Watson, R. B., 14, 20
Watts, M. C., 110, 119, 299
Weatherly, G. L., 185, 187
Webb, C. M., 232, 310–11
Welborn, J. R., 132
Weston, D. P., 262, 276
Wheatcroft, R. A., 62, 236, 262
Whitlach, R. B., 224
Wiebe, P. H., 46, 64, 82, 93, 213–14, 216–17
Wild, R. A., 193
Wilde, W., 62
Willemoes-Suhm, R. von, 14, 15, 20
Williams, A. B., 217
Williams, A. J., III, 62, 159
Williams, S. L., 159
Willis, B. L., 192
Wilson, B. R., 79
Wilson, E. O., 202
Wilson, G. D., F., 203
Wilson, R. R., 204, 222,
Wimbush, M., 185–86
Wirsen, C. O., 213
Wirth, U., 138
Wishner, K. F., 203–4, 263, 265–66, 268
Wolf, W., 185
Wolff, T., 61, 159
Woodard, T. M., 154, 157
Woodin, S. A., 269
Wright, S. H., 51

Yasumasu, I., 194
Yingst, J. Y., 63, 275
Yonge, C. M., 162, 178
Yoshizawa, H., 32
Young, C. M., 10, 18, 32–33, 59, 93, 142, 144, 146, 153–54, 159, 166, 181–82, 186–87, 193–96, 216, 307–9
Yund, P. O., 192, 200

Zajac, R. N., 224
Zezina, O. N., 227
Zhang, X., 210
Ziebis, W., 47

SUBJECT INDEX

Abundance, species, 224–25, 227, 244, 255, 267, 269, 274, 277–78, 280
Abundance, larval, *see* Larval abundance
Abyssal, 56–57, 62, 99, 101, 103, 112, 126, 137, 139, 153, 169, 179, 185, 187, 189, 229, 277, 299
Accessory cells, 138
Accessory sex glands, 76
Adaptation, 27, 44–69, 76, 94, 98, 111, 139, 152, 213, 306, 309
Adaptive radiation, 119
Aegean Sea, 2
Africa, 277, 280
Agassiz, Alexander, 7–9
Agassiz trawl, 160, 228, 230
Age, 169
Age structure, 223–38, 276
Aggregation, larval, 212
Aggregation, reproductive, 166, 179, 188, 193–94
Albert I of Monaco, 14
Albumen, 83, 99
Albumen gland, 89
Allozyme variation, *see* Genetic variation
Alvin, DSV, 78, 225, 266
Amphiblastula larva, 18, 20
Androdioecy, 16
Animal remains, 159, 261, 262, 279, 310, 158, 174–75, 225; *see also* Whale remains
Antarctic, 6, 7, 13, 26, 29–34, 100, 308
Arabian Sea, 280
Aragonite, 100
Arctic, 4, 29–31, 100
Asexual reproduction, 29
Ashworth, Henry, 13, 15
Atlantic Ocean: eastern, 67, 250; equatorial, 14; north, 54, 118, 120–21, 125–26, 132, 139, 187, 194, 245–46, 248, 250–54, 276, 310; northeast, 13–14, 26, 31–33, 50, 60, 150, 166, 182, 223, 232, 236, 263, 27; northwest, 60, 62, 128, 225, 226, 250, 266, 279, 297, 307; south, 120, 246
Atlantis II, R. V. 266
Auricularia larva, 11–12
Australia, 13, 152
Autonomous vehicle, 306
Azoic sediment trays, *see* Sediment tray experiments

Back-arc basin, 2, 213
Bahamas, 139, 144, 146, 153, 166, 187, 188, 194, 276
Banyuls Sur Mer, 11
Barent Sea, 47
Basin, oceanic, 58, 185, 202, 230, 250, 263
Bathyal, 56–57, 99, 101, 103–5, 111, 120, 126, 133, 137, 139, 142, 144, 146,

Bathyal (*Continued*)
148, 179, 188–89, 227, 277, 263, 299
Bathymetric range, 107, 132, 299
Bay of Biscay, 11, 47, 150, 276
Benthic boundary layer, 45, 47, 66, 94, 203, 225
Benthic storm, 159, 185
Benthos, 159, 219, 262
Bering Sea, 280
Biogeography, 45, 64, 67, 69, 110, 125–26, 132, 134, 202; island, 202, 204
Biomass, 184, 236–37, 262–63
Bioturbation, 47
Blake Plateau, 185, 187
Body size, 261–62, 270, 276
Box core, 160, 267
Brazil, 152
Brooding, 4–8, 14, 18, 27–28, 30–34, 40, 42–43, 63, 79, 180, 182, 184, 189, 224, 246, 257, 262, 273–76
Buoyant egg, *see* Egg buoyancy
Buoyant plume, *see* Plume, hydrothermal
Byssal drifting, 64

California, 162
Canada, 243–44, 249–50, 258
Canyon, submarine, 47, 62, 266
Cape Cod, 250
Cape Fear, 269, 307
Cape Hatteras, 250, 256–57, 266, 269, 275
Cape Lookout, 266, 269, 307
Cape Verde Islands, 11
Capsule gland, 89
Carbon, total organic, 263, 267
Carbon flux, 265, 267, 273
Carcass, *see* Animal remains
Caribbean, 308
Cellular automaton model, 284–302, 309
Cephalic tentacle, 86–88
Challenger, HMS, 2, 4–5
Challenger office, 5
Challenger Reports, 4–6, 19–20
Chemosynthesis, *see* Primary production
Chlorophyll a, 263, 265, 268
Chukchi Sea, 280
Chun, Carl, 9, 11–12
Coastal fauna, 118, 133, 134
Cold methane/sulfide seeps, 63, 76–94, 308
Colonization, 45, 65–67, 110, 119–20, 204, 213, 224, 226, 284, 287, 289, 291, 309
Community, deep-sea, 125, 133, 202, 263
Community dynamics, 311
Community structure, 118, 119, 134, 244, 262–63, 277, 279, 310
Competency, metamorphic, 93, 102–14, 206
Competition, 65, 119, 236–37, 286, 299, 309
Competitive exclusion, 284, 287–89, 291–92, 294–96, 298, 300–302
Complemental male, 16
Continental rise, 47, 62, 185, 246, 250, 307
Continental shelf, 186–87, 257, 277
Continental slope, 57, 181, 185, 229–30, 243, 246, 250, 258, 261, 266–67, 269, 307
Continuous reproduction, *see* Temporal breeding patterns
Convection cell, 218
Copulation, 83, 86, 87, 92, 180, 182; *see also* Fertilization, internal
Copulatory organs, 76–77; *see also* Penis
Coral reefs, 119
Coriolis acceleration, 206–8
Coronate larvae, 14
Coupled map lattice, 286
Cross seamount, 187
Cues, reproductive, 180
Currents, 45, 62–63, 66, 109, 124, 152, 184–86, 192, 204, 206–9, 216–17, 224–25, 266, 311
Current speed, *see* Currents
Cyclic events, 46
Cyprid larva, 16, 19–20

Demersal larval development, 31–32, 43, 93–94, 98, 110, 131
Demographic model, 223, 227, 233–36
Demography, 64
Density, water, 185, 216
Density dependence, 213
Depth, 120–21, 126–27, 129, 184, 204, 208, 229, 244, 249, 266, 297, 299–300, 302, 307
Developmental mode, 53–58, 61, 64, 67–69, 76, 98–99, 110–12, 118, 120, 124–25, 131–32, 134, 256, 275, 280, 285, 297–99, 302; *see also* Demersal larval development, Direct development; Lecithotrophic larval development; Mixed development; Planktotrophic larval development
Developmental rate, 34, 110, 214, 230, 236
Diet, 158, 162, 169; *see also* Feeding; Gut contents
Digestion, 236; *see also* Feeding
Digestive efficiency, 236
Digestive tract, 163, 165–66, 172; *see also* Gut contents
Dioecious, *see* Gonochorism
Direct development, 3–4, 7–8, 10, 14, 16, 20, 26–29, 32, 44–45, 55–56
Dispersal, 226, 284–302, 306–8
Dispersal distance, 132, 285, 279
Dispersal potential, 125,

132–33, 138, 285, 287, 289, 291, 293, 297, 299
Dissolved organic material, 51, 60, 93, 131
Distribution, 98; of benthic stages, 98; of larvae, 22, 219; *see also* Bathymetric range; Biogeography; Vertical migration
Disturbance, 119, 224–25, 276, 279, 284–85, 288–302, 307, 309, 310
Diversity, species, 105, 118–34, 202, 224, 225, 227, 269, 274, 277, 280, 288, 298–99, 285, 307, 309; gradient, 133; local, 121, 127–30, 132; regional, 121, 124, 127–29
Dominance, 262, 269, 276
Dormancy, 306
Downwelling, 185
Dumping, ocean, 224; *see also* Pollution, organic
Duration of planktonic life, *see* Larval life, duration of
Dwarf male, 16, 18

East Pacific Rise, 62, 66, 68, 79, 82–89, 93, 152, 264
Echinopluteus larva, 11–12, 19, 20
Eddy, 185, 201, 206, 210, 213, 217, 219
Eddy diffusion, 59
Edinburgh, 4, 13
Egg, 92, 245; *see also* Oocyte; Oocyte structure
Egg buoyancy, 8, 31, 32, 34, 114, 306
Egg capsule, 78, 91, 99, 109, 111, 124, 297
Egg concentration, 190
Egg mass, 124, 154
Egg size, 7–8, 10, 16, 20, 28, 32, 54, 76–77, 79, 86, 92, 131, 138, 144, 153, 162, 173, 182, 229, 246, 249, 307
Egg structure, *see* Oocyte structure
Embryo, 124, 182
Emigration, 203, 210, 255
Endocytosis, 148
Energetics, 114, 131, 132, 174, 226; *see also* Feeding
England, 227
Environmental variability, 285, 302, 306, 312
Ephemeral habitats, 61–64, 68–69, 76–77, 92, 110, 203, 214, 226
Epibenthic sledge, 120, 160, 228–30
Epifauna, 266
Epipelagic zone, 98
Escanaba Trough, 90
Estuary, 269
Eurybathal species, 44
Euspermatozoa, *see* Sperm
Eutrophication, 276
Evolution, 69, 98–99, 110, 204, 213, 214, 285; *see also* Extinction; Genetic variation; Natural selection; Speciation
Explorer Ridge, 93
Extinction, 111, 119, 285
Extra-embryonic layers, 82, 87
Eyes, *see* Sensory structures, larval

Fecal pellet, 263
Fecundity, 16, 32, 44, 45, 51–53, 61, 68–69, 76–77, 82, 93, 138, 162, 223, 226, 229, 238, 284, 288, 291–96, 311
Feeding: carnivory, 50, 172–73, 169; deposit, 47, 50–51, 94, 158, 162–63, 168, 224, 236–38; larval, 34, 53–54, 59–60, 69, 77, 93–94, 99, 102, 112–13, 131, 257; omnivory, 169; scavenging, 50, 158, 159; suspension, 47, 50, 60, 94, 202, 204, 213; *see also* Diet; Gut contents
Fertilization: internal, 83, 86–89, 91–92, 94, 152, 179–80; external, 139, 152, 179–96; mechanism of, 152; success, 166
Fish remains, *see* Animal remains
Fission, *see* Asexual reproduction
Floating egg, *see* Egg buoyancy
Florida, 181
Florida current, 187
Florida escarpment, 63
Flow, *see* Currents
Flow velocity, *see* Currents
Food falls, 262, 276, *see* Animal remains; Macrophyte detritus; Whale remains
Food flux, 307, 309, 311
Food limitation, 34, 152, 204, 213, 236, 262, 307
Foraging behavior, 168
Foraminiferal sand, 266
Free vehicle, 224
Fresh-water, 27
Friction velocity, 185–86
Fugitive species, 284, 287, 289

Galapagos Rift, 62, 66, 79–81, 85, 90–91, 93
Gambia Basin, 120, 122, 125
Gamete, *see* Egg; Sperm
Gamete structure, *see* Oocyte structure; Sperm structure
Gametogenesis, 17, 48–50, 52, 86, 138, 223, 230, 234, 256–57, 309; oogenesis, 48, 150; rate of, 44, 68; spermatogenesis, 48, 150
Gametogenic strategy, 138
Gay Head-Bermuda Transect, 120, 187
Geographic range, 285, 299
Gene flow, *see* Genetic exchange
Generation time, 226
Genetic distance, 154
Genetic differentiation, 308
Genetic exchange, 45, 58, 64, 66–68, 94, 98, 202, 204, 210, 212, 213, 285, 302

Genetic structure, 202, 219
Genetic variation, 66–67, 110, 219, 308
Genital openings, 7–8, 13, 16
Genital papillae, 181
Genital pinnules, 13
German Plankton Expedition, 13
German South Polar Expedition, 13
Golgi complex, 146
Gonad index, 169
Gonad structure, 137–54; Cnidaria, 150; Echinodermata, 13, 139–40, 146, 148–50; Echiurida, 18; Mollusca, 152; Nemertea, 18; Vestimentifera, 148, 150; *see also* Ovary structure
Gonochorism, 77, 79, 82, 86–88, 250
Gonophore, 17
Gorda Ridge, 88–90, 93, 89
Greenland, northeast, 3, 27, 29–30, 32–33, 40
Greenland Sea, 280
Growth, 63, 76, 77, 91, 124, 224–27, 231–34, 237–38, 251, 255, 276
Growth curve, 232–34
Growth rate, *see* Growth
Guaymas Basin, 93
Guaymas Transform Fault, 93
Guiana Basin, 120, 123, 125, 128
Gulf Stream, 266–67
Gully, 266
Gut contents, 161–68, 172, 174
Gyre, oceanic, 185

Habitat area, 202
Habitat complexity, 34
Hadal, 99, 101
Hatching, 100
Hawaii, 139, 144, 153
Heavy metals, 92, 94
Hebrides, 229
HEBBLE site, 185
Hensen, Victor, 13
Herdman, William, 15, 17–18
Hermaphroditism, 16, 77, 79, 194
Hertwig, Richard, 15, 17
Heterotrophs, planktonic, 60
Histology, gonad, 137–38, 230
Hoek, P. P. C., 15–17
Hot spot, 213
Hydrocarbon seep, 102, 109; *see also* Cold methane/sulfide seeps
Hydrodynamic model, 184–86, 189–92, 216, 219
Hydrogen sulfide, 213
Hydrothermal vents, ix, 2, 50, 58, 62, 64–68, 76–94, 98, 102, 105, 108–10, 112, 148, 173, 186, 201–3, 205, 213–19, 226, 238, 308
Hypodermic impregnation, 152

Ibero-Moroccan Gulf, 67
Immigration, 202, 210, 225, 310
Inbreeding, 213
Indian Ocean, 13, 62
Indirect development, *see* Lecithotrophic larval development; Planktotrophic larval development
Infauna, 243–58, 261, 266, 279, 310
Internal fertilization, 76
Interspecific interactions, 285
Intertidal zone, xi, 132, 187, 310
Intramesogleal cyst, 150
Ireland, 229, 244, 256
Iteroparity, 44, 61, 69, 138, 306

Japan, 12, 89, 146, 153
Japanese subduction zone, 63
Johnson-Sea-Link submersible, 78, 267
Juan de Fuca Ridge, 62, 80–81, 88, 91, 93
Juvenile, 99–100, 111, 236, 245, 247, 251, 255, 257, 262, 273; planktonic, 214
Juvenile size, 34

Kaikata Seamount, 89
K-selection, 113

Larvae, deep-sea planktonic, 1, 3–4, 6, 10–21, 31–35, 54–60, 63, 65, 82, 93, 98–99, 104–5, 111, 114, 201–3, 208, 213, 216, 217, 285, 312, 308
Larval abundance, 33, 219
Larval behavior, 202–3, 311
Larval dispersal, 28, 57, 64–69, 77–78, 94, 98, 109–11, 114, 119, 124, 130–32, 201–19
Larval distribution, 33, 219
Larval life, duration of, 132, 174, 204, 230, 293, 308; *see also* Dispersal potential
Larval retention, 201, 203, 206, 210, 212, 213, 219
Larval settling velocity, *see* Larval sinking rate
Larval shell, 76, 297; *see also* Prodissoconch; Protoconch
Larval sinking rate, 217
Larval swimming, 214–17
Larval transport, *see* Larval dispersal
Larval trap, 93
Latitude, 118, 120–21, 125–26, 130–32, 134
Lecithotrophic larval development; benthic, 29, 40, 42–43; pelagic, 10, 13, 18, 26–35, 40–43, 44, 55–56, 64–65, 69, 78, 99–100, 102, 106–7, 109–14, 174, 180, 182, 195, 214, 230–31, 237, 275, 308, 311
l'Hirondelle, 14
Life history trait, 223, 227, 243–44, 246, 255–56, 261, 263, 306
Life span, *see* Longevity

Lightning, HMS, 2
Locomotion, larval, *see* Larval swimming
Longevity, 61, 223, 238
Louisiana Slope, 80–81, 89
Lydonia Canyon, 258

Macroevolution, *see* Evolution
Macrofauna, 184, 223, 225, 230, 236–38, 261–63, 266, 268, 277, 280, 311
Macrophyte detritus, 61, 159, 167, 225, 261–62, 267, 279, 310
Madeira Abyssal Plain, 187
Manca, 245, 255, 257
Manganese, 217, 266
Manganese nodules, 65–66
Mangroves, 119
Mariana Trough, 62
Marianas Back-arc Basin, 83
Massachusetts, 243–45
McMurdo Sound, Antarctica, 30–31, 33, 42–43
Mediterranean Sea, 11, 14, 67
Megafauna, 139, 181, 204, 236–37, 244, 266
Megalops larva, 14, 20
Megaplume, 218
Meiofauna, 184, 224, 232, 236–37, 262, 266
Mesh size, 228–29
Mesocosm, 276
Mesopelagic zone, 158
Mesoscale flow, 201–19, 308
Metabolism, *see* Energetics
Metallic sulfides, 62
Metamorphosis, 34, 67, 99–100, 102, 111, 229, 311; delayed, 93, 114
Meteor, R. V., 103–5, 108
Methane, 213
Mexico, 263
Microhabitat heterogeneity, 307
Microvilli, 150
Mid-Atlantic Ridge, 58, 62, 120
Mid-oceanic ridges, ix, 2
Mineral precipitates, 217; *see also* Metallic sulfides
Mining, deep-ocean, 224
Mixed larval development, 45
Monoecy, *see* Hermaphroditism
Monsoon season, 61
Monterey Bay, California, 33, 41–42
Mortality, 93; adult, 65, 223, 309; juvenile, 229, 232, 235, 237, 255, 293, 311; larval, 16, 26, 113, 109, 293–94, 299, 311
Moriadezek Terrace, 47
Mortensen, Theodor, 9, 11–13
Moseley, Henry Nottridge, 5, 11–12
Mucus, 182–83, 195–96; *see also* Egg mass
Murray, John, 4–6, 27
Mutualism, 302

Naples, 11
Natural selection, 58, 66, 139, 152–54, 299
Nauplius larva, 16, 20
Nekton falls, 61
Nemertea, 18
Neoteny, 112, 114
Nepheloid layer, 46–47
New England, 249
Newfoundland, 62
New South Wales, 152
Nidamental chamber, 184
Nonpelagic development, 54, 56, 61, 67; *see also* Direct development; Lecithotrophic larval development
Nonplanktotrophic development, 28, 64, 77–80, 82–84, 86–89, 91–93, 98–99, 108, 110–11, 114, 119, 124, 131, 133, 214, 284–85, 293, 297–299, 302; *see also* Direct development; Lecithotrophic larval development
Nonptenoglossate gastropod, 125
North American Basin, 58, 120, 122, 126, 128–30, 132–33
North American Plate, 68
North Carolina, 247, 251, 256–57, 261–80, 310
Northeast Pacific Ridges, 68
North Pacific Gyre, 310
Norway, 11
Norwegian Sea, 118, 120–22, 126, 128, 129, 130, 133, 280
Nova Scotia, 18, 62, 185
Nurse eggs, 99
Nutrient input, 119; *see also* Particulate detritus
Nutrient limitation, 184; *see also* Energetics; Feeding
Nutritional flexibility, 182; *see also* Feeding

Oligotrophic, 34, 68
Oocyte, 89, 148, 163
Oocyte structure: cortical ooplasm, 148; cortical granules, 163; microvilli, 150; nucleolus, 150; nucleus, 150; oolemma, 150
Oogenesis, *see* Gametogenesis; Ovary structure
Oogonia, 163
Ophiopluteus larva, 13, 19, 229
Opportunistic species, 223–24, 262, 274–75, 277, 280
Oregon, 63
Organic enrichment, 261–80, 309, 310
Orton's Rule, 1–3, 45–46
Outbreeding, 34
Ovary, 88, 140, 148
Ovary structure: follicle cell, 137; follicular inner epithelial cell, 146; nutritive phagocyte, 148, 150; peritoneal cell, 146
Ovicell, 14, 20
Oviduct, 86

Ovum, *see* Egg; Oocyte
Oxygen isotopes in larval shells, 59, 94, 98, 106, 124
Oxygen stress, 278
Oxygen minimum layer, 261, 263, 275

Pacific Ocean, 13, 62, 246; central north, 187; eastern, 261–63, 269, 275, 280; eastern tropical, 187; north, 204, 210; northeast, 79, 146
Pairing, reproductive, *see* Aggregation, reproductive
Paracalymma larva, 56
Parasitism, 50, 302
Paraspermatozoa, *see* Sperm
Particulate detritus: as larval food, 94; flux to deep sea, 45–48, 50, 158–59, 166, 173–75, 257; horizontal transport, 46–47; *see also* Animal remains; Macrophyte remains
Patch dynamics, 289
Pediveliger, 112
Penis, 82–83, 86–89, 92, 112
Perturbation, *see* Disturbance
pH, 92, 94
Pheromone, 194
Philippine Trench, 194
Phospholipid, 194
Photic zone, 102
Photographic sampling, 159
Photosynthesis, *see* Primary production
Phylogenetic constraint, 98, 112, 133, 302, 309
Phylogeny, 112
Phytodetritus, 47, 158–59, 168, 174–75, 231, 236–37, 243, 257, 262–63, 266
Phytoplankton, 102, 113
Pinnacle, 185
Pisces II, 267
Planetary rotation, 217
Plankton, 99, 102; deep-sea, 33, 58, 82, 104, 201, 206, 210, 212, 213, 215, 216; surface, 98, 103–6, 114
Plankton bloom, spring, 236, 257, 310
Planktotrophic larval development, 11–13, 19–20, 26, 28–31, 33–34, 40–43, 44, 55–61, 65–66, 69, 76–80, 89, 91–94, 98–102, 105–14, 118–19, 122, 124, 125, 128–33, 153, 174, 180, 214, 229–30, 284–85, 293, 297–99, 308–11
Plant remains, *see* Macrophyte detritus
Planula larva, 17, 19, 204
Plesiomorphic character, 98, 112
Plume, hydrothermal, 64, 201, 215–16, 226
Plymouth, England, 11
Podocyte, 137, 146
Poecilogony, 99
Polar faunas, 27, 28, 100, 184, 189; *see also* Arctic; Antarctic; Greenland
Pollution, organic, 261–62, 276, 278
Population biology, 262–63
Population bottleneck, 218, 224
Population density, 153, 184, 225, 262–63, 268–69, 286, 307
Population differentiation, 285, 299
Population dynamics, 212, 224, 227, 237–38
Population genetics, 110
Population structure, 249, 255
Porcupine, HMS, 2
Porcupine Sea Bight, 47, 120, 150
Pore-water chemistry, 267
Postlarvae, 224–26, 229–31, 235–37, 311
Precompetent period, 93, 114, 206
Predation: on benthos, 28, 50–51, 119, 172–73, 237, 302; on larvae, 34, 94
Preservation artifact, 245
Pressure, hydrostatic, 152
Primary production: chemosynthetic, 50, 102; photosynthetic, 44, 47–48, 68, 173, 278, 280, 310
Princess Alice I, II, 14
Prodissoconch, 14, 54–56, 77–82, 162, 307
Protandry, 79
Protoconch, 14, 20, 54–55, 57–60, 77–78, 82–94, 98–101, 103–7, 109, 113, 114, 120, 124–25, 131–32, 307
Prouho, Henri, 9, 11
Pseudocopulation, 179, 182
Pseudopopulation, 67
Pyloric caecum, 169, 174

Radiometric dating, 226–27
Rarefaction curve, 184, 267, 269, 271
Reaction-diffusion-advection model, 286
Reaction-diffusion network model, 286
Receptaculum seminis, 82–83, 86–89, 92
Recolonization, *see* Colonization
Recruitment, 45, 63–65, 119–20, 160, 201, 203, 206, 218, 219, 223–38, 243–44, 247, 251, 255–57, 274–76, 311
Red Sea, 13
Remotely operated vehicle, 306
Reproductive effort, 52, 172, 180
Reproductive mode, *see* Developmental mode
Reproductive output, *see* Fecundity
Reproductive pattern, *see* Developmental mode
Reproductive periodicity, *see* Temporal breeding patterns
Reproductive rate, 63, 225, 275–76

Reproductive strategy, 274, 277
Reproductive success, 302
Respiration, 263, 267
Ridge, oceanic, 58, 62, 77, 185, 202, 205, 214, 226, 266
Rockall Trough, 8, 31–33, 40–41, 139, 175, 223, 227–28, 230–32, 234, 238, 244, 256, 308
Rock outcrop, 202
Rossby number, 208–9
Rough endoplasmic reticulum, 146

Sagami Sea, 12
Salinity, 45, 60
Salmon pen, 276
Sampling frequency, 311
San Diego Trough, 162, 187
Scavengers, 61
Scotian Rise, 62
Scotland, 227, 244, 256
Seagrass detritus, 159, 167
Seamount, 66, 185, 201–13, 219, 263, 269, 308
Season, 169
Seasonal reproduction, *see* Temporal breeding patterns
Secondary sexual structures, 77, 82
Sediment community, 223
Sediment stability, 47
Sediment transport, 275
Sediment traps, 46–47
Sediment tray experiment, 65, 224, 234, 276, 309
Seep, *see* Cold methane/sulfide seep; Hydrocarbon seep
Selective pressures, *see* Natural selection
Semelparity, 61, 63, 69, 138
Seminal vesicle, 112
Senegal, 120
Sensory structures, larval; eyes, ocelli, 16, 18, 20, 107, 108; statocyst, 18
Settlement, larval, 66, 228–30, 232, 310, 311; *see also* Metamorphosis
Settlement cues, 102, 311
Settling velocity, larval, *see* Larval sinking rate
Sewage, *see* Pollution, organic
Sexuality, 76, *see* Androdioecy; Gonochorism; Hermaphroditism; Protandry
Sexual maturity, 52, 61, 63–64, 79, 82, 225, 238, 251, 257
Shallow water, 225, 230, 238, 262, 274, 278
Shear, 187
Shelf, continental, 102–3
Shell apex theory, 14, 100; *see also* Prodissoconch; Protoconch
Shell dissolution, 226
Shell growth rings, 227, 232, 234
Sieve, 311
Sinusigerous lip, 54
Size, body, 50, 65, 92, 251
Size-frequency distribution, 218, 223, 226, 231, 236, 247–49, 251–55, 228–30, 232, 267, 270–71, 273–76
Size structure, population, *see* Size-frequency distribution
Slumps, 62
Soft bottom community, 309
South Carolina, 244, 251, 257, 261–80, 310
Space limitation, 204, 213
Spawning, 49, 76, 99, 148, 152, 179–82, 187–88, 190–92, 194
Speciation, 119, 285, 302
Species composition, 307
Species diversity, *see* Diversity, species
Species richness, *see* Diversity, species
Sperm: aflagellate, 92; anthozoan, 150; ascidian, 17–18; bipolar tailed, 141, 142, 154, 179, 194, 196; conical-headed, 139; cylindrical-headed, 139; dimorphic, 137, 142, 154, 179, 194, 196; echinoderm, 139–47, 179–96; euspermatozoa, 89, 141, 154, 196; modified, 139–42; molluscan, 76–77, 82–83, 86–87, 89, 91, 92, 152; paraspermatozoa, 89, 141–42, 154, 196; primitive, 139; sickle-shaped, 153, 195; size, 144, 146; tabloid-headed, 139; vestimentiferan, 148–50
Spermatid, 142, 148
Spermatogenesis, *see* Gametogenesis
Spermatophores, 152, 246
Sperm concentration, 185–86, 189, 192
Spermiogenesis, 148
Sperm motility, 194–96
Sperm release rate, 185, 187
Sperm storage, 180; *see also* Receptaculum seminis
Sperm structure: acrosomal vesicle, 142, 144; acrosome, 137, 139, 141–42, 148; axoneme, 141, 146; centriolar rootlet, 141; centriole, 142, 144; cytophore, 148; distal centriole, 144, 146; excentric mitochondrion, 17–18; flagellum, 139, 141–42, 148; head (nucleus) 137, 139, 141–42, 144, 146, 148, 153, 179, 194–95; lipid droplets, 137, 144, 146, 153, 179, 194; microtubular doublets, 141; microtubules, 141; middlepiece, 137, 139, 142, 144, 146, 150; mitochondria, 139, 141–42, 144, 146, 150; mitochondrial helix, 148; neck region, 148; nuclear envelope, 148; pigment, 137, 142, 179, 194; proximal centriole, 141, 146; subacrosomal material, 142
Spine barbs, echinoid, 154, 182–83
Spreading center, 213

Statocyst, *see* Sensory structures, larval
Stenothermic, 100
Storms, benthic, 62, 68
Straits of Gibraltar, 67
Subadult, 232
Subduction zone cold seeps, 63
Sublittoral, *see* Subtidal zone
Submersible, manned, ix, 181, 224, 267, 306
Subtidal zone, 132, 169
Succession, 224, 226, 289
Sulfides, 92, 94
Surinam, 120
Swimming, larval, 59
Synchrony of reproduction, *see* Temporal breeding patterns

Tadpole larva, 18, 19
Taylor cap, 201, 206–8, 210
Taylor column, 207
Teleconch, 59–60, 78, 88, 100, 106, 114
Teleplanic larvae, 105, 111, 112, 113; *see also* Dispersal, larval
Temperate faunas, 184, 189
Temperature: at hydrothermal vents, 92, 109, 215; control of breeding by, 2–3, 45–46; in deep-sea, 27, 45, 60, 67, 93, 152, 308; effect on developmental rate, 34, 230, 236; polar, 27; surface, 106, 109
Temporal breeding patterns, 45–51, 68; analysis by histological method, 48, 161, 163, 169; asynchronous reproduction, 257; continuous reproduction, 48–50, 68, 76, 87, 91, 94, 223, 231,234, 237, 249, 251, 255, 256, 257; inferred from size distributions, 50; quasi-continuous reproduction, 158, 162, 166–67, 169, 172, 174; reproductive periodicity, 44, 48–51, 77, 138, 169, 223, 255, 279, 308; reproductive synchrony, 180, 249; seasonal reproduction, 3, 33, 48–51, 68, 158–60, 163, 166, 169, 172–74, 179, 193, 223, 229–32, 234, 236–37, 243–44, 246–47, 249, 255–57
Testis, 88, 112
Theel, Hjalmar, 7, 9
Thermocline, 34
Thomson, Charles Wyville, 4–6
Thorson, Gunnar, 4,5
Thorson's Rule, 1, 3–21, 26–35, 179
Tides, 47, 185, 204, 206, 210
Tongue of the Ocean, Bahamas, 187
Topographic features, 68, 124, 201, 206–7, 267
Transport of larvae, 67
Trenches, 47
Triton, HMS, 16
Trophonema cells, anthozoan, 150
Tropical forests, 119
Trough, 185
Turbidity currents, 62, 68, 159, 279
Turbulent diffusion, 185–87

Ultrafiltration, 137
Ultastructure, gamete and gonad, 137–54, 182
United States, Northeast, 162, 248–50, 257, 307
Upwelling, 59, 185, 261, 267, 279

Veliger larva, 44, 58–60, 63, 65, 91, 98–100, 102–5, 108, 114, 124
Velum, 112
Vertical migration: diel (of plankton), 46; ontogenetic (of larvae), 12, 44, 54, 58–60, 67, 69, 98, 102, 105–6, 108–9, 114, 124, 131
Vesicula seminalis, *see* Seminal vesicle
Vicariance, 68
Vitellogenesis, 138, 146, 148, 150; autosynthetic, 148; heterosynthetic, 148
Viviparous development, 33, 42; *see also* Brooding; Direct development
Volcano Seven, 261–80
Voring Plateau, 263
Vortex, 201, 217, 21; anticyclonic, 217; baroclinic, 217; cyclonic, 217

Wake, 210
Walvis Bay, 277
Wasteage, *see* Mortality, larval
Waves, bottom-trapped, 210
Western boundary current, 26
West European Basin, 58, 120, 122
West Florida Escarpment, 80, 88
Whale remains, 62, 110
Whiting, blue, 175
Willemoes-Suhm, Rudolf von, 14–15
Wood, deep-sea, 63, 110, 309, 310
Wood borers, 223, 274

Yolk, 93, 99, 138, 142, 307
Yolk synthesis, *see* Vitellogenesis

Zoogeography, *see* Biogeography
Zooplankton, *see* Plankton